Global Biodiversity

Earth's living resources in the 21st century

UNEP - World Conservation Monitoring Centre

B. Groombridge and M.D. Jenkins

UNEP

WORLD CONSERVATION
MONITORING CENTRE

Aventis *f*oundation

The **UNEP World Conservation Monitoring Centre** was established in 2000 as the world biodiversity information and assessment centre of the United Nations Environment Programme. The roots of the organisation go back to 1979, when is was founded as the IUCN Conservation Monitoring Centre. In 1988 the World Conservation Monitoring Centre was created jointly by IUCN, WWF-International and UNEP. The financial support and guidance of these organisations in the Centre's formative years is gratefully acknowledged.

The Aventis Foundation, based in Strabourg, Germany, was formed in 1996. The Aventis Foundation promotes projects at the interface of culture, science, business and society. The Foundation aims to select projects that are international, interdisciplinary, and looking toward the future. One of its prime aims is to identify the people who will shape tomorrow and to enable them to contribute to sustainable development through their activities in science, politics, and society.

Published by: World Conservation Press, 219 Huntingdon Road, Cambridge, UK. CB3 0DL

ISBN 1 899628 15 0

Disclaimer: The views expressed in this book do not necessarily reflect those of UNEP-WCMC or its collaborators. The designations of geographical entities in this publication and the presentation of the material do not imply the expression of any opinion whatsoever on the part of UNEP-WCMC, the Aventis Foundation, or other collaborating organisations, concerning the legal status of any country, territory, or area, or of its authorities, or concerning the delimitation of its frontiers or boundaries.

Citation: World Conservation Monitoring Centre 2000. *Global Biodiversity: Earth's living resources in the 21st century*. By: Groombridge, B. and Jenkins, M.D. World Conservation Press, Cambridge, UK.

Design and Production

Designed jointly by Hart McLeod of Waterbeach, Cambridge, UK, and WCMC.

Produced by Hart McLeod. Main text set in Novarese 10.5pt.
Printed by G&E 2000, Digital Media Group, Orton Southgate, Peterborough, UK.

Printed on HELLO Silk 100 gsm paper, made from virgin wood fibre from sawmill residues, forest thinnings and sustainable forests in Austria. Pulps used in the production of the above grade are Totally Chlorine Free (TCF) giving zero AOX.

Maps produced using ESRI ArcView GIS 3.1 software, mainly using data provided by conservation GIS. Spatial data are plotted to the Robinson projection; although not an equal area projection, this gives a well-balanced representation of both area and shape for global mapping purposes.

Cover design by Brian Groombridge, created by Chris McLeod.

CONTENTS

TABLES, FIGURES AND MAPS

PRODUCTION TEAM

WCMC The book was planned, written and edited by Brian Groombridge and Martin Jenkins. Advice and additional text were contributed by Valerie Kapos, Charlotte Lusty and Neil Cox. GIS analysis and map production by Corinna Ravilious and Simon Blyth, with technical advice from Jonathan Rhind. Project Manager: Neil Cox. Project Administrator: Julie Reay. Research Assistants: Anna Morton, Janina Jakubowska and Rachel Cook. Other contributions were made by Laura Battlebury, Mary Cordiner, Mark Spalding, Christoph Zöckler and Tim Johnson.

Hart McLeod Design by Chris McLeod, production by Gill McLeod, graphics by Helen Humphreys.

ACKNOWLEDGEMENTS

First and foremost we would like to express our deepest thanks to the Aventis Foundation, without whose generous funding the research and production work for this book could not have been undertaken.

The Owen Family Trust is also acknowledged with much gratitude for the additional financial support which allowed the book to be produced to a high standard and in colour.

We also acknowledge with thanks the generous assistance extended by the following, who allowed use of important material that we believe greatly enhances the value of the book overall.

BirdLife International, of Cambridge, in particular Colin Bibby, Mike Rands and Allison Stattersfield, for allowing our use of spatial data on EBAs and on threatened bird species, as represented in Maps 5.5 and 5.7.

Jonathan Loh, responsible for the WWF Living Planet Report, for kindly approving use of global trend indices from the 1999 Living Planet Report.

Prof. Wilhelm Barthlott of the Botanisches Institut und Botanischer Garten, Abt. Systematik und Biodiversitaet, Meckenheimer Allee, Bonn for kindly allowing use of a map showing contours of global plant species diversity, reproduced with minor design modification in our Map 5.1.

Robert Lesslie of the Department of Geography, Australian National University, Canberra, for allowing us to use data resulting from his global wilderness analysis.

The journal *Science* and Christopher Field of the Department of Plant Biology, Carnegie Institution, Washington for approving our use of a figure representing global net primary production, reproduced with minor design modification in our Map 1.2.

John E.N. Veron, Chief Scientist at the Australian Institute of Marine Sciences, Townsville, Queensland for allowing use of coral generic diversity data incorporated in our Map 6.4.

Gene Carl Feldman, Oceanographer at NASA/Goddard Space Flight Centre, Greenbelt, Maryland, for approving use of material from the SeaWiFS Project of NASA/Goddard Space Flight Center and ORBIMAGE as used in Map 7.1.

We are grateful also to Dave Ward of Goodfellow & Egan Digital Media Group for coping with our idiosyncratic schedule for copy delivery, and to Mark Chapman and Simon Crowhurst of the Department of Earth Sciences, Cambridge University, and Chronis Tzedakis, Department of Geography, Cambridge University, for advice on Holocene sea levels and temperatures.

Life Counts *project*

Life Counts: *a global balance of biodiversity*

Michael Gleich, Dirk Maxeiner, Michael Miersch and Fabian Nicolay.

Life Counts is a popular account of the status of global biodiversity at the beginning of the 21st century. The book, published simultaneously with WCMC's Global Biodiversity will appear first in German and subsequently in English and other languages. The book examines mankind's relationship with living resources at the brink of the next century, looking at the links between conservation and economic development and suggesting methods of sustainable use as alternative solutions to current levels of exploitation.

Current population estimates for many animal and plant species are included to answer questions such as: how many whales are swimming in the world's oceans, how many individual ants are there, how do population sizes of domestic cattle and krill compare, what number of trees and birds exist per person and what does it mean for mankind ? The economic importance of wild animal and plant resources is demonstrated to have been considerably underestimated; for example, pollination services provided by insects are worth billions of dollars to the global economy.

Key aspects of biodiversity are clearly illustrated with many interesting and well designed graphics. *Life Counts* has been prepared by three well known scientific authors and an award-winning graphics designer in close collaboration with WCMC, IUCN and UNEP, and draws upon specially commissioned research by WCMC. Guest authors include UNEP president Klaus Töpfer and other leading individuals from the international conservation community.

ISBN 3-8270-0350-4
Available from: Berlin Verlag, Berlin
Cost: DM 44

FOREWORD

The beginning of a new millennium is a fine opportunity to reflect on humanity's relationship with the rest of life on Earth – its biodiversity. The issue has moved steadily up the agenda over the past fifty years. Are natural ecosystems and species essential to our own success? Are we destroying them and, if so, what will be the consequences?

What we learn from this book is that mankind does nothing fundamentally different from other organisms in using Earth's resources. But with our sophisticated societies and technologies we have become extraordinarily successful at it.

There has been a price to pay. The evidence is persuasive that over the past 70 000 years human beings have time and again colonised new lands and destroyed the living resource base on which they themselves depended. Throughout the New World, including Australia and New Zealand, and on many small islands, mammals and birds, many of them giants of their day, were hunted to extinction, and their habitats burned.

Today extinction rates are between 100 and 200 times higher than normal, and whole biomes are polluted and fragmented. But domestication of crops and animals has saved us from the full consequences of poor stewardship. From the beginnings of cultivation just 13 000 years ago, we now appropriate 40% of the world's natural productivity to our own use, much of it turned into the dozen crops that provide 75% of our food. Croplands and pastures have increased six-fold in 300 years and as a consequence it appears that only one other multicellular species, the Antarctic krill, remotely approaches humanity in global numbers and biomass.

But unlike the Antarctic krill, we don't measure our success simply in terms of survival and reproduction! As reflective beings we seek quality, dignity and harmony in our lives. There is no quality in burnt forests and over-exploited soils; no dignity in polluted wetlands and seas; no harmony in a world without its complement of other creatures. The challenge we face in the new millennium is to reduce our impact on biodiversity by overcoming poverty, reducing human population growth and consumption, and using technology within the environmental limits of sustainability.

This fine book gives us the baseline against which to measure our success in achieving those great goals.

Mark Collins
UNEP - World Conservation Monitoring Centre

INTRODUCTION

The present decade has seen remarkable growth in concern at all levels for wildlife and the environment, and increased appreciation of the links between the state of ecosystems and the state of humankind. This was given impetus by the 1992 United Nations Conference on Environment and Development (the Rio Summit), at which the Convention on Biological Diversity (CBD) was opened for signature, and many subsequent activities worldwide have arisen from efforts to meet the objectives framed by the CBD text.

In that same year, the World Conservation Monitoring Centre (WCMC) and collaborators produced a substantial volume – *Global Biodiversity: status of the Earth's living resources* – giving an extended introduction to the many and varied themes covered by the then unfamiliar term "biodiversity". This book has since found wide application as an educational resource and source-book of biodiversity data.

Now, thanks to the kind and much valued support of the Aventis Foundation, we are pleased to present *Global Biodiversity: Earth's living resources in the 21st century*.

The present volume builds on nearly a decade of further research and analysis by WCMC and the conservation community worldwide. We aim to use the data now available to provide a more comprehensive view, exploring key global issues in biodiversity to greater depth and in a more accessible style. The purpose is twofold: to outline some of the broad ecological relationships between humans and the rest of the biosphere, and to summarise information bearing on the health of the biosphere.

The first part of the book opens with an outline of some fundamental aspects of material cycles and energy flow in the biosphere. This is followed by discussion of the components of biological diversity, and an account of the expansion of this diversity through geological time and the pattern of its distribution over the surface of the earth.

The second part is concerned with material relationships between humankind and biodiversity, covering salient features of human impacts on the environment from the earliest hominines onward, and the uses currently made of elements of biodiversity. In this part we also summarise information on trends in condition of the main ecosystem types and the species integral to them, contributing to objective analysis of the state of the biosphere at the turn of the second millennium.

This book has a companion volume in *Life Counts*, also supported by Aventis, and produced by a team based in Germany.

1

THE BIOSPHERE

The biosphere is the thin and irregular envelope around the Earth's surface, containing all living organisms and the elements they exchange with the non-living environment.

Water makes up about two-thirds of an average living cell, and organic molecules based on hydrogen, carbon, nitrogen and oxygen make up the remaining one-third. These and other elements of living cells cycle repeatedly between the soil, sediment, air and water of the environment and the transient substance of living organisms.

The energy to maintain the structure of organisms enters the biosphere when sunlight is used by bacteria, algae and plants to produce organic molecules by photosynthesis, and all energy eventually leaves the biosphere again in the form of heat. Photosynthetic organisms themselves use a proportion of the organic material they synthesise; net primary production is the amount of energy-rich material left to sustain all other life on Earth.

Humans now appropriate a large proportion of global net primary production, and have caused planetary-scale perturbations in cycling of carbon, nitrogen, and other elements.

THE LIVING PLANET

The defining characteristic of the planet Earth is that it supports life, and has done so for at least 70% of its history (see Chapter 3). The position of the Earth relative to the sun, its size and composition appear to be the main factors that have allowed life to develop here. Most importantly, these factors have combined to ensure the permanent presence of a large amount of liquid water on the planet's surface, and this is the fundamental prerequisite of life.

The part of the planet that supports life is called the biosphere[1]. The non-living biosphere comprises the hydrosphere (the waters), the soil and upper part of the lithosphere (the solid matter that forms the rocky crust of the earth), and the lower part of the atmosphere (the thin layer of gas coating the planet's surface). These domains interact in ways critical to the operation of the biosphere, and are linked in particular by the properties of water as a solvent and medium that fosters the chemical reactions basic to life.

While providing the conditions necessary for life, the structure and composition of the non-living parts of the biosphere have themselves been profoundly affected through time by living organisms. Most obviously, and from the point of view of humans, most importantly, the presence of significant quantities of free oxygen (O_2) in the atmosphere is entirely the product of oxygen-releasing photosynthesis by cyanobacteria starting more than 2000 million years ago. The idea that living organisms do not merely influence conditions in the biosphere but in some way regulate them to maintain the conditions conducive to life has received considerable attention in recent years, chiefly in terms of the *Gaia* hypothesis[2] as first proposed in the 1970s.

The extent of the biosphere

At planetary scale, the biosphere can be pictured as a thin and irregular envelope around the Earth's surface, just a few kilometres deep on the globe's 6371 km radius. Because most living organisms depend directly or indirectly on sunlight, the regions reached by sunlight form the core of the biosphere: ie. the land surface, the top few millimetres of the soil, and the upper waters of lakes and the ocean through which sunlight can penetrate.

The biosphere is not homogenous, because actively metabolising living organisms are sparse or absent where liquid water is absent, eg. permanent ice at the poles and on the very highest mountain peaks, but abundant where conditions are favourable. Nor are its boundaries sharply defined, because although actual organisms are limited in occurrence, dormant forms of life, such as the spores of bacteria and fungi, become passively dispersed virtually everywhere, from polar icecaps to several tens of kilometres above the surface of the Earth (approaching the upper limit of the stratosphere). Recent findings[3] also indicate that surprisingly large, active, bacterial populations occur at depths of up to a few kilometres in the

Figure 1.1

Hypsographic curve

The horizontal baseline in this figure represents the Earth's total surface area of 510 million km². The figure shows that 71% of this surface is covered by marine waters and 29% is dry land. It also shows the mean land elevation and mean ocean depth, and the amount of Earth's surface, in percentage terms, standing at any given elevation or depth.

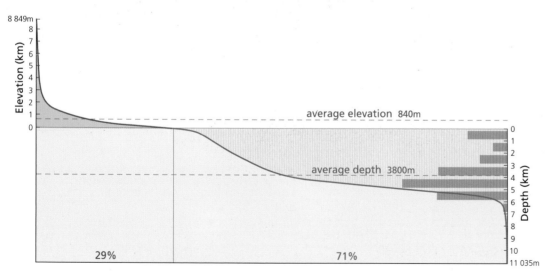

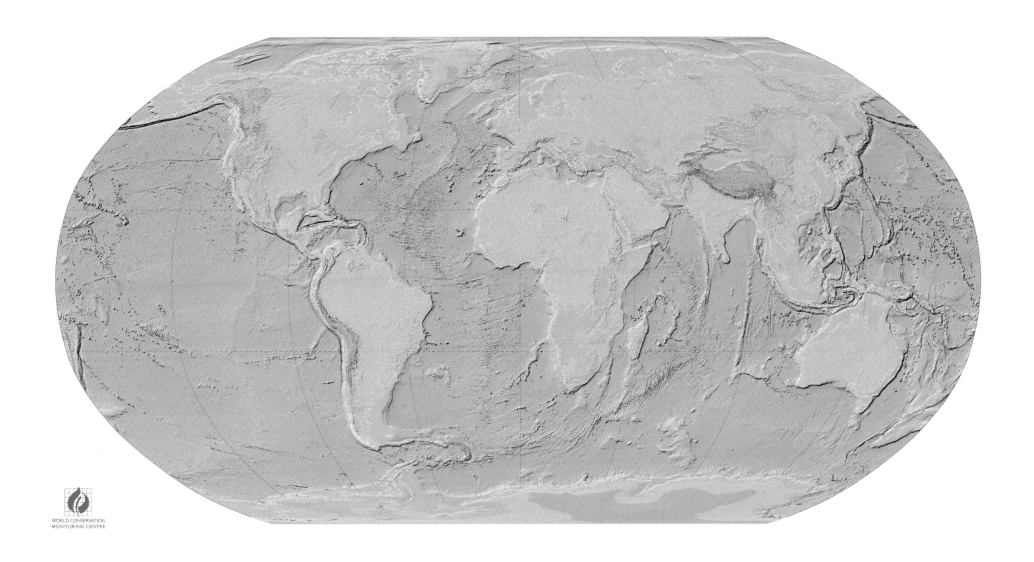

Map1.1

Physical geography of the Earth

This map, best viewed with Figure 1.1, shows the relative areas occupied by dry land and by water, and the general distribution of areas of extreme height or depth.

lithosphere in a range of different sub-surface environments, including terrestrial aquifers, oil-bearing shales, granites and basalts.

The whole of the sea is theoretically capable of supporting active life and comprises therefore the vast majority of the volume of the biosphere (Fig. 1.1). Depending on water clarity, the sunlit (photic) zone may reach just a few centimetres to a few hundred metres in depth, but the marine biosphere is extended into regions of total darkness, down to more than 10 000 metres in the ocean depths, by organisms that subsist on the rain of organic debris falling from the upper waters. Locally on the ocean floor, animal communities exist based on bacteria deriving their energy from hydrogen sulphide emitted from hydrothermal vents. However, the amount of living material in most of the sea – that part of the open ocean below the upper hundred or so metres – is low although still measurable.

The atmosphere plays a vital role in the biosphere, not only in providing a source of essential gases, but also in buffering conditions at ground level, by regulating temperature and providing a shield against excessive ultraviolet radiation. Many organisms, from microscopic bacteria to bats and birds, spend part of their lives suspended in the atmosphere; however, no organism is known that passes its complete life-cycle in the air, and living biomass per unit volume above the earth's solid or liquid surface is extremely low.

Photosynthesis and the biosphere

Life on Earth is based essentially on the chemistry of water and carbon. Indeed, in biochemical terms, living organisms are simply elaborate systems of organic macromolecules dispersed in an aqueous medium. The average cell is about 70% water by weight; the remainder consists very largely of carbon-containing (organic) compounds composed mainly of the four elements hydrogen, carbon, nitrogen and oxygen. These compounds include four major types of large organic molecule – proteins, carbohydrates, lipids, and nucleic acids – and about 100 different small organic molecules. A number of other elements are required in much smaller, though still vital, quantities. These include phosphorus, sulphur, iron and magnesium. All these elements cycle through the biosphere in a variety of forms, both organic and inorganic, following complex and interlinked pathways many of which are yet to be fully elucidated. The engine that drives the organic part of this turnover is photosynthesis – the capture by living tissues of energy from the sun.

Photosynthesis essentially involves the use of energy from sunlight to reduce carbon dioxide with a source of electrons (almost invariably hydrogen) to produce carbohydrates, water and, generally a by-product from the hydrogen donor. In some bacteria the hydrogen donor is hydrogen gas, in others it is hydrogen sulphide, but in cyanobacteria, algae (protoctists) and plants, water is the hydrogen donor and gaseous elemental oxygen (O_2) is the by-product. This is overwhelmingly the predominant and most important form of photosynthesis on the planet, and is described by the following equation:

$$2n H_2O + n CO_2 + \text{light} \longrightarrow n H_2O + n CH_2O + {}_2 n O$$

The initial products of photosynthesis in plants are simple sugars such as glucose. Larger carbohydrate molecules made from glucose include cellulose, the main component of plant cell walls and woody tissues, and starch, a key storage carbohydrate found in roots and tubers. Energy is needed to make the chemical bonds within these organic molecules, and energy is released when the bonds are broken down again. The controlled breakdown of these molecules within cells is the mechanism by which all cells obtain energy to do useful work.

All organisms can break down sugars very directly without the need for oxygen. Many bacteria that live in aerobic conditions only use this method. Virtually all eucaryotes (see Chapter 2) have evolved a more complex additional pathway that requires oxygen but yields much more energy. This latter pathway –

aerobic respiration – essentially reverses the basic photosynthetic reaction shown above.

The major cycling process of the biosphere, therefore, consists of the photosynthetic fixing of carbon dioxide with water to produce organic compounds, in which energy is stored, and oxygen; this is followed by respiration of these compounds, in which the stored energy is released and carbon dioxide and water are produced. Photosynthesis therefore is not only responsible for the vast majority of organic production, but also for the maintenance of free oxygen in the atmosphere, without which aerobic organisms (the great majority of eucaryotic organisms, including humans) could not survive.

Although photosynthesis is the primary engine of the biosphere, in the sense that it injects energy into the system and creates basic organic molecules, production of the full range of organic molecules on which life depends requires additional elements. Of the four key elements, nitrogen is often the one in limited supply, but is an essential component of nucleic acids and proteins. Although the atmosphere consists of 79% nitrogen, this inert gaseous form of the element cannot be used by plants or most other organisms until combined (fixed) with other elements. In the biosphere, atmospheric nitrogen is fixed by a range of bacteria, including cyanobacteria, some free-living soil bacteria, and most importantly by specialised bacteria that live symbiotically in the root nodules of leguminous plants (peas, beans, etc). Some nitrogen is also fixed by lightning in electric storms and, in the modern world, industrially in the production of fertilizer. Fixed nitrogen is made available to plant roots through association with fungi (mychorrhizae) and as nitrogen-fixing organisms decay. From plant roots, it is transported to metabolising plant cells. On death, these steps are reversed, and the fixed nitrogen may be immediately recycled or revert to elemental nitrogen.

PRODUCTIVITY AND THE CARBON CYCLE

Of the solar energy reaching the upper atmosphere of the earth, about half is immediately reflected, most of the remainder interacts with the atmosphere, ocean or land, where it evaporates water and heats air, so driving atmospheric and ocean circulation; much less than 1% of the incoming energy is intercepted and absorbed by photosynthesisers. On land these photosynthesisers are overwhelmingly green plants, although cyanobacteria and algae are also present, the latter particularly in the symbiotic associations with fungi known as lichens, common in areas where green plants are absent or scarce. In aquatic habitats, particularly the sea, virtually all photosynthesis is carried out by cyanobacteria and algae, although green plants (seagrasses) are also present in shallow coastal and inland waters.

Photosynthesisers fix carbon and therefore accumulate organic mass or biomass (often measured in dry form – that is the living or once-living tissues of an organism with the water extracted). These organisms are the *primary producers*. The amount of carbon fixed is referred to as *gross primary production* and is typically measured in grammes of carbon per unit of space (area or volume) per unit of time.

The photosynthetic producers also respire to meet their own energetic needs. Under some circumstances, respiration of photosynthesisers over a given period may balance their carbon fixation, so that there is no net accumulation of organic carbon. More normally, however, there is a surplus of fixation over respiration, so that organic matter is accumulated over time. This accumulation is referred to as *net primary production*. The accumulated matter is available to the vast suite of organisms of all sizes, including humans, that cannot synthesise their own organic compounds from an inorganic base or harness energy from inorganic sources. Such organisms are referred to as *heterotrophs*, while photosynthesisers and a few kinds of bacteria that use other energy sources to synthesise organic compounds are referred to as *autotrophs*.

Food webs

An organic product produced by a photosynthesiser may pass through a number of heterotrophs before finally being broken down again to its inorganic constituents. Conventionally this can be viewed as a food-chain. At macroscopic level, a green plant may be eaten by a herbivore – a grasshopper, say – which is eaten by a lizard, which is itself eaten by a hawk, which dies and is disassembled and partially consumed by animal scavengers, with the remainder decomposed by bacteria and fungi.

In reality, this is an enormous over-simplification. The plant will almost certainly have a complex network of symbiotic fungi at its roots, which make use of some of the gross production of the plant but which also provide it with some essential nutrients. The plant itself may shed leaves which are directly broken down by other fungi, protoctists such as slime moulds, and many forms of bacteria. The grasshopper is likely to be parasitised by a host of smaller organisms, some of which are themselves in turn parasitised. It will also support a host of benign microorganisms in its intestine that are themselves constantly growing and reproducing. The lizard may die and decompose and the hawk may eat the grasshopper directly. The overall pattern of feeding relationships thus forms a web of immense complexity in any but the simplest ecosystems.

Each organism in the food web respires, releasing energy which is eventually dissipated in the form of heat, carbon dioxide and water. At each stage, therefore, some carbon is returned to the inorganic part of the carbon cycle. In addition, all living organisms produce waste products, some of which are incompletely metabolised organic compounds. Heterotrophic organisms are also not completely efficient in their appropriation of the organic material they consume, so that some proportion of this is excreted as waste product. These organic wastes are theoretically available to other organisms in the food web. The assimilation efficiency of heterotrophic organisms may be anything from 20% (in the case of some terrestrial herbivores) to 90% (in the case of some carnivores), with the remainder excreted.

Of the amount assimilated, a high proportion is expended as respiration, with the remainder available to add biomass, ie. to enable the organism to grow and reproduce. The proportion available to add biomass is dependent on the organisms involved as well as a range of other factors. It can be as low as 10% or less and as high as 50% or more. This proportion is a measure of the *net growth efficiency* of the organism.

For purposes of ecological analysis, particularly involving productivity estimates, the *gross growth efficiency* is the most commonly used measure. This is simply the product of the assimilation efficiency and the net growth efficiency of a particular heterotroph and is a measure of the proportion of food consumed by that organism that, after excretion and respiration, is ultimately available for its growth. As a very coarse generalisation, a value of 10% is widely used, although it is acknowledged that in terrestrial herbivores the figure is likely to be lower and in planktonic communities and terrestrial carnivores it is likely to be higher. Using the figure of 10% in the example above, for every kilogramme of plant matter eaten by the grasshopper, the latter would add 10 grammes to its body weight. When the grasshopper was eaten by the lizard, this would add 1 gramme to the lizard's body weight, and when the lizard was eaten by the hawk, this would add 0.1 grammes to the hawk's weight. This explains why at the species level, so-called higher predators are rarer than herbivores and in any given area have a lower biomass, while the biomass of primary producers exceeds that of all heterotrophs combined.

Measures of local and global productivity

Primary productivity varies enormously, both spatially and temporally, at all scales. Most obviously, under natural conditions productivity effectively ceases every night. Seasonal

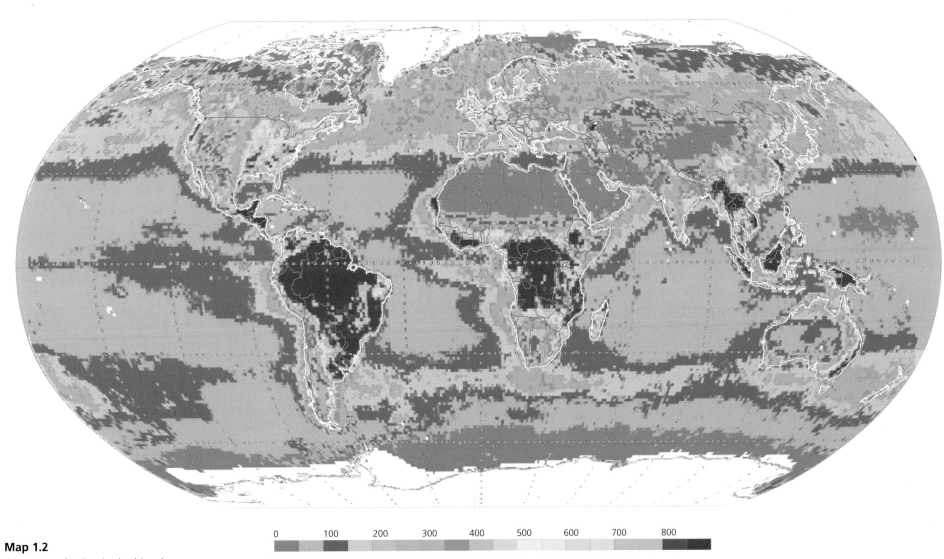

Map 1.2

Primary production in the biosphere

This map shows global spatial variation in annual net primary production (NPP), in g Carbon/m^2/year, calculated from an integrated model of production based on satellite indices of absorbed solar radiation.

Source: Reproduced by permission with modification from Field, C.B., Behrenfeld, M.J., Randerson, J.T. and Falkowski, P. 1998. Primary production of the biosphere: integrating terrestrial and oceanic components. *Science* 281(5374):237-240.

© 1998 American Association for the Advancement of Science.

Table 1.1
Global annual net primary
production
Source: modified from Field *et al.*[4].
Note: ocean data averaged 1978-1983, land
1982-1990, units in petagrams
(1 Pg = 10[15] g).

Biosphere units	NPP (x 10[15] g C)
Ocean	48.5
Terrestrial	56.4
tropical rainforest	17.8
deciduous broadleaf forest	1.5
broadleaf and needleleaf forest	3.1
evergreen needleleaf forest	3.1
deciduous needleleaf forest	1.4
savanna	16.8
perennial grassland	2.4
broadleaf shrubs with bare soil	1.0
tundra	0.8
desert	0.5
cultivation	8.0

variations in most parts of the world are also marked. Productivity, both gross and net, is difficult to measure so that estimates at all scales are subject to considerable uncertainty.

On land one major source of uncertainty is below ground productivity: in natural ecosystems, less than 20% of plant production is typically consumed by herbivores. The remainder enters the soil system, either through the plant roots or as leaf litter. Measuring this portion of terrestrial productivity – probably over 80% of the total – is particularly problematic. In the past there has been a marked tendency to underestimate it. Similarly in the sea it had long been assumed that the nutrient-poor waters of the open ocean were extremely unproductive. It is now known that large populations of extremely small photosynthesising organisms – the so-called picoplankton – form the basis of a surprisingly productive ecosystem in these regions.

Current best estimates of global productivity, based on measurements in particular ecosystems and extrapolation from these using estimates of global extent of those ecosystems, indicate that net primary production on land is of the order of 45-65 x 10[15] g C per year (that is 45-65 thousand million tonnes).

The current best estimate for the sea is around 51 x 10[15] g C per year (that is 51 thousand million tonnes). This gives a global estimate of annual net primary production in the order of 100 x 10[15] g C, or one hundred thousand million tonnes. Gross primary production is estimated to be about twice this.

However, global measures using a somewhat different technique, involving assessments of relative concentration of oxygen isotopes, have indicated that annual gross primary production on land may be greater than 180 x 10[15] g C, while that in the sea is around 140 x 10[15] g C. This would give a global figure of over 320 x 10[15] g C, implying global net primary production of more than 160 x 10[15] g C. This global figure is 60% higher than those global estimates based on summation of individual ecosystem measurements.

Another approach[4] has used a comprehensive set of satellite indices of photosynthetic activity in the ocean and on land, combined with a model of primary production, to generate a more integrated global estimate of NPP. There is considerable spatial and temporal variability, but on average annual NPP on land amounts to around 56 x 10[15] g C, while that in the sea is around 48 x 10[15] g C (see Table 1.1).

The carbon cycle and global biomass estimates

Carbon fixation by photosynthesis forms one crucial step in the carbon cycle[5]. Once fixed, the carbon will remain for a greater or lesser period within living tissues, that is, form part of the planet's biomass. As all cells and individual organisms have a limited lifespan, eventually the carbon will rejoin the non-living carbon pool (see Table 1.2). It may, however, remain as organic carbon compounds for a far greater period than it remained part of the biomass; eg. the woody tissues of Palaeozoic forests were formed several hundred million years ago but remain, fossilised, as a source of coal and oil. Eventually all carbon will recycle through the inorganic pool, as carbon dioxide in the atmosphere, the soil, or dissolved in the sea, or as inorganic carbon compounds (carbonates) in rocks or dissolved in the sea.

The great majority of carbon at any one time lies within the lithosphere, around 80% as carbonate and the remainder as organic carbon compounds. A large proportion of this carbon is effectively inaccessible to the biosphere in the short term but itself participates in the overall carbon cycle, mainly through tectonic activity. As seafloor crust is gradually consumed along subduction plate margins, carbon sediments are taken into the Earth's mantle and later released as carbon dioxide by volcanic and hydrothermal activity. Although the loss of carbon to the mantle is extremely slow, without volcanic activity tectonic processes would eventually exhaust the available carbon pool.

Most of the carbon incorporated in living organisms is associated with green plants, and almost all of this is in the form of cellulose-rich woody tissues. The total terrestrial animal biomass appears to be insignificant in comparison, probably more than two orders of magnitude less. The cyanobacteria and algae that are the primary producers in the ocean are estimated[4] to amount to only 0.2% of the biomass of all primary producers globally, although they generate in the region of half the global NPP (they cycle organic material much more rapidly than land plants, which also sequester large amounts in woody tissues).

The rôle of diversity in the biosphere

The biota play the pivotal rôle in the major biogeochemical cycles, with different groups of organisms (eg. nitrogen-fixing bacteria and photosynthesising plants) mediating different processes. At a very fundamental level, therefore, at least some biological diversity is necessary to maintain the biosphere as it currently operates. However, just how much diversity is needed, and how much, if any, redundancy is built into the system, remain unclear. Indeed, the relationship between biological diversity and a whole suite of ecological measures including stability, resilience and productivity remains incompletely understood[6]. However there is an increasing volume of theoretical and experimental work[7,8] indicating that diversity may play an important rôle in long-term ecosystem functioning.

Human influence on the biosphere

There are now believed to be six billion humans on the planet. A significant proportion of net primary production is diverted away from natural ecosystems to support these. That proportion has been calculated as between 20% and 40% of the terrestrial global total. The former estimate takes account of the global agricultural and natural production used by humans but not the indirect losses, eg. waste or forest loss incidental to extraction[9]; the latter attempts to evaluate all the additional effects of land degradation, pollution and permanent change in land cover[10]. A detailed analysis in one European country[11] suggested that around 50% of NPP is appropriated by humans.

Over enormous areas of the Earth's surface, humans have replaced complex and species-rich natural habitats with simplified modified habitats specialised for agricultural production. Clearance by fire, burning of fuelwood and charcoal, soil cultivation, and fossil fuel use all increase movement of organic carbon into the atmosphere. Global cycling of nitrogen, phosphorus and sulphur has also been perturbed. Application of industrially produced fertilizer has doubled the rate at which

Table 1.2

Estimated global carbon budget and biomass totals
Source: modified from Schlesinger[5].
Note: Biomass figure on land refers to plants.

Total carbon content on earth:		10^{23} g
Amount buried in sedimentary rocks:	organic	1.6×10^{22} g
	carbonate	6.5×10^{22} g
Active carbon pool near surface of which:		$40\,000 \times 10^{15}$ g
Dissolved inorganic carbon in sea		$38\,000 \times 10^{15}$ g
Atmospheric CO_2		750×10^{15} g
Organic carbon in soil		1500×10^{15} g
Biomass on land		560×10^{15} g
Biomass in the sea		$5\text{-}10 \times 10^{15}$ g

nitrogen in fixed form enters the terrestrial cycle, and industrial processes have doubled movement of sulphur from the lithosphere into the atmosphere. Increasing levels of nitrogen and phosphorus lead to shifts in nutrient availability which can cause radical change in natural communities, and sulphur is a major contributor to acidification phenomena.

That human activities may have profound local impacts on natural biota is indisputable. What is now becoming clear is that these activities may also have planet-wide impacts, particularly on climate. Analysis of atmospheric samples trapped in polar ice cores indicates that present-day concentrations of atmospheric carbon dioxide and methane are unprecedented in the past 420,000 years[12]. Although their absolute concentration in the atmosphere is low (CO_2 around 360 and CH_4 at 1.7 parts per million by volume) these two gases play an extremely important rôle in determining atmospheric temperature. It is indisputable that the rise in these gases is a result of human activities and clearly, therefore, these activities are having some impact on global climate. The extent of this impact, particularly when compared with natural climatic fluctuations, remains a subject of great controversy.

References

1 Hutchinson, G.E. 1970. The Biosphere. *Scientific American* 233(3):45-53.
2 Lenton, T.M. 1998. Gaia and natural selection. *Nature* 394(6692):439-447.
3 Parkes, R.J. 1999. Oiling the wheels of controversy. *Nature* 401: 644.
4 Field, C.B., Behrenfeld, M.J., Randerson, J.T. and Falkowski, P. 1998. Primary production of the biosphere: integrating terrestrial and oceanic components. *Science* 282(5374):237-240.
5 Schlesinger, W.H. 1997. *Biogeochemistry: an analysis of global change.* 2nd edition. Academic Press. San Diego, USA and London, UK.
6 Schulze, E.D. and Mooney, H.A. (eds) 1993. *Biodiversity and Ecosystem Function.* Springer, Berlin.
7 Grime, J.P. 1997. Biodiversity and ecosystem function: the debate deepens. *Science* 277: 1260-1261.
8 McGrady-Steed, J., Harris, P.M. and Morin, P.J. 1997. Biodiversity regulates ecosystem predictability. *Nature* 390: 162-165.
9 Wright, D.H. 1990. Human impacts on energy flow through natural ecosystems, and implications for species endangerment. *Ambio* 19(4):189-194.
10 Vitousek, P.M., Ehrlich, P.R., Ehrlich, A.H. and Matson, P.A. 1986. Human appropriation of the products of photosynthesis. *BioScience* 36:368-373.
11 Haberl, H., Erb, K-H., Krausmann, F., Loibl, W., Schulz, N. and Weisz, H. 1999. Colonizing landscapes: human appropriation of net primary production and its influence on standing crop and biomass turnover in Austria. IFF-*Social Ecology Papers*, No.57. Institute for Interdisciplinary Research of Austrian Universities, Vienna. Also see website: http://www.cloc.org/conference/presentations/in4/npp-abstract.htm.
12 Petit, J.R. *et al.* 1999. Climate and atmospheric history of the past 420,000 years from the Vostok ice core, Antarctica. *Nature* 399(6735):429-436.

Suggested introductory source

Hutchinson, G.E. 1970. The Biosphere. *Scientific American* 233(3):45-53.
Also see other papers in the same special issue on the biosphere, probably still the best introduction to most of the concepts.

2

THE DIVERSITY OF ORGANISMS

Biological diversity may be addressed at many levels, from genes to ecosystems, but for most practical purposes the diversity of organisms is central, and species diversity appears the most useful general measure.

Systematics aims to define species and sort them into a hierarchy of named groups congruent with the branching pattern of evolution. There is no single definition of what a species is, no foolproof method to recognise one, and species-level taxonomy can change with new data and new approaches. For better known organisms, reasonable consensus allows useful assessment of species diversity in an area or the world. About 1.75 million of the species that exist have been described and named, but the majority remains unknown. The global total might be ten times greater, many of these being undescribed insects.

The top levels in the taxonomic hierarchy are intended to identify the major branches in evolution, each holds a group of species sharing basic patterns of form and function, and so provide another indicator of organismal diversity. The deepest division is between bacteria (prokaryotes) and other organisms (eukaryotes), divided into protoctists, animals, fungi and plants. All species known are accommodated in about 100 major groups (phyla).

BIODIVERSITY

Biological diversity is an imprecise term that may refer to diversity in a gene, species, community of species, or ecosystem; it is often contracted to *biodiversity*, and used broadly with reference to total biological diversity in an area or on the Earth as a whole[1-3]. Pest resistance in different rice varieties, or the number and kinds of species present in an area of forest, or the changing quality of natural grassland, are all aspects of biodiversity, but whatever the context, the diversity of organisms is central.

The diversity of organisms may be addressed at two levels, these representing the lowest and the higher levels of the taxonomic hierarchy. First, *species diversity*, or the number and variety of individual species, and second, *taxonomic diversity*, or the number and variety of major groups into which species may be placed. Of the higher taxonomic levels, that of phyla (singular: phylum) is perhaps the most significant, this being the topmost taxonomic category in regular use, with one phylum for each lineage of species sharing a fundamental similarity in form, organisation, and mode of life. While approaching two million species are known on Earth, and ten times this total may exist, as yet unknown, the known species can in current opinion[4] all be accommodated in around 100 phyla.

For most practical purposes, species are the appropriate targets for biodiversity research and management action, and species diversity, rather than diversity at other levels, such as genes or ecosystems, is the most generally useful measure of biodiversity. Although genes provide the blueprint for construction of organisms, and their diversity is clearly fundamental, genes are expressed only through the form, function and differential survival of organisms. Ecosystems are essentially manifestations of the interactions of organisms with each other and the non-living environment. Neither genes nor ecosystems can be manipulated or managed without attention being given to the requirements of organisms. Species are the entities in nature that adapt and evolve, that occupy ecological space, and that become extinct.

EVOLUTION AND SYSTEMATICS

A basic principle in evolution is that just as new individuals arise from ancestral individuals, so new populations arise from existing populations, and ultimately new species arise from existing species. The chief mechanism by which this occurs is believed to be reproductive isolation. Most obviously, an existing single population of interbreeding individuals may become geographically split into two or more separate populations through, for example, in the case of terrestrial species, sea-level rise dividing a single large island into a number of smaller islands. The genetic makeup of these isolated populations will diverge, mainly through natural selection acting on them, but probably also through other mechanisms. This genetic divergence will be manifested in various ways, physically, physiologically and behaviourally. If the period of isolation continues for long enough, the populations will diverge enough that they can be regarded as separate species. Each one of these species may itself in turn give rise to other species in due course, although some will die out without giving rise to any progeny. The surviving descendant species may themselves give rise to two or more new species and so on through the long march of evolutionary time. The result of this is a branching tree-like structure – a phylogenetic tree – rooted in the distant past (rarely in plants and animals, but probably more frequently in bacteria and protoctists, two separate branches may combine into one). If it is assumed that all life on Earth had a common origin in the distant past (see Chapter 3), then all existing organisms form the topmost extremities of a vast and unimaginably complex single phylogenetic tree.

The rôle of systematics[5,6] is twofold. First, is to name the immense variety of different sorts of organisms that exist. Second, is to try to elucidate the relationships between all these different organisms, that is effectively to work out exactly where they stand in the phylogenetic tree. Because the true evolutionary events that generated the overall phylogenetic tree are lost in history, the relationships between organisms have to be inferred from the evidence to hand. The most important forms of evidence are the characters of organisms, both living and fossil. These characters may be genetic, morphological, biochemical or behavioural. In systematics, individuals are grouped on the basis of similar characters held in common.

Methods to reconstruct phylogeny generally use two working assumptions: that species sharing a large number of characters are likely to be related, and that species sharing some uniquely complex and specialised feature are likely to be more closely related than species not possessing this feature. Given a classification congruent with phylogeny, the taxonomic hierarchy becomes a device to store information on hypotheses about evolutionary history.

The traditional output of systematics has consisted of species descriptions or revisions, or lists of species in a given group (which if formally named is a *taxon*) with hypotheses of their evolutionary relationships, or checklists of all species in some higher taxon in a country or even the world. Systematics provides the basic framework for the whole of biology, and is a fundamental discipline of biodiversity studies. The term *taxonomy* is a near synonym of systematics, but tends to imply emphasis on classification and names.

GROUPS AND NAMES

In the current system for naming species (nomenclature), each has a two-part scientific name (binomial), based on Latin or latinised Greek, comprised of the genus name (eg. *Vipera*) and specific epithet (eg. *berus*). In technical works, and to clarify in which sense a name is being used, the author of the specific epithet (eg. Linnaeus, 1758) may be given after the binomial. By convention, both parts of the binomial are italicized when printed, and the author name is shown in parentheses if the species was originally put into a different genus. Similar species (eg. the European adder *Vipera berus* and asp viper *Vipera aspis*) are grouped together in the same genus (*Vipera*), similar genera in families (Viperidae), families in orders (Serpentes), orders in classes (Reptilia), and classes in phyla (Craniata or Vertebrata) up to the highest level, the kingdom, of which five are generally recognized at present (bacteria, protoctists, animals, fungi, plants). An organism can only be assigned to a single species, genus, family, etc., and the taxonomic system forms a hierarchy with each lower taxonomic level being nested entirely within each increasingly inclusive higher level.

Groups such as mammals or snakes, that because of shared unique characters are considered to contain all the living descendants of a common ancestor, are called *monophyletic* groups. Groups with no shared unique characters but only unspecialised or non-unique characters in common, are termed *paraphyletic* groups. These typically are groups of related species left over after one or more clearly monophyletic lineages that evolved within the group have been recognised and named. Familiar examples are fishes (the craniate vertebrates without the unique features of tetrapods) and reptiles (the amniote vertebrates without the unique features of birds or mammals). The kingdom Protoctista is such a group. Groups defined on characters that appear to have evolved more than once are called *polyphyletic* groups. Although the goal of most systematists is to recognise only monophyletic groups in order to be able to retrieve evolutionary relationships from a classification, many paraphyletic groups, such as 'fishes' and 'reptiles', continue to be very widely used in everday practice, and it will never be possible to establish a completely resolved evolutionary classification.

Because hypotheses about relationships are always subject to revision as new information becomes available, or existing data are reinterpreted, the taxonomy of species is not fixed.

Although rules of nomenclature exist, the fact that species names are liable to change can cause confusion, for example when, as is commonly the case, conservation legislation uses a name no longer current.

SPECIES CONCEPTS AND DIVERSITY ASSESSMENT

Despite the importance of 'the species', there is no unequivocal and operational definition of what they are and how they can be recognised[3,5,7]. There are at least half a dozen different definitions of what a species is, often differing in subtle and mainly theoretical ways, but much of the existing body of systematic knowledge has been built up around elements of the *biological species concept*. This defines a species as a population of organisms that actually or potentially interbreed in nature, and that are reproductively isolated by morphological, behavioural or genetic means, from other such groups. It is, however, applicable only in organisms where sexual reproduction is the norm.

In most real cases, especially where all the systematist has to hand is a collection of preserved specimens, whether criteria concerning reproductive isolation are met or not cannot in fact be tested, but an experienced worker will come to hold some particular level of morphological or other difference as deserving of species status. Where there is good evidence from fieldwork and geographic data attached to specimens that two somewhat similar populations occur in the same locality (are sympatric) but maintain their differences, they may be presumed not to interbreed, and will be treated and named as species. The magnitude of the differences between them can then become a benchmark against which other putative species populations may be assessed.

Different taxonomists will often use different criteria for the same group of organisms, so that one specialist may regard a group of fundamentally similar populations as a single species, whereas another will treat each smaller recognisable population as a separate species. In the latter case, often termed the *phylogenetic species concept*, the assumption is that each distinct lineage once established on its own evolutionary course is *de facto* a separate species.

Different characters and criteria are used to classify species in different groups of organisms. For example, defining species of fungi and species of bird rely on very different taxonomic characters, and demand narrowly-specialised taxonomists. Some organisms are difficult or logically impossible to accommodate in any species concept involving criteria predicated on the existence of sexual reproduction. Many higher plants, for example, have individuals that reproduce vegetatively instead of sexually, so that lineages consist of genetically identical clones; others are prone to hybridisation. Dissimilar bacteria can readily receive genetic material, through direct entry of genes from the fluid environment, or from viruses or other bacteria[4]. Strains of bacteria are usually defined on the reaction of the cell wall to particular cytological stains (eg. Gram-negative and Gram-positive groups) and on biochemical properties of colonies, rather than on clues to reproductive isolation, and the biological species concept simply cannot be applied to them.

Such factors mean that even if the species is the basic currency in which biodiversity is counted, the value of 'one species' is not equivalent across all groups of organisms. The use of different species concepts by different systematists can make a very large difference to the number of species recognised in a group, and to complications in nomenclature (these will both affect the outcome of biodiversity inventory, an important application of systematics).

It is not usual, however, to have a large number of taxonomists working on exactly the same group of organisms, so that while the species level taxonomy of organisms is in a continual state of flux, it is not subject to radical and wholesale change. The key point appears to be that units broadly corresponding to biologist's concept of 'the species' do indeed exist in nature, and they to an extent define themselves through their reproductive behaviour. It has thus been possible to reach some measure of consensus on species-level classifications, particularly amongst well-studied groups of larger organisms such as terrestrial vertebrates. This makes it possible to estimate and make comparisons, often coarse but usually meaningful, of the number and kinds of species in different sites, areas or countries.

NUMBERS OF LIVING SPECIES

From a practical point of view it is more important to know how many species, and which ones, occur in some spatially restricted area, such as a protected area or a country, than in the world overall. However, proper evaluation of each local situation requires some knowledge of the wider context, and where the goal is maintenance of global biodiversity in the face of increased risk, it is clearly important to have, if not an accurate count, certainly a sound appreciation of the full baseline range of diversity. This requires both an estimate of the number of known valid species, and an estimate of the number of unknown species, neither of which is readily available.

Although the goal of systematics is to recognise and name species, and to maintain an ordered body of information on names and associated biological data, there is no master catalogue of all known species. Developing such a resource has only become feasible with advances in communication and information technology during the present decade. However, while many systematic data, in the form of checklists and museum catalogues, are now available in digital form over the Internet, and more will become so (see eg. Species 2000[8]), a harmonised catalogue in this format of all known species remains a distant prospect.

The number of known species can be estimated by collating data from systematists and the taxonomic literature; although many species names are synonyms (ie. different names inadvertently applied to the same species), this can be done with reasonable precision for more familiar and well-reviewed groups of species. Recent calculations of this kind suggest that around 1.75 million of the species that exist have been discovered, collected, and later named by systematists[2,9].

Any estimate of how many undiscovered and hence undescribed species are likely to exist in any given group, and in the biosphere overall, involves substantial uncertainty[2,9]. In taxonomic groups where individuals are readily visible, popular or economically important, and subject to sustained systematic attention, eg. mammals and birds, the number of known species is certainly very close to the total number of species in the group that exist. On average around 25 and 5 new species of mammals and birds respectively have been described in recent years[9], and changing systematic opinion on which populations should be regarded as separate species and which should not, rather than completely new discoveries, is the major source of change in the number of named species.

The converse applies to groups whose individuals are small, difficult to collect, obscure and of no popular interest, eg. many groups of invertebrate animals. In some cases, where new sampling and collection methods have been used, unexpectedly large numbers of new species have been found (eg. tropical forest canopy insects and marine sediment nematodes), and if findings from such local work are extrapolated to global level the total number of species calculated to exist is many orders of magnitude greater than the number actually known. Expert opinion suggests that in many taxonomic groups the number of known species is likely to be significantly less than the total number of species that exist, and in some cases to form only a small proportion of the possible total. Most undescribed species are likely to be insects, such as tropical forest beetles; many will be fungi, and many forms will be bacteria.

Frequently there are so few systematists actively working on a

Table 2.1

Estimated numbers of described species, and possible global total.

Notes: this table presents recent estimates of the number of species of living organisms in the five kingdoms recognised[4], and in some selected groups within them. Vertebrate classes are distinguished because of the general interest in these groups. The described species column refers to species named by taxonomists. These estimates are inevitably incomplete, because new species will have been described since publication of any checklist and more are continually being described; most groups lack a list of species and numbers are even more approximate. The estimated total column includes provisional working estimates of described species plus the number of unknown and undescribed species; the total figure may be wildly inaccurate. Only a small selection of animal phyla is shown, but the total figures in the bottom row are for all species in the five kingdoms.

Sources: data mainly from[2,9]; vertebrates from individual sources indicated.

Kingdoms	Phyla	Described species	Estimated total
Bacteria		4 000	1 000 000
Protoctista		80 000	600 00
Animalia			
	Craniata (vertebrates) total	52 000	55 000
	Mammals[11]	4630	
	Birds[12]	9946	
	Reptiles[13]	7400	
	Amphibians[14]	4950	
	Fishes[15]	25 000	
	Mandibulata (insects & myriapods)	963 000	8 000 000
	Chelicerata (arachnids etc)	75 000	750 000
	Mollusca	70 000	200 000
	Crustacea	40 000	150 000
	Nematoda	25 000	400 000
Fungi		72 000	1 500 000
Plantae		270 000	320 000
TOTAL		1 750 000	14 000 000

group that the number of named species appears to be limited mainly by the rate at which collected specimens waiting on museum shelves can be studied and described, and changing opinion on which populations are separate species is insignificant.

Recent estimates of the numbers of known and possibly existing species in the world biota are given in Table 2.1. These are mostly big numbers, and the fossil record suggests that overall diversity has been increasing for some 600 million years up to the very recent past, but the numbers themselves are without significance except in a wider context. The context is partly defined by issues such as the following. What effects are exerted on global biodiversity by exponential increase in human numbers? Have species recently become extinct at a faster than average rate? If so, do accelerated losses matter to humans, or to the persistence of the biosphere?

THE DIVERSITY OF LIVING ORGANISMS

If the total number of species that exist represents the first key parameter of global biodiversity, the diversity of different *kinds* of species is arguably the second. As noted above, defining species involves some subjectivity, but this is far greater when defining groups of different kinds of species. These may be defined on the basis of some highly distinctive combination of form, organisation, biochemistry, and mode of life; ie. not variations on a theme, like species within a genus or families within a class, but entirely different themes.

Exactly what constitutes a different theme is of course debatable, but in general very highly distinctive groups are placed in one of the higher taxonomic categories (ie. kingdom, phylum, or class, in current systems), and if the taxonomic hierarchy is to a degree congruent with the broad branching pattern of past evolution, then each such group should consist of all or some of the members of one evolutionary lineage of organisms. Clearly, the number of phyla, for example, depends on subjective systematic opinion as to where the line between kinds of species is drawn, and opinion changes with the research tools and the data available, and with different attitudes to classification and the value of different evidence.

Until van Leeuwenhoek observed microorganisms through a primitive microscope in the late 17th century, humans had been aware only of organisms visible to the naked eye (macroscopic) and regarded all organisms as either plants or animals. At the end of the nineteenth century, with improved cytological techniques and new views on evolution, a third kingdom of organisms (Protista) was recognised for bacteria lacking an

Table 2.2
Key features of the five kingdoms of living organisms
Source: after Margolis and Schwartz[4].

Bacteria	Prokaryotic microorganisms. Reproduce asexually by cell splitting, or produce genetic recombinants without any fusion of cells by accepting genes from other bacteria, or from the fluid medium, or through viruses, independent of cell reproduction. Metabolically uniquely versatile; key mediators of major biogeochemical cycles. Permeate the entire biosphere, including other organisms, although dominant only in exceptional habitats.
Protoctista	Mainly microorganisms. Possess the features of eukaryotes, but lack the characteristics of fungi, animals or plants. Extraordinary variation in life cycle and morphology. Early evolution probably based on symbiotic relationships between different kinds of bacteria forming lineages of composite organisms resulting in the protoctist grade of organisation. Include photosynthetic algae (formerly classed as plants) and heterotrophs (formerly called 'protozoa').
Animalia	Multicellular, mainly macroscopic, eukaryotes. Reproduce through fertilization of an egg by a sperm, the fertilized egg (now diploid, ie. a duplicate set of chromosomes) is called a zygote and (except sponges) this forms a characteristic hollow multicelled *blastula* from which the embryo develops. All heterotrophic.
Plantae	Multicellular macroscopic eukaryotes. The fertilized egg develops into a multicelled embryo different from blastula of animals. Alternate spore-producing generations and egg or sperm-producing generations. Virtually all are terrestrial photosynthetic autotrophs.
Fungi	Mainly multicellular, micro- to macroscopic eukaryotes. Fungi develop directly without an embryo stage from resistant non-motile haploid (one set of chromosomes) spores that can be produced by a single parent. Sexual reproduction also results in haploid spores. Most consist of network of threadlike hyphae. Heterotrophs, vital to decomposition processes; form mycorrhizal symbioses with plants, facilitating exchange of soil nutrients.

organised cell nucleus (Monera). Contemporary views further refine the distinction between major groups, but differ to some extent from each other. For example, some workers[10] attach highest significance to analysis of differences in ribosomal RNA (ribonucleic acid) sequence, resulting in a tripartite division of living organisms: two containing bacteria (Archaebacteria and Eubacteria), and one containing all eukaryotes (Eukaryota), among which fungi, animals and plants form insignificant and scarcely discernible clusters. Others[4] have attempted to evaluate a broad range of biological characters in addition to molecular sequence data, and recognise five kingdoms: one containing bacteria (both archaebacteria and eubacteria), one containing protoctists (Protoctista – essentially the eukaryotes that are not fungi, animals or plants), and separate kingdoms Fungi, Animalia, and Plantae (see Table 2.2). Although these systems differ in approach and in taxonomic outcome, the differences are mainly of emphasis. Both make clear the enormous range of diversity within bacteria; both stress the difference between prokaryotes (bacteria) and eukaryotes (all other organisms).

The prokaryote-eukaryote division appears to be the most fundamental evolutionary discontinuity between living organisms[4]. In prokaryotes, the genetic material is free within the cell. In eukaryotes, the genetic material is linked to proteins and organised into chromosomes that are packed within a membrane-bounded cell nucleus. There are several other profound differences. One of the most important is that in eukaryotes the enzymes needed to extract energy from organic molecules are organised into discrete membrane-bounded organelles (mitochondria) within the cell, and in the eukaryotes that photosynthesise (plants, some protoctists) the pigments and enzymes needed to fix solar energy are also in discrete organelles (chloroplasts) within the cell. There is good evidence that both mitochondria and chloroplasts represent vestiges of former symbiotic bacterial cells that more than a billion years ago became fully incorporated within other cells to form eukaryotes at the protoctist grade of organisation.

The only known 'organisms' that are not cells, or assemblages of cells, are viruses. They exist on the very boundary of most definitions of life. Consisting only of nucleic acids and protein, they are much smaller than the smallest bacteria, they can only replicate inside other living cells, and they are totally inert outside other cells, when they can survive for years in a crystallized state. Each type of virus may be more closely related to the organism in which it grows than to other viruses[4]. They are not discussed elsewhere in this book.

Aiming to provide an overview of global organismal diversity, and following the most comprehensive recent synthesis[4], each of the five kingdoms of living organisms is very briefly characterised in Table 2.2. The remainder of the chapter consists of an outline of key features of all the 96 phyla recognised.

The symbols associated with each phylum name indicate whether the species occur in marine, inland water or terrestrial habitats. Where more that one symbol is shown, this does not mean that species are equally distributed between them. In some cases, the text notes the principal habitat. For parasitic forms the symbol refers to the host habitat.

marine   freshwater   terrestrial

References

1 World Conservation Monitoring Centre. 1992. Groombridge, B. (ed). *Global biodiversity: status of the Earth's living resources*. Chapman & Hall, London.
2 United Nations Environment Programme. 1995. Heywood, V. (ed). *Global biodiversity assessment*. Cambridge University Press, Cambridge.
3 Gaston, K.J. (ed.). 1996. *Biodiversity: a biology of numbers and difference*. Blackwell Science Ltd., Oxford.
4 Margulis, L. and Schwartz, K.V. 1998. *Five kingdoms. An illustrated guide to the phyla of life on earth*. 3rd edition. W.H. Freeman and Company, NY.
5 Vane-Wright, R.I. 1992. Systematics and diversity. In, World Conservation Monitoring Centre. *Global biodiversity: status of the Earth's living resources*. Pp. 7–12. Chapman & Hall, London.
6 Minelli, A. 1993. *Biological systematics: the state of the art*. Chapman & Hall, London.
7 Vane-Wright, R.I. 1992. Species concepts. In, World Conservation Monitoring Centre. *Global biodiversity: status of the Earth's living resources*. Pp. 13–16. Chapman & Hall, London.
8 Home page of Species 2000 project. See: http://www.sp2000.org/default.html
9 Hammond, P. 1992. Species inventory. In, World Conservation Monitoring Centre. *Global biodiversity: status of the Earth's living resources*. Pp. 17–39. Chapman & Hall, London.
10 Woese, C.R., Kandler, O. and Wheelis, M.L. 1990. Towards a natural system of organisms: proposal for the domains Archaea, Bacteria and Eukarya. *Proceedings of the National Academy of Sciences, USA* 87:4576-4579.
11 Wilson, D.E. and Reeder, D.M. (eds). 1993. *Mammal species of the world: a taxonomic and geographic reference*. 2nd edition. Smithsonian Institution Press, Washington D.C. and London.
12 Sibley, C.G. 1996. *Birds of the world* 2.0. Thayer Birding Software, Cincinnati.
13 http://www.embl-heidelberg.de/~uetz/LivingReptiles.html.
14 Duellman, W.E. 1993. *Amphibian species of the world: additions and corrections*. Special Publication No. 21, University of Kansas Museum of Natural History, Lawrence, Kansas.
15 Eschmeyer, W.N., Ferraris, C.J., Mysi Dang Hoang, and Long, D.J. 1998. *A catalog of the species of fishes*. Vols 1–3. California Academy of Sciences, San Francisco. Also see: http://www.calacademy.org/research/ichthyology/species/.

Suggested introductory sources

Gaston, K.J. and Spicer, J.I. 1998. *Biodiversity: an introduction*. Blackwell Science Ltd., Oxford.
Margulis, L. 1998. *The symbiotic planet: a new look at evolution*. Weidenfeld & Nicolson, London.

Euryarchaeota
Methanogens and halophils

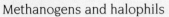

Diversity	Size	Nutrition	Mode of life
	Microscopic	Methanogens are chemoautotrophs halophils are photosynthetic	Free-living or symbiotic, inhabiting the intestines of animals

Euryarchaeota share similarities in ribosomal RNA sequence but consist of two very different groups. Methanogens cannot tolerate oxygen (are obligate anaerobes) and free-living forms tend to occur in swamps, bogs and estuary sediments; many others live in the gut of herbivorous animals, from termites to cows. They are chemoautotrophs that obtain energy by reducing CO_2 and oxidising H_2 to produce CH_4 (methane) and H_2O. They are responsible for liberation of organic carbon from sediments into the atmosphere where it can be reused, involving around 2 billion tonnes of methane annually. Halophils live in extremely salty or highly alkaline environments such as soda lakes worldwide; they respire oxygen.

Crenarchaeota
Thermoacidophils

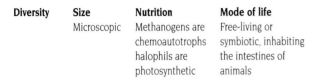

Diversity	Size	Nutrition	Mode of life
	Microscopic	Chemoautotrophs or heterotrophs	Free-living

Bacteria adapted to hot, acidic sulphur-rich environments often found in hot springs and around submarine vents. *Pyrolobus* grows at temperatures of 113°C. *Sulfolobus* tolerates temperatures up to 90°C and may die if the temperature drops below 55°C; it also tolerates highly acid conditions (pH of less than 1, or stronger than concentrated sulphuric acid).

Proteobacteria
Purple bacteria

Diversity	Size	Nutrition	Mode of life
High but imprecisely known	Microscopic	Includes virtually all nutritional modes known	Major parasites and symbionts; some free-living aquatic forms

An enormous and extremely varied group of bacteria including many disease-causing forms (eg. *Salmonella*, a cause of food-poisoning, and *Neisseria*, which causes gonorrhea) and symbionts such as *Escherichia coli*. Proteobacteria show a great range of physical structure and metabolic activity; the group includes heterotrophs, chemotrophs, chemoheterotrophs, chemolithoautotrophs, photoautotrophs, photoheterotrophs, methylotrophs, hydrogen-oxidizers, and sulphide-oxidizers. Many are facultative aerobes, respiring oxygen when this is available but able to survive by respiring eg. nitrogen (N_2) or sulphate (SO_4^{2-}) when not. Responsible for a significant proportion of atmospheric nitrogen fixation.

Spirochaetae
Spirochaetes

Diversity	Size	Nutrition	Mode of life
12 genera	Microscopic	Heterotrophic	Free-living or symbiotic, some parasitic

Spiral-shaped bacteria occurring in marine and freshwater habitats, including deep muddy sediments, and in animals, where many are major parasites or symbionts. Some respire gaseous oxygen, others are poisoned by it. *Treponema pallidum* causes syphilis and yaws, and *Leptospira* causes leptospirosis.

Cyanobacteria
Blue-green bacteria and Chloroxybacteria

Diversity	Size	Nutrition	Mode of life
Many thousands of types	Microscopic but relatively large	Photosynthetic	Free-living

Photosynthesising bacteria, present in a great variety of habitats. Until recently called 'blue-green algae' and considered to be plants. These bacteria dominated the landscape in the Proterozoic eon between 2600 and 545 million years ago. *Prochlorococcus* occurs at the base of the photic zone throughout the world's oceans and may be one of the commonest bacteria. Many fix atmospheric nitrogen. Form reef-like stromatolites in some shallow water marine environments.

Saprospirae
Fermenting gliders

Diversity	Size	Nutrition	Mode of life
	Microscopic	Heterotrophic	Free-living, or symbiotic in animals

Anaerobic fermenters restricted to anoxic environments, but requiring organic compounds. Some free-living, some (eg. *Bacteroides*) inhabit intestinal tract of vertebrates, including humans, in enormous numbers.

Chloroflexa
Green nonsulphur phototrophs

Diversity	Size	Nutrition	Mode of life
Three genera	Microscopic	Photosynthetic	Free-living

Anaerobic filament-forming bacteria known from sulphur-rich habitats such as hot springs. Whilst these forms are typically photosynthetic, *Chloroflexus* can also grow heterotrophically in the dark.

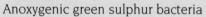

Chlorobia
Anoxygenic green sulphur bacteria

Diversity	Size	Nutrition	Mode of life
	Microscopic	Photosynthetic	Mainly free-living, some symbiotic with other bacteria

Phototrophic obligate anaerobes, inhabiting sunlit sulphide-rich habitats, particularly anaerobic muds. Some are tolerant of extremely high or low temperatures and salinities. Most use hydrogen sulphide or sodium sulphide in photosynthesis, instead of water, releasing sulphur instead of oxygen. Others form symbiotic associations with oxygen-respiring heterotrophic bacteria.

Aphragmobacteria
Mycoplasmas

? ?

Diversity	Size	Nutrition	Mode of life
	Microscopic	Heterotrophic	All symbionts, some parasitic

Very small bacteria, lacking a cell-wall, widespread in insect, plant and vertebrate tissues. Normally benign, but pathogenic in some conditions, and responsible for some forms of pneumonia and tick-borne diseases (eg. E*hrlichia*).

Thermotogae
Thermophilic fermenters

Diversity	Size	Nutrition	Mode of life
	Microscopic	Heterotrophic	Free-living

Recently discovered obligate anaerobic bacteria known from submarine hot vents, terrestrial hot springs and subterranean oil reservoirs. Highly heat tolerant, living at temperatures of 50°C to 80°C. Ferment sugar and other organic compounds.

Pirellulae
Proteinaceous-walled bacteria and their relatives

Diversity	Size	Nutrition	Mode of life
	Microscopic	Heterotrophic	Mostly aquatic in freshwaters, some symbionts, some parasitic

Diverse bacteria with proteinaceous cell walls, mostly obligate aerobic heterotrophs living in freshwaters. *Chlamydia* is parasitic, inhabiting animal cells and with apparently no independent means of producing energy. C. *psittaci* causes psittacosis, C. *trachomatis* causes trachoma blindness.

Actinobacteria
Actinomycetes, actinomycota and their relatives

Diversity	Size	Nutrition	Mode of life
	Microscopic	Heterotrophic	Some free-living, some symbionts

A large and diverse group of heterotrophic unicellular rod-shaped bacteria (coryneforms), and filamentous, multicelled bacteria (actinomycetes) originally regarded as fungi. Some form pathogenic lesions on skin, others are found in leaf litter; some of the latter can break down cellulose. *Frankia* is a nitrogen-fixing symbiont in plants. *Streptomyces* produces streptomycin and other antibiotics.

Deinococci
Heat- or radiation-resistant bacteria

Diversity	Size	Nutrition	Mode of life
	Microscopic	Heterotrophic	Free-living

Spherical, heterotrophic, obligate or facultative aerobic bacteria highly resistant to heat (*Thermus*) or radiation (*Deinococcus*). Most metabolise sugars. *Thermus aquaticus*, isolated from hot springs in Yellowstone National Park, USA, is the source of Taq polymerase used in the Polymerase Chain Reaction technique.

Endospora
Endospore-forming and related bacteria

Diversity	Size	Nutrition	Mode of life
	Microscopic	Heterotrophic	Many symbionts and parasites

A very large, important and varied group of heterotrophic bacteria, some obligate anaerobes, others facultative or obligate aerobes. Most form endospores (propagules within the parent cell resistant to heat and desiccation). Some can break down lignin and cellulose, others are fermenters, breaking down sugars to produce compounds such as lactic acid and ethanol. Some, such as *Streptococcus* are associated with infections.

Archaeoprotista
Amitochondriates

Diversity	Size	Nutrition	Mode of life
	Single-celled	Mostly heterotrophic	Free-living in aquatic habitats, and symbionts, often parasitic, in animals

Anaerobic and lacking mitochondria. Many forms are parasitic or symbiotic in the intestines of animals, eg. wood-eating termites and cockroaches. *Giardia* causes giardiasis in humans.

Granuloreticulosa
Foraminifera and reticulomyxids

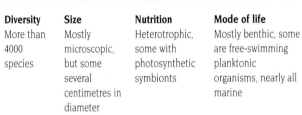

Diversity	Size	Nutrition	Mode of life
More than 4000 species	Mostly microscopic, but some several centimetres in diameter	Heterotrophic, some with photosynthetic symbionts	Mostly benthic, some are free-swimming planktonic organisms, nearly all marine

Foraminifera have multi-pored shells (tests) composed of organic matter reinforced with minerals (sand or calcium carbonate). Important in marine food webs. Many marine sediments are composed largely of foraminifera and fossil species, about 40 000 of which are known, are important in stratigraphy. Some of latter, eg. *Nummulites*, up to 10 cms diameter. Reticulomyxids lack shells and form soft reticulate masses.

Rhizopoda
Amastigote amoebas and cellular slime-moulds

Diversity	Size	Nutrition	Mode of life
Around 200 species	Single- or multicellular	Heterotrophic	Mainly benthic in aquatic habitats, or in water film on land; some amoebas are parasitic

Often abundant in soil, where cyst-forming types highly resistant to desiccation. *Entamoeba histolytica* is responsible for some forms of amoebic dysentery in humans. Some amoebas construct a coating (test) from detritus and these have a fossil record from Palaeozoic times; some fossil acritarchs (see Chapter 3) may represent testate amoebas. Cellular slime moulds typically exist amid decaying vegetation, on logs, or bark, and feed by enveloping bacteria and protoctists. The reproductive form of slime moulds is an aggregation of cells each formerly having independent existence. Key experimental organisms in studies of cell communication and differentiation.

Microspora
Microsporans

? ?

Diversity	Size	Nutrition	Mode of life
About 800 species	Single-celled	Heterotophic	Intracellular parasites of animals

Anaerobic and lacking mitochondria. Frequently form large single-cell tumors in host animals; some highly pathogenic, some harmless. *Nosema* causes pebrine, a disease of silkworm larvae.

Xenophyophora
Xenophyophores

Diversity	Size	Nutrition	Mode of life
42 known species	Sometimes several centimetres diameter	Heterotrophic	Benthic marine forms

Little-known bottom living marine protoctists from deep sea and abyssal regions. Make shells (tests) from detritus (eg. foraminiferan shells, sponge spicules). Xenophyophores are the most abundant macroscopic organisms in some deep-sea communities, with several individuals per square metre. Some acritarch fossils (see Chapter 3) may have been xenophyophores.

Myxomycota
Plasmodial slime-moulds

Diversity	Size	Nutrition	Mode of life
500 species	Microscopic cells but macroscopic, up to several centimetres, in plasmodial form	Heterotrophic	Free-living organisms in damp terrestrial habitats

Similar in some respects to cellular slime moulds (Rhizopoda). Myxomycotes have a sexual stage, and the plasmodium that develops from the zygote is multinucleate. Fruiting stage develops in drier conditions. Feed by enveloping bacteria and protoctists growing on decaying vegetation. Key organisms in studies of cell motility. *Cercomonas* is a small swimming amoeboid organism that does not form plasmodia.

Dinomastigota
Dinoflagellates

Diversity	Size	Nutrition	Mode of life
4000 species	Single-celled, up to 2 mm, occasionally colonial	Some are heterotrophic, others photosynthetic	Mostly free-living marine plankton

Typically planktonic; some symbiotic with or live on marine animals or seaweed, some occur in freshwaters. Many adopt very different forms at different life stages. *Gymnodinium microadriaticum* is the most common intracellular photosynthesising symbiont in corals. Some produce powerful toxins and are important cause of fish mortality (eg. *Pfeisteria piscicida*) and may form toxic 'red tides' (eg. *Gonyaulax tamarensis*). Ciguatera poisoning in humans is caused by accumulations of dinoflagellate toxins in fishes and marine invertebrates. Many species (eg. *Noctiluca*) are bioluminescent.

Ciliophora
Ciliates

Diversity	Size	Nutrition	Mode of life
10 000 species	Mostly microscopic and single-celled	Mostly heterotrophic	Mostly free-living, some symbionts or parasites

Although most are unicells, a few multi-cellular forms resembling slime-moulds exist. Ciliates feed on bacteria or absorb nutrients from the surrounding medium. Entodiniomorphs live as symbionts in the stomachs of ruminants; the parasite *Balantidium* sometimes causes disease in humans. The free-living *Paramecium* and *Stentor* are well-studied and much used in research and education.

Apicomplexa
Sporozoa

Diversity	Size	Nutrition	Mode of life
4600 species	Single-celled	Heterotrophic	Symbiotic with or parasitic on animals

Many of these spore-forming protoctists are bloodstream parasites with very complex life-cycles. Coccidians are the best known group because infection often causes serious or fatal intestinal tract infection. Many, eg. *Eimeria*, infect livestock; *Isospora hominis* is the only direct coccidian parasite of humans. *Plasmodium* causes malaria, probably at present the most important single infectious disease affecting humans.

Haptomonada
Prymnesiophytes

Diversity	Size	Nutrition	Mode of life
	Mostly single-celled	Photosynthetic	Aquatic, with free-living and resting stages

Most are marine, some occur in freshwaters. Two distinct life stages: a motile, golden-coloured alga and a resting, coccolithophorid stage, covered in distinctive calcareous plates (*coccoliths*). Coccolithophorids are important in calcium carbonate sediments and in stratigraphic studies. Some are endosymbionts of radiolaria (Actinopoda).

Cryptomonada
Cryptophyta

Diversity	Size	Nutrition	Mode of life
	Single-celled	Some heterotrophic, others photosynthetic	Most are free-living

Cosmopolitan in moist areas. Most cryptomonads are flattened elliptical free-swimming cells in freshwater. Marine species may form blooms on beaches, others are intestinal parasites. Heterotrophs ingest bacteria and protoctists. Some of the photosynthetic forms possess yellow and red pigments in addition to chlorophyll, and some also contain blue-red phycocyanin pigments. Some form colonies of non-mobile cells embedded in a gel-like matrix.

Discomitochondria
Flagellates, zoomastigotes

Diversity	Size	Nutrition	Mode of life
	Mostly unicellular	Generally heterotrophic, most euglenids are photosynthetic	Mainly free-living in a wide range of aquatic and terrestrial habitats

All formerly regarded as protozoan animals, and in medical literature are commonly termed flagellates. Most feed on bacteria or absorb nutrients directly from surroundings; some, ie. euglenids, are usually photosynthetic. Some are symbiotic or parasitic, the latter including organisms (*Trypanosoma*) responsible for sleeping sickness and Chagas disease.

Chrysomonada
Chrysophyta

Diversity	Size	Nutrition	Mode of life
	Most single-celled, some form large branching colonies	Photosynthetic	Free-living, mainly in freshwaters

A large and diverse group of algae with golden-yellow pigments. The silicoflagellates are a component of marine plankton and extract silica from sea water to form shells.

Xanthophyta
Yellow green algae

Diversity	Size	Nutrition	Mode of life
About 600 species	Single-celled or colonial	Photosynthetic	Free-living mostly freshwater algae

Free-swimming unicells, or highly structured multicellular or multi-nucleated organisms, with gold-yellow xanthin pigments. Often form scum in pond water and margins. Typically form pectin-rich cellulosic cell walls; cysts often rich in iron or silica.

Eustigmatophyta
Green eyespot algae

Diversity	Size	Nutrition	Mode of life
Nine genera known	Single-celled	Photosynthetic	Free-living algae, mostly freshwater

Planktonic algae with yellowish-green pigments, typically at the base of freshwater food webs. A few multicellular forms are known.

Diatoms

Diversity	Size	Nutrition	Mode of life
Around 10 000 living species	Single-celled, some colonial	Mostly photosynthetic, some saprophytes	Mostly free-living

Widely distributed in the photic zone of marine and inland waters worldwide. Some occur in moist soils. Diatoms have distinctive paired tests or shells of organic material impregnated with silica extracted from surrounding water. Very important basal components of marine and freshwater food webs.

Phaeophyta
Brown algae

Diversity	Size	Nutrition	Mode of life
900 species	Macroscopic plant-like organisms, mostly few centimetres, some-times much larger	Photosynthetic	Most live anchored to the substrate on rocky coasts

Most widespread in temperate regions, where they usually dominate the intertidal zone. Generally fixed but some, eg. *Sargassum*, form large floating mats far out to sea. The largest protoctists: Pacific giant kelp (*Macrocystis pyrifera*) sometimes to 65 m length. Brown algae are major primary producers in inshore environments and also provide habitat for a large number of macroscopic marine organisms.

Labyrinthulata
Slime nets and thraustochytrids

Diversity	Size	Nutrition	Mode of life
Eight known genera	Colonies up to a few centimetres long	Heterotrophic	Colonial marine protoctists

Slime nets consist of a complex colonial network of cells that move and grow within an extracellular slime matrix of their own making. *Labyrinthula* grows on eel grass (*Zostera*) where possibly pathogenic.

Plasmodiomorpha

? ?

Diversity	Size	Nutrition	Mode of life
29 species	Microscopic	Heterotrophic	Obligate intracellular symbionts, mainly of terrestrial plants, some parasitic

Zoospores occur in soil and infect the host; a plasmodium with many cell nuclei but no dividing walls develops within the host cell. Most species do not appear to harm their hosts but *Plasmodiophora brassicae* causes club-root disease of brassicas and *Spongospora subterranea* powdery scab of potatoes.

Oomycota
Oomycetes

Diversity	Size	Nutrition	Mode of life
100s of species		Heterotrophic symbionts	Mostly in freshwaters or soil, some parasitic on land plants

Feed by extending threadlike hyphae into host tissue where release digestive enzymes and absorb nutrients. Familiarly known as water moulds, white rusts and downy mildews. Many oomyctes are very important crop pests, eg. *Phytophthora infestans* causes potato blight; *Saprolegnia parasitica* attacks freshwater and aquarium fishes. Formerly regarded as fungi.

Hyphochytriomycota
Hyphochytrids

Diversity	Size	Nutrition	Mode of life
23 species		Heterotrophic	Present in freshwaters and soil moisture, saprophytic or parasitic

Feed by extending threadlike hyphae into host tissue, typically algae or fungi, or into organic remains, eg. insect or plant debris, where digestive enzymes are released and nutrients absorbed. Formerly regarded as fungi.

Haplospora

Diversity	Size	Nutrition	Mode of life
33 species	Single-celled	Heterotrophic	Unicellular symbionts living in the tissues of marine animals

Life history incompletely known, but characterised by production of spores into water or host tissue. Many are benign symbionts, and exist in multinucleate plasmodium form, but several are parasitic and damage host tissues. Host animals include molluscs, nematodes, trematodes and polychaetes. Often found as parasites of parasites, eg. within trematode parasites of oysters. Formerly regarded as sporozoans.

Paramyxa

Diversity	Size	Nutrition	Mode of life
6 species	Microscopic	Heterotrophic	Obligate symbionts living within the cells of marine invertebrates

Characterised by production of multicelled spores within host tissue. Live within annelids, crustaceans, molluscs, and probably other groups of marine invertebrates. Formerly regarded as sporozoans.

Myxospora
Myxosporidians

Diversity	Size	Nutrition	Mode of life
1100 species	Infected tissue may have growths of several centimetres in diameter	Heterotrophic	Multicellular symbionts, mostly parasites of fishes but also of marine and freshwater invertebrates

Myxosporidians penetrate the host integument and travel to the intestine where amoeboid forms carried to target organs are released. Many form large plasmodial masses attached to internal organs. Hosts include sipunculans and freshwater oligochaete worms. Most appear benign, including the fish symbionts, but some are important pathogens, eg. *Myxostoma cerebralis* causes twist disease of salmon. Formerly regarded as sporozoans.

Rhodophyta
Red algae

Diversity	Size	Nutrition	Mode of life
4100 species	Macroscopic plant-like organisms, up to 1 metre in size	Nearly all photosynthetic; a few are symbionts on other red algae	Virtually all are marine, a few species are freshwater or terrestrial

Red algae occur attached to substrate on beaches and rocky shores worldwide. Most abundant in tropics. Many forms become encrusted with calcium carbonate; calcified red algae have a fossil record from the early Palaeozoic. Agar jelly is extracted from red algae, and other extracts are used in food manufacture. Along with the Phaeophyta (brown algae) the largest and most complex protoctists.

Gamophyta
Conjugating green algae

Diversity	Size	Nutrition	Mode of life
Many thousand species	Multicellular forms are macroscopic	Photosynthetic	Freshwater algae

Multi-cellular filament-forming or unicellular green algae found in freshwaters. Many contribute to algal blooms and pond scum. Filamentous forms include *Spirogyra*. Desmids consist of paired cells joined at a narrow bridge through which their cytoplasm is continuous.

Actinopoda
Radiolarians

Diversity	Size	Nutrition	Mode of life
More than 4000 species		Heterotrophic but most hold symbiotic photosynthetic haptomonads	Mostly marine, although the Heliozoa is mainly freshwater

Relatively large, generally unicellular protoctists with radial symmetry. Some form large colonies in which many individuals are embedded in a jelly-like matrix. Some occur in open ocean waters, some are benthic. Many have siliceous skeletons with spines or oars used for swimming. Most acantharians include photosynthetic grass green haptomonad or yellow or green algae symbionts.

Chlorophyta
Green algae

Diversity	Size	Nutrition	Mode of life
16 000 species	Range from single-celled to macroscopic green seaweeds	Photosynthetic	Diverse mostly marine and freshwater algae; a few symbiotic with other organisms

Chlorophytes include unicellular and complex multicellular species as well as forms with many nuclei sharing the same cytoplasm. Major primary producers, they are estimated to fix over one billion tonnes of atmospheric carbon annually. Symbiotic forms include *Platymonas* in the flatworm *Convoluta roscoffensis*. Some forms are resistant to at least periodic desiccation. Some early form of chlorophyta almost certainly gave rise to plants.

Zoomastigota
Zoomastigotes

Diversity	Size	Nutrition	Mode of life
	Single-celled, some colonial	Heterotrophic	Some are free-living in marine and freshwater environments, others are symbionts in the intestines of vertebrates

Many feed by ingesting bacteria. Parasitic forms occur in the intestine of aquatic vertebrates, eg. the opalinids, found in frogs and toads. One group of colonial forms, the choanomastigotes, may be ancestral to the sponges.

Chytridiomycota

Diversity	Size	Nutrition	Mode of life
1000 species	Microscopic	Heterotrophic	Decomposers or parasites in freshwater or moist soils

Feed by extending threadlike hyphae into living hosts or dead material. Simplest forms grow entirely within the cells of their hosts. Some are associated with plant diseases, eg. *Physoderma zea-maydis* causes brownspot in maize. Cell walls of chitin, some with cellulose also. Chytrids may be ancestral to fungi.

Placozoa
Trichoplax

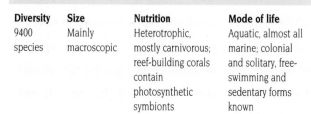

Diversity	Size	Nutrition	Mode of life
1 species	Up to 1 mm	Heterotrophic	Very small marine animal

Discovered in a seawater aquarium in 1883, and since reported in shallow marine water and marine research stations. *Trichoplax adhaerens* is the least complex of all living animals, consisting of a few thousand cells but no distinct tissues; little is known of its life history.

Porifera
Sponges

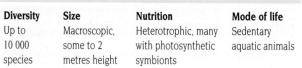

Diversity	Size	Nutrition	Mode of life
Up to 10 000 species	Macroscopic, some to 2 metres height	Heterotrophic, many with photosynthetic symbionts	Sedentary aquatic animals

The vast majority of sponges is marine, about 100 species are freshwater. Filter-feeders (one Mediterranean form passively captures crustaceans and digests them externally). Many include photosynthetic symbionts, eg. cyanobacteria, and brown, red or green algae. Sponges have simple structure with no tissues or organs and are generally supported by calcareous or siliceous spicules or fibrous, proteinaceous matrix.

Cnidaria
Cnidarians, hydras

Diversity	Size	Nutrition	Mode of life
9400 species	Mainly macroscopic	Heterotrophic, mostly carnivorous; reef-building corals contain photosynthetic symbionts	Aquatic, almost all marine; colonial and solitary, free-swimming and sedentary forms known

A diverse phylum of radially symmetrical animals, including sea anemones, jellyfishes and corals. Specialised stinging cells called cnidoblasts are diagnostic. Largest individuals (eg. lion's mane *Cyanea*) may have tentacles many metres long. Reef-building corals are of major importance in clear water coastal shallows in tropics and subtropics. Coral reefs often highly diverse and of great economic value. Photosynthetic symbionts (dinomastigotes) of reef corals occur in polyp tissue at density up to 5 million/cm^2; these require sunlight and limit reef growth to upper part of the photic zone. Non-reef coral without symbionts range down to 3000 m.

Ctenophora
Comb jellies

Diversity	Size	Nutrition	Mode of life
100 species	Typically around a centimetre, largest up to 2 metres length	Heterotrophic	Free-swimming marine organisms

A small phylum of translucent, soft-bodied predators. Widespread in marine waters and possibly the most abundant planktonic animals between 400 and 700 m depth. Their fragility makes them difficult to collect and study.

Platyhelminthes
Flatworms

Diversity	Size	Nutrition	Mode of life
20 000 species	Often a few millimetres, tapeworms to 30 m in length	Heterotrophic	Free-living or symbionts, many parasitic; found in freshwater, marine and terrestrial environments

Flatworms, flukes and tapeworms. Free-living soil flatworms most abundant in tropics, aquatic forms mainly temperate. Some can survive in environments low in O$_2$ by oxidising hydrogen sulphide. Parasitic forms include flukes such as *Schistosoma*, the cause of schistosomiasis, and tapeworms, obligate parasites of vertebrate gut. Many flatworm parasites have complex life cycle with infective larvae and intermediate hosts.

Gnathostomulids
Jaw worms

Diversity	Size	Nutrition	Mode of life
80 described species	Average length around 1.5 mm	Heterotrophic	Free-living marine worms

A small phylum of translucent, benthic worms capable of surviving in sediments very low in O$_2$ and high in hydrogen sulphide. Graze on bacteria, protoctists and fungi in marine sediments. Have been found at several hundred metres depth. Population densities may exceed 6000 per litre of sediment, outnumbering nematodes.

Rhombozoa
Rhombozoans

Diversity	Size	Nutrition	Mode of life
65 species	Up to 5 mm	Heterotrophic	Worm-like internal parasites or symbionts of benthic cephalopod molluscs

Mostly found in the kidneys of squid and octopus in temperate waters.

Orthonectida
Orthonectida

Diversity	Size	Nutrition	Mode of life
20 species	Microscopic	Heterotrophic	Worm-like internal parasites or symbionts of marine invertebrates

Recorded from echinoderms, nemertines, annelids, molluscs and flatworms. Less benign than rhombozoans; may affect host reproduction.

Nemertina
Ribbon worms

Diversity	Size	Nutrition	Mode of life
900 species	Macroscopic, from 0.5 mm to 30 metres in length, mostly small	Heterotrophic	Mostly free-living predatory marine worms

Characterised by the slender anterior proboscis, used for predation, defence and locomotion. Abundant in the intertidal zone, some forms are pelagic. Freshwater and terrestrial forms are known, and some species are symbionts or parasites. *Malacobdella* is a filter-feeder, inhabiting the mantle cavity of clams.

Nematoda
Nematodes

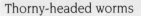

Diversity	Size	Nutrition	Mode of life
25 000 species described	From 0.1 mm to 9 m in length	Heterotrophic	Free-living or parasitic worm-like animals

Possibly the most abundant animals living on Earth, found in virtually all habitats and in many other organisms. Free-living forms are key to decomposition and nutrient cycling. Many species are important parasites of plants and animals, including humans (eg. filariasis). Nematodes have provided important research animals in genetics and cell differentiation.

Nematomorpha
Nematomorphs

Diversity	Size	Nutrition	Mode of life
240 species	Ranging from 10 to 70 cm length	Heterotrophic	Adults are free-living and usually aquatic; all are endoparasitic at some stage

A small phylum of leathery, unsegmented, worm-like animals. Occur widely in aquatic or moist terrestrial habitats. Eggs hatch into minute motile larvae which enter host and metamorphose into immature worms. These burst out, killing the host, when near water or during rain. Hosts include annelids and arthropods. Rarely found in humans, where appear non-pathogenic.

Acanthocephala
Thorny-headed worms

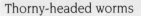

Diversity	Size	Nutrition	Mode of life
More than 1000 species	Between 1 mm and 1 metre length	Heterotrophic	Parasitic worms that lack a free-living stage

Adult individuals anchor themselves to the gut wall of vertebrates. Infection generally occurs after an intermediate invertebrate host is ingested. Thorny-headed worms appear to alter host behaviour so as to increase probability of host being ingested by predator, and so transfer parasite to further host. Humans are seldom parasitised.

Rotifera
Rotifers

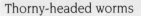

Diversity	Size	Nutrition	Mode of life
2000 species	Mostly microscopic, some to 2 mm	Heterotrophic	Mostly free-swimming in freshwaters

Mainly freshwater, also in moist habitats on land; about 50 species occur in benthic and pelagic marine habitats. Rotifers are the most abundant and cosmopolitan of the freshwater zooplankton. Mostly free-living; many live on other invertebrate organisms, many are endoparasites of invertebrates. Most free-living rotifers reproduce parthenogenetically.

Priapulida
Priapulids

Diversity	Size	Nutrition	Mode of life
17 species	Range between 0.5 mm and 30 cm	Heterotrophic	Exclusively marine, free-living worm-like animals

Found in sand or mud, from intertidal pools to abyssal depths, and from tropical waters to the Antarctic. Approximately half of the described species are part of the marine meiobenthos (ie. small bottom or sediment-living species between about 0.5 and 1 mm).

Loricifera
Loriciferans

Diversity	Size	Nutrition	Mode of life
100 species	Microscopic	Heterotrophic	Benthic marine species

Widespread, probably cosmopolitan part of interstitial fauna. Life history incompletely known. Adults are sedentary on sand or gravel, sometimes ectoparasites; the larvae are believed to be free-living and mobile. Protective plates cover the abdomen, into which the neck and head with mouth cone can be retracted.

Kinorhyncha
Kinorhynchs

Diversity	Size	Nutrition	Mode of life
150 species	Up to 1 mm in length	Heterotrophic	Free-living marine animals

Kinorhynchs are cosmopolitan in muddy bottom habitats, including estuaries and the intertidal zone, and to a depth of approximately 5000 m. Some species are commensal with hydrozoans, bryozoans and sponges.

Gastrotricha
Gastrotrichs

Diversity	Size	Nutrition	Mode of life
400 species	Average length 0.5 mm	Heterotrophic	Free-living worm-like animals, mainly marine

The ventral side is ciliated and often glued to substrate; exposed surfaces bear bristles or scales. Most occur in subtidal or intertidal sediments where part of the marine meiobenthos; freshwater forms most abundant in small still waters. Important scavengers of dead bacteria and plankton.

Entoprocta
Entoprocts

Diversity	Size	Nutrition	Mode of life
150 species	Very small macroscopic animals, up to 1 cm	Heterotrophic	Mostly sessile colonial marine organisms

Widely distributed in shallow coastal waters. One freshwater species known. Colonies permanently attached by stalks, horizontal stolons and basal discs to solid substrate, algae or other animals. Often form conspicuous mat-like growth on seaweed and rocks. Filter-feeders, consuming diatoms, desmids, other plankton and detritus. *Loxosomella* is free-living, and moves by somersaulting basal disc over tentacles.

Chelicerata
Chelicerates

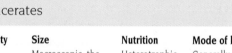

Diversity	Size	Nutrition	Mode of life
75 000 species	Macroscopic, the largest species of Pycnogonida have a leg span of almost 80 cm	Heterotrophic	Generally free-living and in most habitats

A very large and diverse arthropod phylum characterised by claws (chelicerae) on the anterior pair of appendages, and sharing other features (eg. segmented bodies, chitinous exoskeleton, jointed appendages) with insects and crustaceans. Most chelicerates are in Arachnida (more than 75 000 species); others are horseshoe crabs (Merostomata) and sea spiders (Pycnogonida). Arachnids are ubiquitous on land, with a few freshwater species; the group includes ticks and mites, some of considerable importance as vectors of disease in humans and livestock. Merostomata include *Limulus*, superficially unchanged since the Silurian (see Chapter 3). Pycnogonids range from the shallows to the deep ocean (6800m) and from pole to pole.

Mandibulata
Mandibulates

Diversity	Size	Nutrition	Mode of life
950 000 described species	From near microscopic to many centimetres	Heterotrophic	Most are free-living terrestrial species

An exceptionally large and diverse arthropod phylum distinguished by a pair of crushing mandibles. Jointed chitinous exoskeleton and a single pair of antennae. Includes insects (Hexapoda), centipedes and millipedes (Myriapoda), symphyla and pauropods. The largest group of animals, Hexapoda contains some 950 000 described species and may number in millions. Insects range up to 30 cms length (ie. giant stick insect *Pharnacia serratipes*). Social insects such as termites, ants, some bees and wasps can form large colonies. Many insects are very important crop pests or disease vectors; others are beneficial because of crop pollination or pest control.

Sipuncula
Peanut worms

Diversity	Size	Nutrition	Mode of life
150 species	A few millimetres to 0.5 m in length	Heterotrophic	Exclusively marine, worm-like, burrowing or crevice-dwelling organisms

Benthic species, mainly in shallow warm marine habitats, most abundant on rocky shores, also present in polar regions and down to 7000 m in the abyssal ocean. Ingest diatoms and other protoctists or organic debris. Used locally as human food in the Indo-Pacific and China.

Annelida
Annelids

Diversity	Size	Nutrition	Mode of life
15 000 species	From 0.5 mm to 3 m	Heterotrophic	Segmented worms, mostly free-living in soils and sediments; some parasitic

A large phylum including polychaetes (9000 species), oligochaetes (6000), and leeches (500). Most are active predators and scavengers. Polychaetes include free-living and tube-dwelling marine species, mainly benthic but some pelagic. Oligochaetes occur in freshwater, estuaries and deep sea, but are most numerous on land where earthworms are very important to soil structure. Leeches are mainly free-living predators of vertebrates and invertebrates in freshwaters or water film on land. Formerly more widely used for medicinal purposes (*Hirudo medicinalis*).

Crustacea
Crustaceans

Diversity	Size	Nutrition	Mode of life
40 000 species	Mostly macroscopic	Heterotrophic	Mostly free-living in aquatic and humid terrestrial habitats; some parasitic

A very large and very diverse arthropod phylum distinguished by having two pairs of antennae. Species occur in virtually all habitats. Includes crabs, crayfish, prawns, barnacles, copepods, brine shrimp, water fleas, woodlice, etc. Size ranges between 0.25 mm (*Alonella*) and 2.8 m (*Macrocheira* claw span). The predominant arthropods in most freshwaters. Widespread in all marine habitats, from pelagic waters to ocean depths at 5000 m, and in moist terrestrial situations. Numerous parasitic and commensal forms exist; pentastomids sometimes parasitise humans. Crustaceans such as krill (*Euphausia superba*) form key components of the marine food web. There are numerous important crustacean fisheries.

Echiura
Spoon worms

Diversity	Size	Nutrition	Mode of life
140 species	From a few millimetres to 40 cm	Heterotrophic	Exclusively free-living marine organisms

Echiurans live in U-shaped burrows in marine sediments, rock crevices, and mangrove, with some forms extending to abyssal depths around 10 000 m. Echiurans have a flexible proboscis which may extend to 1.5 m from the bulbous unsegmented body. Cilia move food items down the proboscis to the mouth.

Pogonophora
Beard worms

Diversity	Size	Nutrition	Mode of life
120 species	From 10 cm to 2 m length	Heterotrophic, some with chemoautotrophic symbionts	Sessile benthic marine worms

Pogonophora live in fixed upright chitin tubes secreted in sediments, shell, or decaying wood on the ocean floor. Most abundant in cold deep waters, shallow polar seas or (the vestimentiferans) around hot submarine vents with a high hydrogen sulphide and methane content. Greatest diversity in the western Pacific. Adult pogonophorans have no gut and probably absorb nutrients directly from tentacles. Vent-living forms derive nutrients and energy from the oxidation of hydrogen sulphide through symbiotic chemoautotrophic bacteria, which can occur at densities of 1 billion per gram of body tissue.

Mollusca
Molluscs

Diversity	Size	Nutrition	Mode of life
70 000 species.	Range from near microscopic to several metres	Heterotrophic	Mainly free-living species present in most habitat types; some parasitic

A large and highly diverse phylum, with species occurring in benthic and pelagic marine waters, in freshwaters of all kinds, and on land, from forests to deserts. Includes snails, slugs, mussels, chitons, octopods, squid and others. Most species are free-living, although some parasitic or commensals. Many, eg. bivalves, are sedentary as adults. Size reaches maximum, approaching 20 m, in the giant squid *Architeuthis*. Many important as food source. Some forms act as intermediate hosts to parasites (eg. *Schistosoma*) that cause serious human disease; other species can cause significant damage to crops and constructions (eg. *Dreissena*). Venom of some marine gastropods is of medical interest. Monoplacophorans, the most primitive molluscs, first seen alive in the 1970s but abundant in Palaeozoic (see Chapter 3).

Phoronida
Phoronids

Diversity	Size	Nutrition	Mode of life
14 species	From 1 mm to 50 cm	Heterotrophic	Sedentary filter-feeding marine worms

Most inhabit leathery chitinous tubes encrusted with sand or shell fragments; some burrow in mollusc shells or rock. From coastal shallows to 400 m depth. Filter-feed on plankton and detritus. Cosmopolitan but not abundant; half the known species occur on Pacific coast of N America.

Onychophora
Velvet worms

Diversity	Size	Nutrition	Mode of life
100 species	From one to 20 cm	Heterotrophic	A small phylum of free-living terrestrial worm-like animals

Require high humidity levels to counter water loss through thin chitinous cuticle. Many occur in forest, some in caves. All carnivorous. Walk slowly on 14-43 pairs of stumpy legs. Two geographic groups exist, one mainly warm northern hemisphere, the other southern hemisphere. Many tropical species are viviparous.

Bryozoa
Ectoprocts

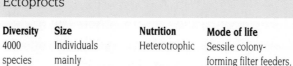

Diversity	Size	Nutrition	Mode of life
4000 species	Individuals mainly microscopic, colonies to 0.5 m diameter	Heterotrophic	Sessile colony-forming filter feeders, mainly marine

Marine bryozoans mainly intertidal, but also on seafloor to considerable depths. About 50 freshwater species known, with jelly-like colonies on plant surfaces in slow streams. Large colonies (0.5 m in diameter) derived from the asexual budding of zooids (<1 mm in length) may contain several million individuals. Marine forms contribute to reef diversity.

Brachiopoda
Lampshells

Diversity	Size	Nutrition	Mode of life
350 species	From 2 mm to 10 cm	Heterotrophic	Benthic, mainly sessile, marine animals

Cosmopolitan. Present between the intertidal zone and 4000 m depth. Usually occur cemented to surface by stalk (pedicle); some species free-living on or in marine sediment. Unlike molluscs, brachiopods have a lophophore, a specialised surface for gas exchange and food collection, and have dorso-ventral symmetry instead of lateral symmetry. Previously very diverse, especially during the Paleozoic era (see Chapter 3). Some 30 000 extinct species have been described, and *Lingula*, with fossils from 400 Mya, may be the oldest genus with living species.

Tardigrada
Water bears

Diversity	Size	Nutrition	Mode of life
750 species	Mainly microscopic	Heterotrophic	Free-living animals, mostly in moist terrestrial and freshwater habitats

Widely distributed from pole to pole. All are aquatic; land species live in the water film on mosses, forest litter and other habitats. A few marine species. Move on four pairs of stumpy legs. Mainly ingest liquid food obtained by piercing protoctists, animals or plants. The Mesotardigrada (genus *Thermozodium*) inhabit hot springs. Can survive extreme desiccation with low metabolism or in encysted form, and when dormant (cryptobiotic) may be tolerant of exceptionally high or low temperatures, approaching absolute zero.

Chaetognatha
Arrow worms

Diversity	Size	Nutrition	Mode of life
70 species	From 0.5 and 15 cm	Heterotrophic	Worm-like planktonic marine predators

Common plankton in open seas, especially abundant in warm seas down to 200 m. Detect prey, mainly copepods, by vibration sensors, and can inject neurotoxins. Important to marine fisheries as a source of food for fishes.

Hemichordata
Acorn worms

Diversity	Size	Nutrition	Mode of life
90 species	Adults between 2.5 and 250 cm	Heterotrophic	Sedentary benthic marine species

A small phylum of soft bodied benthic marine worm-like animals. Adults mostly sedentary and live burrowed in soft sediment of shallow seas, or in secreted tubes. Sexual and asexual reproduction occurs, colonies may be formed by budding. Hemichordates were previously classified as chordates, and resemble them in having ciliated gill slits in pharynx.

Echinodermata
Echinoderms

Diversity	Size	Nutrition	Mode of life
7000 species	A few centimetres to near 2 m	Heterotrophic	Mostly free-living benthic marine species

Invertebrates with five-part radial symmetry, an internal calcium carbonate skeleton, and a water vascular system. Includes starfish, sea urchins, sea cucumbers, and others. Mostly benthic in intertidal or subtidal habitats. Sea lilies extend to 10 000 m, and sea cucumbers in places make up nearly entire animal biomass at these abyssal depths. Viviparous forms exist. Some, eg. dried sea cucumbers (trepang), used as human food.

Cephalochordata
Lancelets

Diversity	Size	Nutrition	Mode of life
23 species	From 5 cm to 15 cm	Heterotrophic	Free-living filter-feeding marine animals

Lancelets occur in estuary sediments and shallow sandy seafloors, and live with the head protruding in order to screen out small plankton and organic materials. They make up a small phylum of chordates with a cartilaginous rod dorsal to the gut (notochord), a dorsal hollow nerve cord, and persistent gill slits in the pharynx, but without an internal bony skeleton or cerebral ganglion. They are the closest living relatives of vertebrate animals. Used as human food in some areas.

Urochordata
Sea squirts

Diversity	Size	Nutrition	Mode of life
1400 species	From 1 mm to 2 cm	Heterotrophic	Small marine filter-feeding animals

Adults may either be benthic and sedentary (Class Ascidiacae, tunicates) or pelagic and free-swimming (Class Larvacea); Thaliacea or salps also free-swimming. All are ciliary filter-feeders. Urochordata have a dorsal hollow nerve cord, a cartilaginous rod dorsal to the gut (notochord) and gill slits in the pharynx at some stage, these being features of chordates.

Craniata
Craniates or vertebrates

Diversity	Size	Nutrition	Mode of life
52 000 species	From about 1 cm to 35 m	Heterotrophic	Free-living species present in most habitat types

This group contains the vertebrates; a very large and very diverse phylum of chordates, all of which, unlike the acraniate chordates (urochordates, cephalochordates), have a brain enclosed within a skull (cranium). The majority have a bony internal skeleton. A small group of mainly marine species (lampreys and hagfish) lack jaws and are grouped in Agnatha in contrast to other vertebrates, all of which possess jaws (Gnathostomata). About half of all described vertebrate species are fishes: the Chondrichthyes (sharks and rays), Osteichthyes (bony fishes), the lungfishes and coelacanths. These make up around 25 000 species in total. The tetrapods (the four-limbed non-fish vertebrates) include amphibians, reptiles, birds and mammals. Although not so versatile as bacteria, vertebrates between them extend from the air above the highest mountain to abyssal ocean depths, from sand desert to tropical forest, and from hot springs to polar ice and subzero waters. The vertebrates include the most familiar animals, and, with molluscs and crustaceans, most of those of direct nutritional importance to humans.

Zygomycota
Zygomycotes

Diversity	Size	Nutrition	Mode of life
1100 species	Generally microscopic or little larger, but hyphae may extend considerable distance through soil	Heterotrophic	Mainly terrestrial, in soil and leaf litter

Most are saprobic (saprophytic), secreting digestive enzymes into organic material and absorbing nutrients released. Many are parasites on protoctists, small animals, plants or other fungi. Zygomycotes include about 100 species that form mycorrhizal associations, in which fungal partner contacts or enters plant roots and assists inflow of nutrients and absorbs organic plant substances in exchange. Most vascular plants probably have such relationships with fungi and these may be critical in nutrient-poor soils. A few occur in freshwaters.

Basidiomycota
Basidiomycotes

Diversity	Size	Nutrition	Mode of life
22 250 species	Generally microscopic or somewhat larger, but hyphae extend considerable distance, and fruiting bodies to 10 cm or more	Heterotrophic	Mainly terrestrial, in soil and leaf litter, or on trees and other organisms

A large phylum including typical mushrooms, puffballs, stinkhorns, and rusts and smuts (some of which cause economically important plant diseases). All are characterised by microscopic club-shaped spore-producing reproductive structures, typically borne in great numbers on a basidiocarp – the familiar mushroom. Largest known mushroom specimen to 146 cm wide and 54 cm high. Many basiomycotes form mycorrhizae with trees and shrubs; as with zygomycotes, the fungi move basic nutrients from soil to plant and plant carbohydrates move into the fungus. Many ascomycotes fulfil a key ecological role in breaking down organic material, and transfer of inorganic nutrients and water from soil to plants. Several species are valued wild food items. A few freshwater forms are known.

Ascomycota
Ascomycotes

Diversity	Size	Nutrition	Mode of life
30 000 species	Mainly microscopic but reproductive body up to several centimetres	Heterotrophic	Mainly terrestrial, in soil and leaf litter, or on and in other organisms

A large diverse phylum distinguished from other fungi by the microscopic sac-like spore-producing reproductive structure (ascus). Includes yeasts, morels, truffles, blue-green moulds and lichens. Thread-like hyphae form network (mycelium) through substrate. Most are either free-living, many are parasitic. Several hundred forms occur in freshwaters. More than 10 000 are the heterotrophic components of lichens, which are joint organisms formed by ascomycotes with either photosynthetic green algae or cyanobacteria. As with other fungi, ascomycotes fulfil a key ecological role in breaking down organic material, and transfer of inorganic nutrients and water from soil to plants. Some are important in food preparation (eg. baker's yeast). Many cause important diseases of plants and animals, including humans, while others are sources of key medicinal substances such as penicillin and similar antibiotics.

Bryophyta
Mosses

Diversity	Size	Nutrition	Mode of life
10 000 species	Low growing plants	Photosynthetic	Terrestrial, mainly in moist habitats and wetlands

A large phylum of non-vascular plants (ie. lacking specialised xylem and phloem transport tissue) comprised of mosses and the genus *Takakia*. Often conspicuous in cold or cool temperate habitats, particularly tundra, where mosses the dominant plants, and also in heathland, bogs, woodland, waterlogged areas and freshwater margins. Most diverse in moist tropical habitats. Many mosses well adapted to withstand desiccation; some occur in warm arid regions. Peat moss (*Sphagnum*) contributes to the development of new soils.

Hepatophyta
Liverworts

Diversity	Size	Nutrition	Mode of life
6000 species	Low growing plants	Photosynthetic	Terrestrial, in moist habitats

Non-vascular plants typically found in moist habitats growing on woodland floor, shaded stream banks, waterfalls, or rocks; often epiphytic and often occur with mosses (Bryophyta). Widespread in cold temperate regions, present in Antarctica, but species diversity highest in tropics. Often among first plants to colonise burned or newly exposed substrates.

Anthocerophyta
Horned liverworts

Diversity	Size	Nutrition	Mode of life
100 species	Low growing plants	Photosynthetic	Terrestrial, in moist habitats

A small group of non-vascular plants of moist habitats, typically on woodland floor or water margins. Present worldwide in temperate and tropical regions. Among first colonists of bare substrates, including rocks. Some species have associated nitrogen-fixing cyanobacteria.

Lycophyta
Club mosses

Diversity	Size	Nutrition	Mode of life
1000 species	Mainly low growing herbaceous plants	Photosynthetic	Terrestrial, in moist and dry habitats

Small seedless evergreen vascular plants found in temperate and tropical habitats, typically on forest floor in temperate regions although most tropical species are epiphytic. A few occur in arid areas. Lycophytes were prominent in Palaeozoic plant communities before evolution of flowering plants; although all living species are small, trees up to 40 m height were dominant in Carboniferous coal forests. Some similarity to mosses and conifers but unrelated to either.

Psilophyta
Whisk fern

Diversity	Size	Nutrition	Mode of life
10 species	Small herbaceous plants	Photosynthetic and symbiotic with fungi	Terrestrial

A very small group of vascular plants, the only ones lacking both roots and leaves. Similar to earliest simple leafless land plants of late Silurian and Devonian times, 400 Mya, and conceivably direct descendants of them. Present as epiphytes or ground-living species with a restricted range in subtropics and temperate areas. These plants have a mycorrhizal association (also seen in earliest fossil forms) with fungal hyphae that increase the flow of soil nutrients to the non-photosynthetic plant cells.

Sphenophyta
Horsetails

Diversity	Size	Nutrition	Mode of life
15 species, one genus	Herbaceous plants	Photosynthetic	Terrestrial

A small phylum of seedless vascular plants, with jointed ridged stems and tiny scale-like leaves. Found in moist or disturbed areas, including urban areas and roadsides, more typically in moist woods and wetland margins, also in salt flats. Historically consumed as food in Europe and North America; poisonous to livestock. As with lycophytes, sphenophytes were diverse and abundant in Devonian and Carboniferous forests, with tree-like forms to 15 m high.

Filicinophyta
Ferns

Diversity	Size	Nutrition	Mode of life
12 000 species	From a few centimetres to 25 m	Photosynthetic	Terrestrial, a few in freshwater

A diverse phylum of vascular plants, the most species-rich group of plants lacking seeds, widespread from cold temperate areas to the tropics. Mainly in moist areas, such as forest floor and stream margins; species diversity highest in tropics, where many forms epiphytic and some species grow as trees to 25 m height. The aquatic *Azolla*, a very small floating fern, has symbiotic nitrogen-fixing cyanobacteria. Several food, medicinal and other products are derived from ferns. Ferns, especially tree ferns, were diverse and very abundant in Devonian and Carboniferous times.

Cycadophyta
Cycads

Diversity	Size	Nutrition	Mode of life
145 species	From shrubs to small trees of 18 m height	Photosynthetic	Terrestrial

A small phylum of seed-bearing, often palm-like, vascular plants restricted to the tropics and subtropics, where present in a range of habitats, from moist forest to deserts and coastal mangroves. Diverse in the Cretaceous. Cycads are gymnosperms, ie. seeds do not become enclosed in a fruit. Many are insect pollinated, often by beetles. All species have symbiotic nitrogen-fixing cyanobacteria. Cycads provide a variety of materials, including thatch, food, medicines, and ornamental plants. Cycad starch for bread requires special treatment to destroy potentially fatal toxins.

Ginkgophyta
Ginkgo

Diversity	Size	Nutrition	Mode of life
One species	To 30 m height	Photosynthetic	Terrestrial

A vascular seed-bearing tree, characterised by the fan-shaped leaf with bifurcating veins and the fleshy exposed ovule, now restricted as a wild species to steep forest in southern China. A wide diversity of ginkgophytes, of which *Ginkgo biloba* is the only survivor, existed during the Mesozoic. Now widely planted for ornamental purposes. It is a gymnosperm, ie. seed not enclosed in a fruit. Leaf extract used as a traditional food and medicine in East Asia.

Coniferophyta
Conifers

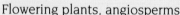

Diversity	Size	Nutrition	Mode of life
550 species	Shrubs or large trees, up to 100 m height	Photosynthetic	Terrestrial

Conifers are cone-bearing gymnospermous vascular plants, with needle-shape leaves, mostly evergreen trees. They form extensive forests at high latitudes in the northern hemisphere, and also occur more locally, often on arid mountains; also common in the tropics and in temperate southern forests, where *Araucaria* is widespread. In *Sequoiadendron* and *Sequoia*, conifers include the largest living plants. Many species in mountainous and northern areas have characteristic symbiotic mycorrhizal fungi. Conifers provide timber, paper pulp, and ornamental plants, and some have food or medicinal value.

Gnetophyta
Gnetophytes

Diversity	Size	Nutrition	Mode of life
70 species	Small trees, shrubs or vines	Photosynthetic	Terrestrial

A small phylum of vascular seed plants, distinguished from other gymnosperms by having vessels for water transport similar to those of flowering plants. The three living genera differ greatly from each other. Some plants of the genus *Gnetum* in tropical moist forest grow to 7 m in height; some *Welwitschia*, a unique low-growing conebearing plant of southwest African deserts, may be 2000 years old.

Anthophyta
Flowering plants, angiosperms

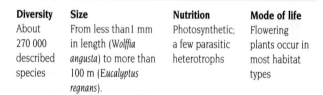

Diversity	Size	Nutrition	Mode of life
About 270 000 described species	From less than 1 mm in length (*Wolffia angusta*) to more than 100 m (*Eucalyptus regnans*).	Photosynthetic; a few parasitic heterotrophs	Flowering plants occur in most habitat types

An extremely diverse, geographically cosmopolitan, phylum of vascular seed plants, distinguished by flowers, and fruits that enclose the fertilised seeds. The great majority of species is terrestrial, in virtually all habitat types. Many occur in or around lakes, rivers and wetlands, and seagrasses occur subtidally in shallow marine waters. Two main groups are distinguished, according to whether the germinating seed has one or two seed leaves: Monocotyledones and Dicotyledones. Monocots include palms, lilies, and the economically vital grasses; most monocots are herbaceous and woody forms lack special tissue that secondarily adds width to the trunk. Dicots form the larger group. Success of anthophytes appears linked to coevolution with animals, in particular with specialised modes of pollination and seed dispersal. All major food and medicinal plants, and hardwood timber trees, are found in this phylum.

3 BIODIVERSITY THROUGH TIME

Two fundamental patterns can be distinguished in the fossil record. On one hand, new groups of organisms appear, diversify and generally persist for very long periods of time. On the other hand, most such groups and presumably all species eventually cease to exist. Global biodiversity has been greatly reduced during several periods of radical environmental change when it appears that the majority of multicellular species then living became extinct.

Despite the prevalence of extinction and the disappearance of many groups, the fossil record shows an erratic but relentless increase in biological diversity, exponential in overall rate through the Mesozoic and Cenozoic and reaching a peak around the end of the Tertiary. Evidently, the great wealth of biodiversity existing now on Earth is the result of a modest net excess of originations over extinctions during the 3500 million year evolution of life.

THE FOSSIL RECORD

Knowledge of the history of diversity through geological time is based on analysis of the fossil record. Fossils, as the preserved parts or casts or imprints of dead organisms, are just mineralised bone or stone, but when interpreted in a biological context they provide the only direct evidence of the 3500 million year history of life on the planet.

The fossils discovered and described by palaeontologists represent more than a quarter of a million species, virtually all of them now extinct[1], but these are believed to make up only a very small fraction of all the species that have ever existed. For example, the fossil record of marine animals is far more comprehensive than that of terrestrial forms, but the marine sample is estimated to represent only about two percent of all the marine animals that have lived[2]. The fossil record overall may represent as little as one percent, or less, of all the species that have existed[1]. Clearly, statements about broad patterns in the evolution of life, and the ascendancy or extinction of groups of organisms, thus rest on a very narrow base of tangible evidence.

For most kinds of organism, exceptional circumstances are required if a dead individual is to become preserved and found. With some exceptions, microscopic or soft-bodied organisms rarely leave discernable traces of their existence, and larger organisms are usually decomposed, disassembled and never discovered. Terrestrial fossils are often of individuals that must have been preserved by the smallest of chances, perhaps sudden

burial during a natural disaster, or as a result of the body falling into a rock crevice out of reach of scavengers.

Macroscopic animals with hard skeletons that lived in shallow marine environments, where their remains could be buried by sediment, petrified, and later exposed in uplifted rock strata, are by far the most likely to be both preserved and found. The fossil record from the past 600 million years is thus dominated by molluscs, brachiopods and corals[1].

All else being equal, the probability of an individual being preserved will rise greatly the more widespread and abundant the species is, and the longer it persists through time. Conversely, there is a very low probability that any individual of a numerically rare or restricted range species with a short persistence time will die in circumstances conducive to fossilisation and subsequently be found. Factors such as these mean that even the relatively better known groups are certain to be very incompletely known. The plant and animal species now living include a substantial number of rare or local species and it is difficult to imagine many of them being represented in the fossil record of our time as recovered in the future[3].

Because the fossil record gives an incomplete and biased view of the past history of life, the reconstruction of that history has been the subject of great debate. It is, however, generally accepted that the record can give a reasonable insight into past diversity in terms of taxonomic richness[4], particularly at higher taxonomic levels. It is far more difficult to derive other, more ecologically based, measures of diversity from it, as these

Figure 3.1

The four eons of the geological timescale.

Arrows indicate approximate age of oldest confirmed fossils of the groups named.

Source: modified from Margulis and Schwartz[7].

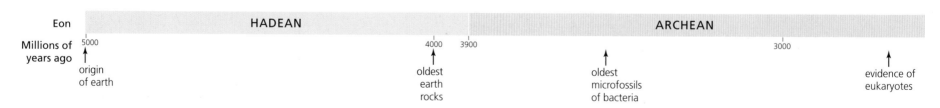

Eon	HADEAN		ARCHEAN	
Millions of years ago	5000	4000 3900	3000	
	↑ origin of earth	↑ oldest earth rocks	↑ oldest microfossils of bacteria	↑ evidence of eukaryotes

require the reconstruction of palaeoenvironments, a far more contentious exercise than palaeotaxonomy.

PATTERNS OF DIVERSIFICATION

The early history of life

In its earliest history, there was no life on the Earth. Now, there are about 1.75 million different species of all kinds known, and estimates suggest that at least 14 million, and possibly many more millions, may exist in total, mostly unknown and undescribed. Self-evidently, life has both arisen and diversified.

The planet Earth is nearly 5000 million years old, and the oldest known rocks are about 4000 million years in age. The first clear evidence of organisms in the fossil record is of filaments and spheroids in rocks about 3500 million years old. These are believed to be the remains of procaryotes (bacteria), indicating that recognisable cellular life must have arisen at the latest at this time, and probably well before this. Stromatolites also first appear at around this time. These rock domes built up of multiple layers of sediment and bacteria occur very locally today in extreme environments, such as salt ponds, but are the most abundant fossils known for the next 3000 million years of life on Earth, for much of which time bacteria were the only known life forms. The available evidence indicates that diversity was very low for most of this period and that many forms apparently persisted for unimaginably long periods of time (hundreds of millions of years). Indeed some of these archaic fossil bacteria are outwardly indistinguishable from modern bacteria,

indicating that some life forms may conceivably have remained largely unchanged almost since the beginning of life on Earth.

The next major step in the evolution of life was the development of eucaryotic organisms. Biochemicals characteristic of eukoryotes have been found in shales 2700 million years old[19], far earlier than the first fossils. The microfossils known as acritarchs, which are recorded as far back as 1200 Mya (million years ago) in the Proterozoic, are almost certainly the cysts or resting stages of eukaryotic protoctists, probably of planktonic marine algae. Depending on when eucaryotes arose, previous life on Earth consisted solely of bacteria for between 800 million and 2000 million years.

Diversity in acritarchs, and the rate at which different forms replaced one another in the record, were low until around 1000 Mya when both species number and species turnover increased markedly[5].

Radiations around the early Phanerozoic boundary

For many years it was assumed that animals originated in the Cambrian period at the base of the Phanerozoic, some 545 million years ago (the Phanerozoic is the eon of time characterised by presence of animal fossils; it includes the Palaeozoic, Mesozoic and Caenozoic). This is now known not to be the case, as a wide range of fossil animals, including recognisable arthropods and possibly echinoderms, is now known from about 100 million years before the Cambrian. Most fossils from this time, however, appear completely unrelated to extant

PROTEROZOIC | PHANEROZOIC

2600   2000   1000   545   0

oldest protoctist fossils

oldest animal fossils

oldest fungi fossils

oldest plant fossils

forms, and consist mainly of enigmatic frond- and disc-shaped soft-bodied animals: the so-called Ediacaran fauna.

The lower Cambrian marks a dramatic change from this early fauna, with the sudden appearance in the fossil record of a wide range of animals, many with calcareous skeletons. It is generally accepted that this represents a genuine explosion of diversity which took place over only a few million years, and is not an artefact of the fossil record. The lower Cambrian thus represents the most important period of high-level diversification in the history of animal life on Earth. These archaic invertebrates had by the end of the Cambrian period, around 500 Mya, established all the basic body plans seen in extant animals, and many others besides. Each such basic line-age is recognised taxonomically at phylum level, and the range of morphological diversity was higher at 500 Mya than at any time before or since. As many as 100 different animal phyla may have existed during the Cambrian[6] including every well-skeletalised animal phylum living today (except perhaps the Bryozoa), whereas in the latest synthesis all extant animals are placed in 37 phyla[7].

Plants and animals began to extend into terrestrial habitats during the first half of the Palaeozoic, with the first fossil material known from the late Silurian, around 400 Mya. At this point approximately 90% of the history of life to the present had already passed. Fossils suggest low diversity for the next 100 million years until the later Devonian period. No new animal phyla appeared with the colonisation of land, millions of years after the initial Cambrian radiation of animal phyla.

Diversity of marine animals in the Phanerozoic

The overall pattern of diversity (assessed as numbers of families) shows a possible early peak around the start of Phanerozoic time, followed by a plateau of somewhat higher diversity extending through most of the Palaeozoic era, and then, after the end-Permian mass extinction (see below) a steady increase in diversity over remaining geological time[4].

Although the number of phyla has decreased markedly since the Cambrian, diversity at all lower taxonomic levels has either increased overall or in a few cases remained more or less level. The number of orders of marine animals present in the fossil record climbed steadily through the Cambrian and Ordovician, levelling off towards the end of the Ordovician to a figure of between 125 and 140, which has been maintained throughout the Phanerozoic.

The diversity of marine families represented in the fossil record shows a similar pattern of increase through the Cambrian (possibly falling during the latter half of the period) and Ordovician, levelling off at around 500, a figure which was maintained until the late Permian mass extinction. This extinction event resulted in the loss of around 200 families, but diversity increased subsequently to the modern level of around 1100 families, with a number of temporary reversals during

Figure 3.2

Periods and eras of the Phanerozoic.

This is an expanded version of the most recent segment of the geological timescale shown in Figure 3.1. The Phanerozoic is the eon of time extending from the base of the Cambrian, some 545 million years ago, to the present, and to which the entire fossil record was formerly thought to be restricted.

Source: modified from Margulis and Schwartz[17], Cambrian base date from International Subcommission on Cambrian stratigraphy website[17].

Period	Cambrian	Ordovician	Silurian	Devonian	Carboniferous
Era	PALAEOZOIC				
Millions of years ago	545 490	438	408	360	286

Figure 3.3

Animal family diversity through time.

The lines plotted represent the number of families in the fossil record.

Notes: The blue line essentially represents marine invertebrate animals. Although a small number of vertebrate groups, notably fishes and a few tetrapod species, is included, these make up a very small proportion of the total marine family diversity shown. The curve for fishes includes an increasing proportion of freshwater forms through the Cenozoic. Tetrapods are amphibians, reptiles, birds and mammals.

Sources: marine animals, modified from Sepkoski[2]; fishes and tetrapods, modified after Benton[16]; insects, modified from Labandeira and Sepkoski[8].

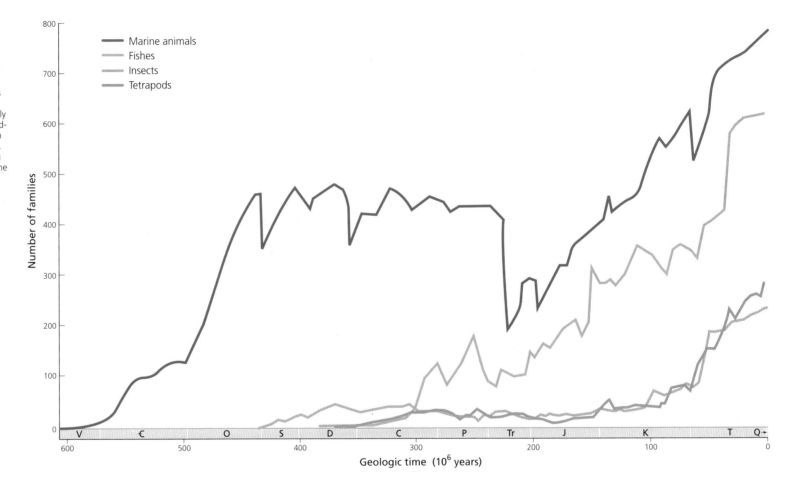

minor extinction events. The trend in number of species in the fossil record is even more extreme. From the early Cambrian until the mid-Cretaceous, the number of marine species remained low; since then, that is during the past 100 million years, it has increased dramatically, perhaps by a factor of 10.

Diversity of terrestrial organisms in the Phanerozoic

Although low to moderate peaks and troughs are evident in the record, the overall pattern of family diversity in terrestrial organisms shows a continuing rise from the Silurian to the present, thus differing somewhat from the pattern shown by marine animals[4].

It is generally accepted that vascular terrestrial plants first arose in the Silurian, although some palaeobotanists argue for a Late Ordovician origin. Diversity increased during the Silurian, and then more rapidly during the Devonian, owing to the first appearance of seed-bearing plants, leading to a peak of over 40 genera during the late Devonian. Diversity then declined

slightly, but started to increase markedly during the Carboniferous, with 20 families and more than 250 species in the mid-Carboniferous record of the northern hemisphere. Following this, diversity increased only slowly until the end of the Permian. There was a marked decrease in diversity at the end of the Permian, coinciding with or preceding the mass extinction of animal species, followed by a rapid rebound to previous levels. Diversity then continued increasing slowly, reaching around 400 species in the early Cretaceous. Starting at the mid-Cretaceous, diversity began increasing at an accelerating pace.

This overall pattern masks important changes with time in the composition of the flora, most notably in the relative importance of the three main groups of vascular plant: the pteridophytes, gymnosperms and angiosperms. The Silurian and early Devonian are marked by a radiation of primitive pteridophytes. During the Carboniferous, more advanced pteridophytes and gymnosperms developed and underwent extensive diversification. Following the late Permian extinction event, pteridophytes were largely replaced by gymnosperms (although ferns remain abundant) and these became the dominant group until the mid-Cretaceous. The dramatic increase in plant diversity since then is entirely due to the radiation of the angiosperms which first appeared in the lower Cretaceous.

Colonisation of land by animals has occurred many times; although the oldest body fossils of terrestrial animals date from the early Devonian, it is generally accepted that the primary period of land invasion by animals was the Silurian.

The overwhelming number of described extant species of terrestrial animals are insects. Their fossil record is more extensive than might be expected, but had been little studied until recently. Data on insect diversity at family level have been collated, based on nearly 1300 families[8]. This analysis shows a very slow increase in families from the first appearance of

Figure 3.4

Plant diversity through time.

Notes: Pteridophytes are ferns Filicinophyta and allies, gymnosperms are conifers Coniferophyta and allies, angiosperms are flowering plants Anthophyta. Note changing numerical dominance of each group over time. The figure represents species in the fossil record instead of families as in Figs. 3.3 and 3.6, and it relates to the northern hemisphere record.

Source: modified from Kemp[14], after Niklas.

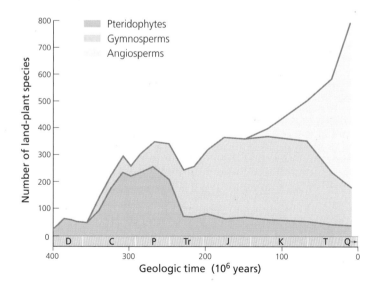

insects in the Devonian, a rise in numbers during the Carboniferous, and a steeply increasing rise throughout the Mesozoic to the Tertiary. The explosion in insect diversity had previously been attributed to ecological opportunities provided by the expansion of flowering plants, but the insects are now known to have begun their ascendancy some 150 million years before the flowering plants[8].

The fossil record of vertebrates includes around 1400 families, with tetrapods somewhat outnumbering fishes. The bird record is much less substantial than that for other groups, probably because their light skeletons have been less frequently preserved.

Terrestrial vertebrates first appear in the fossil record in the late Devonian. Diversity remained relatively low during the Palaeozoic, with around 50 families, and may have declined during the early Mesozoic. From the mid-Cretaceous the number of families started to increase rapidly, reaching a Recent peak of around 340. Diversity of genera follows this overall pattern in a more exaggerated form. It appears that periodic increase in the number of tetrapod families is mainly a result of lineages becoming adapted to modes of life not already followed by other organisms, ie. by adopting new diets or new habitats[9].

PATTERNS OF EXTINCTION

If living species represent between two and four percent of all species that have ever lived[10], almost all species that have lived are extinct, and extinction can be presumed to be the ultimate fate of all species.

Numerous estimates have been made of the lifespan of species in the fossil record; these range from 0.5 My to 13 My for groups as varied as mammals and microscopic protoctists. Analysis of 17 500 genera of extinct marine microorganisms, invertebrates and vertebrates, suggests an average lifespan of 4 million years in these groups[1]. Given this average lifespan, at

a very gross estimate, the mean extinction rate would be 2.5 species per year if there were around 10 million species in total. However, because of bias inherent in the fossil record, such life-span estimates are likely to relate to widespread, abundant and geologically longer-lived species; in effect, the extinction-resistant species, and so not represent the biota as a whole[11]. Most species will therefore survive for less than four million years, and real extinction rates at any given time will be correspondingly higher. Nevertheless, even if background extinction rates were ten times higher than this, extinctions amongst the 4000 or so living mammals would be expected to occur at a rate of around one every 100 years, and amongst birds at one every 50 years.

Major extinctions in animals

In general the Precambrian fossil record is too incomplete to allow detailed analysis of extinction rates. However, there is good evidence of a major loss of diversity during the Vendian period in latest Precambrian times, around 550 Mya, when the entire Ediacaran fauna disappeared (along with many acritarchs). Another wave of extinction affected archaeocyathid sponges, molluscs and trilobites during the lower Cambrian some 530 Mya.

By far the most severe marine invertebrate mass extinction was in the late Permian (250 Mya). At that time, the number of families of marine animals recorded in the fossil record declined by 54% and the number of genera by 78-84%. Extrapolation from these figures indicates that species diversity may have dropped by as much as 95%. Other major extinctions in marine invertebrates occurred at the end of the Ordovician (440 Mya) (see Fig. 3.3), when around 22% of families were lost, and during the late Devonian and late Triassic (21% and 20% respectively). Around 15% of marine families disappeared at the end of the Cretaceous.

The vertebrate fossil record, especially for terrestrial

tetrapods, is much less amenable to analysis of extinction rates than the invertebrate record chiefly because it is less complete and less diverse. However, studies indicate that fishes have been subject to at least eight important extinction events since their recorded origin in the Silurian, while tetrapods have experienced at least six such events since their appearance in the late Devonian. Some of these events coincide with each other and with those recorded for marine invertebrates; in particular, the five major mass extinction events outlined above are paralleled by losses in vertebrate diversity. The most significant is the late Permian event, which, in terms of percentage loss, is the largest recorded extinction both for fishes (44% of families disappearing from the fossil record) and tetrapods (58% of families disappearing). The late Cretaceous event was more significant for tetrapods than for other groups, with at least 30 of the 80-90 families then in the fossil record disappearing at this time. These families were, however, virtually confined to three major groups which suffered complete extirpation – the dinosaurs, plesiosaurs and pterosaurs. Most other major vertebrate taxa were almost completely unaffected.

Major extinctions in vascular plants

Fewer major extinction events have been distinguished in the plant fossil record than in the animal record[12]. Plant extinction rates (based on analysis of families and genera) do vary with time, but in general, periods of elevated plant extinction appear to be more protracted than animal extinction events and do not usually coincide with them. It had been argued that these periods may be more to do with competitive displacement by more developed plant forms, or with gradual climatic change, than with any sudden catastrophic events. However, such generalisations are being modified as evidence for mass extinction has emerged following advances in stratigraphic sampling and dating.

Recent evidence has revealed that a fourfold increase in atmospheric carbon dioxide around the Triassic-Jurassic boundary is correlated with a more than 95% turnover in the megaflora (ie. leaf fossils etc., as opposed to pollen or spores)[13]. The end-Cretaceous catastrophe appears to have had a major influence on the structure and composition of terrestrial vegetation and on the survival of species. Data from fossil leaves suggest that perhaps 75% of late Cretaceous species became extinct, although data from fossil pollens indicate a lower though still significant level of extinction. During the Tertiary there are two other periods of widespread enhanced extinction rates, during the late Eocene and from the late Miocene to the Quaternary, although in the latter, extinction of taxa at generic level and above appears to have been mainly regional rather than global.

Mass extinctions

The very many species extinctions represented in the fossil record are not distributed evenly through time, nor do they occur randomly. In palaeontology much attention has been devoted to mass extinction periods, during which some 75-95%

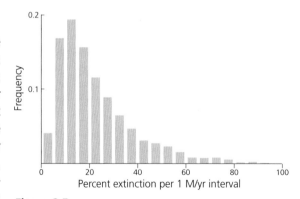

Figure 3.5
Frequency of percent extinction per million year period.
See text for explanation
Source: modified from Raup[1].

Figure 3.6

Number of family extinctions per geological interval and cumulative diversity of living families through the Phanerozoic.

Source: from data made available by M. Benton at University of Bristol Earth Sciences website.[18]

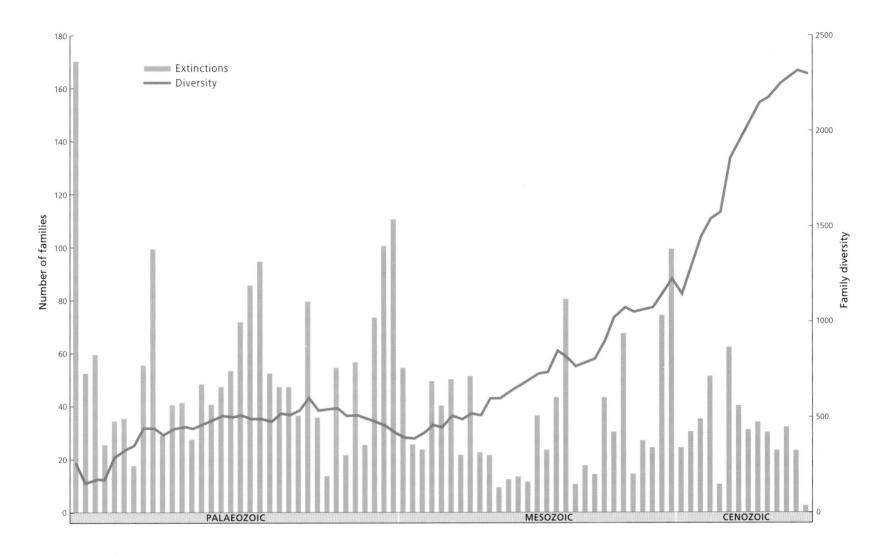

of species then living became extinct during geologically very short periods of time, in some cases possibly as little as a few hundred thousand years or even less. Five such phases (based chiefly on extinction of marine species) are recognised during Phanerozoic time, late in each of the Ordovician, Devonian, Permian, Triassic and Cretaceous periods (Table 3.1).

Although each of the 'Big Five' mass extinctions had a very profound effect on then contemporary life, they are not isolated peaks standing out from a constant, very low, background rate. Rather, the extinction rate has varied continuously throughout the Phanerozoic with periods of more or less elevated rates of extinction, of which the most extreme can be characterised as mass extinction events. A frequency plot (Fig. 3.5) of the percentage of species becoming extinct in each one million year interval of the 600 million year record of animal life provides an indication of this variation in extinction intensity[1]. The plot takes the form of an even right-skewed curve, with no substantial discontinuity between very rare periods of very high extinction rate (>60% of species extinct per one million years) and the most frequent rate (10-15%); the mean intensity is 25% of species extinct per 1 My interval. This suggests that mass extinctions may arise from causes not qualitatively different from those associated with extinctions at other times. It should also be remembered that because of the extremely long duration of the Phanerozoic during which species have been constantly becoming extinct, and the very short duration of the mass extinction events themselves, the latter account for only a small percentage (estimated at around 4%) of all extinctions.

The precise causes and timespans of each of the mass extinctions have been the subject of much debate and study[14,15]. It is now widely accepted that the late Permian mass extinction was a long-term event, lasting for 5-8 million years. It appears to have been associated with geologically-rapid global physical changes (including the formation of the supercontinent Pangaea), climate change, and extensive, tectonically-induced marine transgression and increased volcanic activity. The late Cretaceous extinction is probably the best known, but in terms of overall loss of diversity is also the least important. There is strong evidence that this extinction event was associated with climate change following an extra-terrestrial impact, although this remains somewhat controversial. The late Ordovician event appears to be correlated with the global Hirnantian glaciation, with three separate episodes of extinction spread over only 500 000 years. In all cases, however, the ability to determine accurately the timing and periodicity of extinction is heavily dependent on the completeness of the fossil record and the reliability and precision of stratigraphic analysis.

A mass extinction period is typically followed by a five to ten million year phase of very low diversity, with a handful of species dominant in fossil faunas and floras. When diversity again increases the biota may be very different in composition from those preceding. In several instances, groups previously showing low diversity have radiated and spread widely following the demise of groups previously dominant; eg. ray-finned fishes diversified following loss of placoderms; quillworts and seed ferns diversified for a time after loss of glossopterids; and the mammals radiated after loss of many terrestrial reptile groups. Reef organisms provide a classic example: reef communities have been lost several times during the Phanerozoic but with each reappearance the main reef-building forms have been different[15].

The present day diversity of living organisms is the result of a net excess of originations over extinctions through geological time. Figure 3.6 shows the number of family extinctions in each interval of geological time, and the cumulative diversity of families, both marine and terrestrial.

Table 3.1

The principal mass extinctions in the Phanerozoic fossil record.

Source: summarised from Kemp[14], Hallam and Wignall[15].

Period	Date (Mya)	Species loss (%)	Biotic change	Possible causes
Late Ordovician	440	85%	The last but largest of several extinction events during the Ordovician. More than 25% of marine invertebrate families lost. Entire class Graptolithina reduces to a few species; acritarchs, brachiopods, conodonts, corals, echinoderms, trilobites, all much reduced.	Cooling; Warming Marine regression Marine transgression & anoxia
Late Devonian	365	80%	Mass extinctions came at the end of prolonged period of diversity reduction. Rugose corals lost >95% of shallow water species, stromatoporoid corals reduced by half and reefs disappeared, brachiopods lost 33 families, ammonoids and trilobites severely affected. Fishes suffer only major mass extinction; all early jawed fishes (placoderms) disappear and most agnatha. First major crisis in plants; diversity greatly reduced, but spread of first tree, the gymnosperm *Archaeopteris*.	Marine transgression & anoxia
End Permian	250	95%	The most severe extinction crisis, metazoan life came within a few percent of total extinction. Tabulate and rugose corals terminated, complex reefs disappear (return after 8 My gap); echinoderms almost wiped out; worst crisis in history of foraminifera; severe extinction in ammonites, brachiopods, bryozoa, molluscs. Some losses in early ray-fin fishes. Major loss of terrestrial vertebrates (75% of families) and insects (8 of 27 insect orders extinct). Mass extinction in plants, large plants including peat-forming trees lost (spread of small conifers, lycopods and quillworts), sudden unprecedented abundance of fungal spores at end of period.	Volcanism Warming Marine transgression & anoxia
End Triassic	205	80%	Mass extinction in marine invertebrates, especially brachiopods, cephalopods and molluscs, also mass disappearance of scleractinian corals and sponges. Several seed fern families lost; some land vertebrates lost, but evidence for mass extinction questionable.	Marine regression
End Cretaceous	66	75%	Radical change in planktonic foraminifera; 85% of calcareous nanoplankton lost, also all ammonites, belemnites and many bivalves; losses in echinoids and corals. Many marine reptiles extinct (ichthyosaurs, plesiosaurs, mosasaurs); significant losses in freshwater and terrestrial vertebrates, including last dinosaurs (high turnover throughout dinosaur history, end Cretaceous unusual in that no replacements emerged). Mass extinctions in plants, highest (possibly 60% species loss) in angiosperms, lowest in ferns.	Impact of large meteor Volcanism Cooling Marine regression

References

1. Raup, D.M. 1994. The role of extinction in evolution. *Proc. Natl. Acad. Sci. USA* 91:6758-6763.
2. Sepkoski, J.J. 1992. Phylogenetic and ecologic patterns in the Phanerozoic history of marine biodiversity. In, Eldredge, N. (ed). *Systematics, ecology and the biodiversity crisis*. Pp. 77-100. Columbia University Press, NY.
3. Jablonski, D. 1991. Extinctions: a paleontological perspective. *Science* 253:754-757.
4. Benton, M.J. 1995. Diversification and extinction in the history of life. *Science* 268:52-58.
5. Knoll, A.H. 1995. Proterozoic and early Cambrian protists: evidence for accelerating evolutionary tempo. In, Fitch, W.M. and Ayala, F.J. (eds). *Tempo and mode in evolution: genetics and paleontology 50 years after Simpson*. Pp. 63-68. National Academy Press.
6. McMenamin, M.A.S. 1989. The origins and radiation of the early metazoa. In, Allen, K.C. and Briggs, D.E.G. (eds). *Evolution and the fossil record*. Pp. 73-98. Belhaven Press, a division of Pinter Publishers Ltd., London.
7. Margulis, L. and Schwartz, K.V. 1998. *Five kingdoms. An illustrated guide to the phyla of life on earth*. 3rd edition. W.H. Freeman and Company, NY.
8. Labandeira, C.C. and Sepkoski, J.J. 1993. Insect diversity in the fossil record. *Science* 261:310-315.
9. Benton, M.J. 1999. The history of life: large databases in palaeontology. In, Harper, D.A.T. (ed.). *Numerical palaeontology: computer-based modelling and analysis of fossils and their distributions*. John Wiley & Sons, Chichester and New York.
10. May, R.M., Lawton, J.H. and Stork, N.E. 1995. Assessing extinction rates. In, Lawton, J.H. and May, R.M. (eds). *Extinction Rates*. Pp. 1-24. Oxford University Press, Oxford.
11. Jablonski, D. 1995. Extinctions in the fossil record. In, Lawton, J.H. and May, R.M. (eds). *Extinction Rates*. Pp. 25-44. Oxford University Press, Oxford.
12. Crane, P.R. 1989. Patterns of evolution and extinction in vascular plants. In, Allen, K.C. and Briggs, D.E.G. (eds). *Evolution and the fossil record*. Pp. 153-187. Belhaven Press, a division of Pinter Publishers Ltd., London.
13. McElwain, J.C., Beerling, D.J. and Woodward, F.I. 1999. Fossil plants and global warming at the Triassic-Jurassic boundary. *Science* 285(5432):1386-1390.
14. Kemp, T.S. 1999. *Fossils and evolution*. Oxford University Press, Oxford.
15. Hallam, A. and Wignall, P.B. 1997. *Mass extinctions and their aftermath*. Oxford University Press, Oxford.
16. Benton, M.J. 1889. Patterns of evolution and extinction in vertebrates. In, Allen, K.C. and Briggs, D.E.G. (eds). *Evolution and the fossil record*. Pp.218-241. Belhaven Press, a division of Printer Publishers Ltd., London.
17. International Subcommission on Cambrian Stratigraphy. http://www.uni-wuerzburg.de/palaeontologie/stuff/casub.htm.
18. Benton, M.J. http://palaeo.gly.bris.ac.uk/frwhole/fr2.families
19. Brocks, J.J., Logan, G.A., Buick, R. and Summons, R.E. 1999. Archean molecular fossils and the early rise of eukaryotes, *Science* 285 (5430):1033-1036.

Suggested introductory source

Kemp, T.S. 1999. *Fossils and evolution*. Oxford University Press, Oxford.

4

HUMANS AND BIODIVERSITY

The lineage leading to the human species emerged from an ancestry among the apes, almost certainly in Africa around 5 million years ago. Fossils usually attributed to the genus *Homo* itself date from the late Pliocene, perhaps 2.5 million years ago, and anatomically modern humans appear in the fossil record some 200 000 years or more before present. Humans appear to have been exerting significant impacts on the environment ever since their evolutionary career began.

There is circumstantial evidence that hominids may have made deliberate use of fire more than one million years ago, and might be associated with extensive vegetation burning dating from 400 000 years ago. They are also strongly implicated in the extinction of many large terrestrial mammal and bird species on continents and islands, such as the Americas, Australia and New Zealand, that were colonised later on in hominid history.

Early humans lived at low population density and probably increased in number by moving into new parts of the world. The development of agriculture around 10 000 years ago allowed populations to start increasing in density, but the global human population appears to have continued to grow only relatively slowly until the nineteenth century, when revolutionary developments in agriculture, industry and public health triggered an exponential rise that has continued to the present day. Agriculture is simply a way to channel more of Earth's production into human bodies; humans now use or divert an estimated 40% of net primary production on land, and no other single species approaches humanity in numbers, biomass and the extent of its distribution. Nevertheless, humans are as dependent on other species and their supporting ecosystems as when they first appeared.

USE OF NATURAL RESOURCES

The human species evolved as a natural element of diversity in the living world and it is a simple ecological imperative that humans depend on other species, and the ecosystems that support them, for the basic requirements of existence. The creation of organic compounds by photosynthesising organisms is the point at which the sun's energy enters the biosphere (Chapter 1); humans and other animals are unable to capture energy in this way and must consume and digest either primary producers or other organisms that are themselves dependent on primary producers, in order to obtain these energy rich organic compounds for their own activities. While humans are doing nothing fundamentally different from other animals, with the benefits of society and technology, which serve to increase the rates of resource extraction, they are uniquely successful at it. Self-evidently, humans have not arrived at their extraordinarily dominant position on the planet overnight. The growth of their influence can be traced back several million years, to well before the Pleistocene when a stone-tool-wielding hominid first emerged somewhere in eastern Africa.

HUMAN ORIGINS

Climate during the past two million years

The Earth's climate appears always to have been in a state of flux. Generally the degree of accuracy in our understanding of global climate, and certainly the degree of resolution in the time-scale of climate change, decreases the further back in time we go. It is therefore difficult to compare periods that are recent in geological terms with more distant times. However, it does seem that during the past two million years there have been numerous, intense climate changes that were at least as severe, or perhaps more severe, than any recorded earlier in the Earth's history[1]. It is during this period that hominids very similar to modern humans first appear in the fossil record. By the early Holocene, some 10 000 years ago, technologically sophisticated

humans had spread to all the major land masses except Antarctica, and had evidently started to exert a major, and ever-growing, impact on the biosphere. Indeed there is evidence that such impacts – for example in the extinction of large mammal and bird species – were already being felt considerably earlier than this. However, because the rise of humans coincides with a period of major climatic and ecological fluctuations, it is often difficult to disentangle the effects of the former from the latter, and so the precise nature of these impacts remains controversial.

The Tertiary period, which began some 65 million years ago and ended with the start of the Quaternary (the Pleistocene and Holocene) 1.8 million years ago, is characterised overall by a gradual decrease in global temperature and increase in aridity. Superimposed on this general pattern were many oscillations, occurring on a timescale of thousands of years[2]. These oscillations are believed to be linked to cyclic variations in the Earth's position in its orbit around the sun, known as Milankovitch Cycles. These cycles notwithstanding, the climate during virtually the whole of the Tertiary was notably warmer than at present.

Around 1.8 million years ago, at the very start of the Pleistocene[3], there was apparently rapid global cooling leading to the start of a period dominated by marked climate cycles of around 100 000 years duration. For long periods of each cycle global temperatures were significantly lower than they are today and extensive areas of the northern hemisphere landmasses were covered in ice sheets. Detailed analysis of climate changes over the past 400 000 years or so, particularly through examination of Antarctic ice-cores[4], indicate that each cycle over this period has been broadly similar, with a gradual decline in average temperature, although with many minor and some large oscillations (relatively colder periods being referred to as stadials, warmer ones as interstadials), followed by a short period of intense warming in which temperatures rose from those of fully glacial conditions to those characterising a warm interglacial state, perhaps sometimes over only a few decades.

Mean global temperature during glacial periods was around 6°C cooler than during interglacials, with cooling more pronounced at the poles than the equator. The mid-latitudes and equatorial regions were probably somewhat more arid than they are at present. The large amount of water locked up in the greatly expanded polar ice-caps meant that mean sea level during glacial periods was probably around 100-150 m lower than at present.

Until around 11 000 years ago, each temperature peak was apparently followed by an almost immediate decline as the next glacial cycle began[4]. Overall, it appears that during the past half million years the Earth's climate has been as warm as, or warmer than, today's for only around 2% of the time[1]. It seems likely that this also holds true for the early Pleistocene. The end of the last glacial cycle, around 11 000-12 000 years ago, marked the start of the Holocene. Temperatures similar to today's have prevailed throughout the Holocene, making it by far the longest true interglacial period for at least the past half million years and probably for the last 1.8 million years.

Human origins and dispersal

The origins and early history of humans are among the most controversial subjects in palaeontology. Remains are generally scarce and often open to varying interpretation. Nevertheless, consensus has emerged over the broad outlines. It is likely that the direct ancestors of humans – the hominid line – diverged from the apes in Africa during a cool, dry phase of the Late Miocene, around 6.0-5.3 million years ago. No relevant fossils are known from this time, however, and the evidence for this divergence lies in the form of 'molecular clocks' arising from the comparison of human and ape genetic material[5]. The earliest recorded hominid, *Ardipithecus ramidus*, dates from 4.4 million years ago. Somewhat more recent are various species of *Australopithecus* and *Paranthropus*, all from eastern Africa.

Sometime during the middle to late Pliocene, 2.5 to 1.8 million years ago, the genus *Homo* is thought to have evolved from *Australopithecus* stock[5,6]. Until comparatively recently it had been assumed that early man then remained confined to Africa until less than one million years ago. However, two well-preserved skulls recently found in the Caucasus (southern Georgia) have been dated to about 1.8 million years ago, and fossils and stone tools from tropical and subtropical eastern and southeast Asia have been persuasively dated to around 1.9 million years ago. These finds suggest that populations of *Homo* may have spread through warm parts of southeast Europe as well as Asia within at most a few hundred thousand years of the genus originating in Africa[5]. Other very early hominid remains from Europe date from 780 000 years ago in Spain, and 500 000 years ago in northern Europe[7].

The earliest evidence of hominids outside Africa and Eurasia is much more recent. A human skeleton found in 1974 at Lake Mungo in Australia has recently been dated at 62 000 years BP[8]. This is some 20 000 years earlier than the most generally accepted evidence for colonisation of Sahul – the single land-mass that comprised Australia and New Guinea during most of the Pleistocene.

In the Americas, the oldest good evidence of human presence is that of a coastal settlement in Chile dated around 12 500 ^{14}C years BP (14 000-15 000 calendar years ago)[9], although even this is far from universally accepted amongst archaeologists. There are very controversial and now widely questioned claims for evidence of human settlement much earlier, most notably that dating from 32 000 years BP from Pedra Furada in northeast Brazil[10]. The earliest unequivocal evidence of widespread occupation in the Americas comes from the so-called Clovis hunting culture whose oldest remains are generally dated at around 11 500 ^{14}C years BP (ca 12 000-13 000 calendar years BP)[11]. Settlement of the Caribbean islands including Cuba and

Hispaniola appears to have taken place considerably later, around 6000 years BP[12].

Archaeological evidence indicates that colonisation of the Pacific Islands east of New Guinea began around 4000 years ago, when much of Melanesia and Micronesia were settled. Fiji and Samoa were probably colonised around 3500 years ago, and the outliers of Hawai'i, Easter Island and New Zealand within the last 1500 years[13]. In the Indian Ocean, Madagascar was probably first settled around 1500 years ago[14].

Technology

The earliest evidence of tool manufacture by hominids is from the Gona River drainage in northern Ethiopia where stone tools have been dated to at least 2.5 million years ago. These tools are small (generally less than 10 cm long) and simple, but are already of relatively sophisticated manufacture, suggesting that even older artefacts will eventually be found[15]. They are of essentially the same design as those dated from 2.3 to 1.5 million years ago from other sites in East Africa (eg. Lokalalei in the Lake Turkana basin in northern Kenya and the Olduvai gorge in Tanzania) indicating effective technological stasis for at least one million years (from 2.5 million to roughly 1.5 million years ago)[15]. Collectively these tools are referred to as products of the Oldowan Stone tool industry[15].

Around 1.5 million years ago much more sophisticated and often larger tools including hand-axes and cleavers suddenly appear in the archaeological record in East Africa. These are referred to as Acheulian tools. The Acheulian tool industry spread into Europe, the Near East and India and remained apparently relatively unchanged until around 200 000 years ago, showing similar temporal persistence to the Oldowan industry.

Evidence of tools and artefacts made from organic materials from the Palaeolithic is understandably extremely scarce. In 1995, however, three large wooden implements very similar in design to modern-day javelins and dated to around 400 000 years ago where discovered at Schöningen in Germany. These were found in association with a smaller wooden implement (interpreted as likely to have been a throwing stick), stone tools and the butchered remains of more than ten horses and can be persuasively interpreted as throwing spears used in systematic, organised hunting[16].

There is also intriguing although indirect evidence from Southeast Asia dating to around 900 000 years ago that hominids at this time, at least in this region, were capable of repeated water crossings using watercraft. The evidence is in the form of stone tools dated to this age from the island of Flores in eastern Indonesia. Even at the time of the last glacial maximum (when global sea levels would have been at their lowest), reaching Flores from the Asian mainland would have required crossing three deep water straits with a total distance of at least 19 km. The impoverished – and typically island – nature of the Palaeolithic fauna of Flores would appear to preclude the existence of any now submerged land bridge[17]. The next oldest, again indirect, evidence for the use of watercraft is the colonisation of Sahul, probably sometime around 40 000 to 60 000 years ago. Even at times of lowest sea level, this would have necessitated crossing some 100 km of open sea.

Fire

At some point in their evolutionary history, hominids clearly learnt to control, manipulate and, presumably later, to start fires. Determining even very approximately when this may have happened is difficult, and as with all else to do with human evolution, controversial. This is chiefly because the existence of natural fires caused by lightning strikes and volcanic activity greatly complicates the interpretation of the archaeological and geological record – an association between artefacts and evidence of burning does not necessarily indicate a direct link between the two[18].

The earliest dated associations between artefacts and

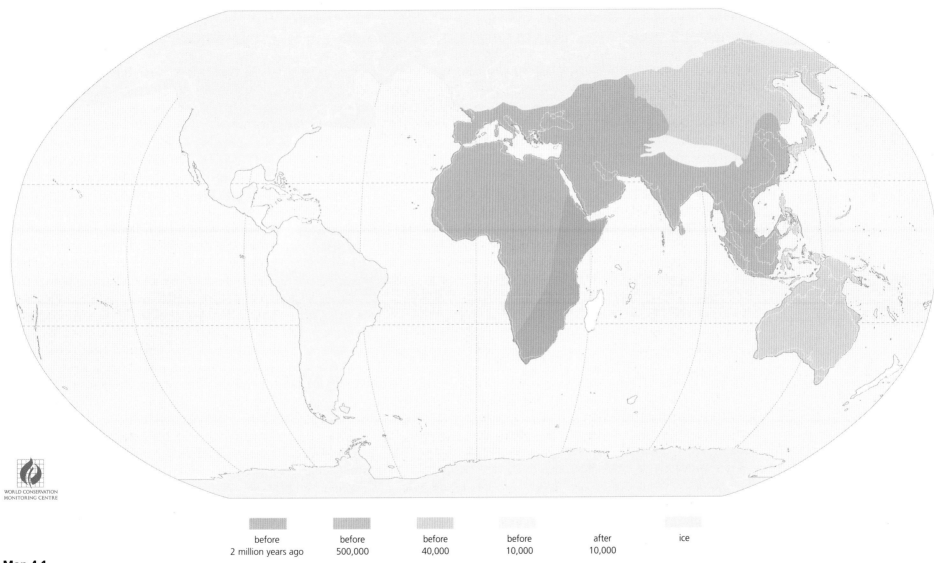

before
2 million years ago

before
500,000

before
40,000

before
10,000

after
10,000

ice

Map 4.1

Early human dispersal

This map presents a highly generalised and in places speculative view of the colonisation of the world by humans from the early Pleistocene onwards. Coastline is shown 150 m lower than today's and, with northern hemisphere ice cover, represents an approximation of that at periods of glacial maximum. Timing of arrival in northeast Eurasia and insular southeast Asia (modern day Philippines and Sulawesi) is unclear.

burning that could be interpreted as deliberate use of fire are from Africa. Stone tools and splintered bones, which can be interpreted as evidence of butchery, have been found associated with clay baked at several hundred degrees for several hours at Chesowanja in Kenya, in deposits around 1.4 million years old[18,19]. The characteristics of the clay are consistent with formation beneath a campfire, but are also consistent with formation around a slow-burning tree-stump which could be associated with a natural bushfire[20]. Charred animal bones and other evidence of human occupation from just over one million years ago have been found at Swartkrans Cave in South Africa, although similar problems of interpretation apply.

Numerous sites in Europe and Asia provide evidence of human occupation and associated fire from the mid-Pleistocene, some 400 000 years ago. The best known of these is at Zhoukoudien in China, although even here the evidence for deliberate use of fire is widely considered equivocal[21,22]. Others are at Torralba-Ambrona (Spain), Terra Amata (France), Westbury-sub-Mendip (England) and Vertesszollos (Hungary)[23].

IMPACTS OF EARLY HOMINIDS ON THE ENVIRONMENT

Charting the course of human evolution is a major challenge. Attempting to assess the impacts on the environment of the hominid line throughout its history is even more problematic. Doing this entails not only the reconstruction of palaeoenvironments, but also demonstrating a causal link between the presence of hominids and a given set of environmental changes. At one level, of course, the latter can never be unequivocally demonstrated as we are dealing with long past events which can never be re-created. However, it is possible to produce plausible interpretations of past events using the best available data.

Biomass burning

Large-scale burning of terrestrial vegetation is undoubtedly one of the major present day impacts by humans on the biosphere, and is believed to constitute around one-third of current annual anthropogenic CO_2 emissions. Evidence of earlier impacts is invariably circumstantial. It is relatively difficult to demonstrate widespread biomass burning in the fossil or subfossil record and even harder to demonstrate a link with human activities – natural fires may be caused by storms or volcanic activity and may be expected to vary in extent, frequency and intensity according to prevailing climatic and ecological conditions. However, the abundance of elemental carbon in marine sediments off Sierra Leone in Africa can be persuasively interpreted as a measure of intensity of biomass burning in sub-Saharan Africa. Analysis of a core from these sediments covering the past million years or so indicates that inferred fire incidence in the region was low until around 400 000 years ago. Since then, five episodes of intense burning of vegetation can be inferred, all except the most recent coinciding with periods when the global climate was changing from interglacial to glacial. The current peak is unique in that it is occurring during an interglacial period. The change from interglacial to glacial climate is generally associated with increased aridity, so that vegetation may be expected to be more vulnerable to fire. It would also be expected that a substantial fuel base in the form of woody biomass would have accumulated during the warmer, wetter, interglacial.

It may well be merely coincidental, but it is intriguing that the period of increased fire incidence in the Sierra Leone core – around 400 000 years ago – coincides with the timing of the first widespread evidence for hominid use of fire, as discussed above.

Species extinctions and humans

A wide range of factors affects the frequency of occurrence of particular species in the fossil record, of which the abundance of that species in life is only one. There is thus not necessarily any direct relationship between the former and the latter, so that deducing past changes in abundance of species from the fossil record is a problematic exercise. Cataloguing, though not dating, extinctions is rather less contentious, although even this may be problematic as evinced by the existence of so-called 'Lazarus' taxa (those presumed extinct that are rediscovered alive) (see Chapter 3).

One of the unusual features of the Quaternary period (the Pleistocene and Holocene) has been the apparently disproportionately high extinction rates in the largest terrestrial species, particularly mammals and birds (Table 4.1). These species are generally referred to as the 'megafauna', often defined as those with an adult mass of 44 kg or more, although the term has not been used consistently.

The extinct American fauna includes such well known genera as the sabretooth cats *Smilodon*, giant ground sloths *Eremotherium*, glyptodonts *Glyptotherium* and mammoths *Mammuthus* as well as a number of scavenging and raptorial birds, including the giant *Teratornis* and *Cathartornis*. Those in Australia include the marsupial equivalents of rhinoceroses (family Diprotodontidae) and lions (*Thylacoleo*), giant wombats (*Phascolonus*, *Ramsayia* and *Phascolomis*), the large emu-like *Genyornis* and the giant monitor lizard *Megalania*.

These extinctions have been followed by a similar series on islands during the Holocene. On New Zealand, several species of moa (giant flightless ratite birds in the family Anomalopterygidae) had become extinct by the end of the seventeenth century. On Madagascar, two endemic hippopotamus species *Hippopotamus lemerlei* and H. *madagascariensis*, the elephantbird *Aepyornis maximus*, and a number of large to very large lemur species, all appear to have died out 500 to 900 years ago. Similarly on the Caribbean islands, a number of large mammals, including several ground sloths (Order Xenarthra, family Megalonychidae) appeared to have survived until human occupation but to have died out at some point since then.

The precise causes of all these extinctions have been the subject of endless debate, which centres chiefly on the rôle of humans. At one extreme lies the 'blitzkrieg' hypothesis, applied particularly to the apparently sudden collapse (ie. over a few hundred years) of the North American megafauna at the hand of humans. At the other are those who maintain that in most, if not all cases, the impact of early humans was negligible and climate change, particularly increasing aridity, was the cause.

Table 4.1

Late Pleistocene extinct and living genera of large animals.

*Recently established date of widespread disappearance of the ratite bird *Genyornis newtoni*[24].

Source: Modified from Martin[25].

	Extinct within last 100 000 years	Still extant	Total genera	%Extinct	Estimated timing of maximum extinction
Africa	7	42	49	14	No peak
North America	33	12	45	73	11 000-13 000 years ago
South America	46	12	58	80	11 000-13 000 years ago
Australia	19	3	22	86	c 50 000 years ago*

Several features of the phenomenon seem to point persuasively to humans having played a pivotal rôle in most, if not all, of these extinctions. The most compelling is their timing. In each case the arrival of humans seems to have preceded the major spate of extinctions (inasmuch as these can be dated), with no or very few such extinctions having been recorded prior to human arrival (compare Map 4.1 and Table 4.1). The cumulative weight of this coincidence, from the colonisation of the Sahul some 60 000-70 000 years ago to the arrival of humans on New Zealand and Madagascar in the past 1500-2000 years is difficult to counter. An exception to this rule is Africa and Eurasia, where megafaunal extinctions were relatively few, and were spread out over the whole of the Pleistocene. In Africa the peak was in the lower Pleistocene (21 genera extirpated from the region between 1.8 and 0.7 million years ago compared with 9 between 0.7 million and 130 000 years ago and 7 later than 130 000 years ago). It is noteworthy that this more gradual, earlier and less extreme pattern of extinctions has typified the region where humans evolved.

It is difficult to formulate an entirely climate-based model of extinction that can account for this asynchronicity – outside Africa and Eurasia, these species survived a series of climatic changes at least as extreme as those they faced at the start of the Holocene. The recent study of *Genyornis newtoni* in Australia[24] seems to indicate a widespread and largely synchronous disappearance from a wide range of habitat types during a period of climatic stability, strongly implying that some other agent was responsible. It is also noteworthy that the fossil and archaeological evidence indicates that, on continents at least, these extinctions were not matched by parallel extinctions of smaller species (for example, as far as is known no insect species at all became extinct in Europe in the entire Pleistocene[1]). If climate change were the cause, it would be expected that these species would be at least as affected as larger ones as in general their opportunities for long-range dispersal and migration are much more limited.

Even if humans are accepted as the major agents in these extinctions, the mechanism or mechanisms involved remains elusive. It seems likely that extensive use of fire, and direct hunting of a fauna that had evolved in the absence of humans, and was therefore unlikely to recognise them as potential predators, may have been sufficient cause to exterminate the large herbivores. The large carnivores and scavengers may then have suffered population collapses owing to the disappearance of their prey base.

Origins of agriculture

All the available evidence suggests that until the end of the Pleistocene, humans and their immediate ancestors and kin depended on hunting and gathering of wild resources for their sustenance. Around the end of the Pleistocene, however, a radical change seems to have taken place. This was the emergence of crop-based agriculture and domestication of livestock, phenomena which appear to have arisen independently in Africa, Eurasia and the Americas.

Study of agricultural origins is a rapidly expanding field, based on study of fossil pollen records, archaeological remains, the genetics of present day crops and their close relatives, and remaining indigenous agricultural systems. The first tangible evidence of cultivated plants concerns grains of rye *Secale orientale*, larger than the wild type, found among the remains of more than 150 plant species in a settlement near Aleppo (Syria); these date from around 13 000 years before present, ie. the latest Pleistocene. Other early plant domesticates, such as squash in Ecuador and rice in China, appear in archaeological records a little later, at the time of warmer postglacial conditions 10 000-11 000 years ago. The first records of domesticated pigs, goats and sheep in the Near East also date from a similar period. The earliest definite indications of domestication of the dog *Canis* are somewhat earlier – from around 14 000 years ago in Oberkassel in Germany[26]. Analysis of 'molecular clocks' appears to show that the dog diverged from its wild ancestor, the wolf *Canis lupus*, far earlier than this – perhaps as much as 350 000 years ago. It is quite possible that domestication of the dog as a guard animal and aid in hunting predates domestication of other animals and plants as food sources by many tens of thousands of years, but in the absence of archaeological evidence this remains speculative.

The development of agriculture began independently in different continents and proceeded at different rates, while early cultivators undoubtedly continued to rely heavily on hunting and gathering from the wild. The adoption of agriculture as an integrated major form of food production appears to have taken place first in the 'fertile crescent' of the Near East, around 10 500 years ago. Wheat *Triticum*, barley *Hordeum*, rye *Secale*, pea *Pisum* and lentil *Lens*, cattle *Bos*, sheep *Ovis*, goat *Capra* and pig *Sus* were all domesticated in this region, and formed the basis of the Neolithic peasant economy that spread rapidly into surrounding areas. The system integrated use of food plants, cereals especially, and domesticated animals, for fertilizer and power as well as food.

Full-scale agriculture in other parts of the world appears to have taken longer to develop, perhaps because of the absence of domesticated animals. Although individual crop plants had been in earlier use, farming systems based on rice cultivation in Asia appear to date from 7500 years ago, and systems based on maize in the Americas became fully productive only some 4300 years ago.

Most of the major crops upon which present day humans depend have been grown continuously since the early or middle Holocene. They have been constantly selected over this period and have developed large amounts of useful genetic variation. Indeed the success of individual crop species over wide geographical areas is partly determined by their flexibility in evolving and sustaining genotypes suitable for local environments. Conventional breeding involves the selection and crossing of desirable phenotypes within a crop in order to create more productive genotypes. The process of harvest, storage and sowing alone may have assisted in the selection of traits such as non-shattering seed heads, uniform ripening of seeds, uniform germination, large fruits and seeds and easy storage. Breeding methods have improved by speeding up the introgression of desired genetic traits into new cultivars: tissue culture allows the crossing of desired genotypes at a cell level and genetic modification involves the incorporation of individual genes directly into genomes.

Domesticated crops have been transported around the world probably since full-scale agriculture began, eg. wheats are recorded in areas outside their presumed centre of origin at least 8000 years before present. Some crops became increasingly widely distributed after the 1500s when European colonists moved out of their home continent. Genetic erosion of agricultural diversity presumably began at the same time, as introduced crops and cultivars displaced indigenous ones.

Agriculture and human numbers

Information on early human population numbers is based heavily on inference from circumstantial evidence, and remains on an uncertain footing even when written historical material become available in some abundance for the past few hundred years. Highly speculative estimates based on extrapolations of population densities of great apes[27] and on studies of contemporary human hunter-gatherers[28] indicate that the global late Pleistocene human population may have been between five and ten million. It seems likely that any increase in human population up to then had been a result of increasing the total area occupied by the species, rather than by any major increase in population density in already occupied areas.

Some authors[27] distinguish three main phases of population change thereafter. First is a cycle of primary increase in Europe, Asia and the Mediterranean brought about by the spread and further development of Neolithic agriculture, which appears to have allowed a great increase in population density. At the start of the Iron Age in Europe and the Near East, some 3000 years ago, the world population was believed to be doubling every 500 years and the total probably reached 100 million around this time or soon after. Growth appears to have slowed to reach near zero by around year 400, possibly because the limits of then current technology had been reached. After the Dark Ages in Europe, a second growth cycle began around the 10th century in Europe and Asia, with numbers rising to a peak of around 360 million during the 13th century, followed by a slight fall. The global population then increased slowly, perhaps with some falls (for example following European colonisation of the Americas in the 16th century, when the large native Indian population appears to have collapsed) until the 19th century, when an increasingly rapid rise began as a result of revolutionary developments in agriculture, industry and public health (Figure 4.1).

Crucially, the rate of global population growth peaked during the late 1960s; it was then at just over 2% per annum, but is now about 1.7%. And the absolute increase per annum has also peaked; it was around 85 million more people per annum in the late 1980s and is now about 80 million. Such trends suggest that the present global total of some 6000 million may not itself double, as all previous totals have done. The medium variant of the current UN long-range forecast suggests the total in 2050 may be 8900 million. Although several countries in Africa have yet to shift to lower fertility rates, thus making a further doubling quite possible, it may be that "children born today may be thinking about their retirement at a time when the global population count will have stabilized – or even begun to decline"[29].

The current human population is very unevenly spread across the land surface of the Earth. While some areas, such as most of Antarctica, the interior of Greenland, and hyperarid hot deserts, have no permanent human presence, in others, human densities may locally reach an extreme of 1000 inhabitants per hectare (eg. in Calcutta and Shanghai). Map 4.2 shows the current density of human populations over the Earth, based on census counts within administrative units of varying size. Among the most striking features of this map are the large areas of very high population density in parts of China and the Ganges-Brahmaputra lowlands, also on Java, and the large area of high population density extending over most of Europe.

Human activities have now made themselves felt throughout the biosphere, but it might be expected that the degree of transformation of terrestrial landscapes would be related to ease of access and proximity to population centres. Settlements, ranging in size from villages to cities of many million inhabitants, are connected by networks of paths, railways and waterways that allows the influence of humans to diffuse far beyond these settlements. Map 4.3 shows results of a GIS-based analysis of the relative distance of points on the Earth's surface

Figure 4.1

Human population

This graph shows the long period of many thousands of years during which the world human population remained small, followed by an exceptionally rapid rise to a total of six thousand million at present.

Source: data from McEvedy and Jones[27], FAO[34].

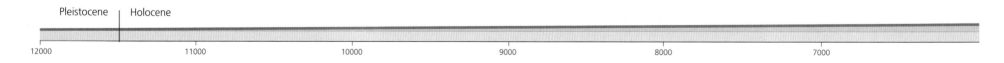

Pleistocene | Holocene

12000   11000   10000   9000   8000   7000

from all such human constructions. Those points most remote can be given a high wilderness value, and conversely, points surrounded by a high density of human structures can be given a low wilderness value. Although human population density was

as such not part of the analysis, there are, unsurprisingly, strong similarities between Maps 4.2 and 4.3.

The unchecked exponential rise in abundance of a single species, to a position of global ecological dominance – that is, in the sense of using a disproportionate share of natural resources, not of having managerial control of ecological processes – is without known precedent in the history of the biosphere. This was not achieved without significant, often adverse, effects on the environment, many of which stem from the agricultural activities required to maintain human numbers. There is no single species in which so many individuals, of such large body size, are distributed so widely over the planet. There appear to be few macroscopic species, the Antarctic krill *Euphausia suberba* being one of these few, in which the number of individuals approaches the size of the present human population. Several different kinds of mostly small or microscopic species together exceed human numbers. Certainly

Table 4.2

Number of individuals and biomass in selected organisms

Notes: Calculations based on data in sources cited after group names. Biomass is estimated dry weight standardised on 30% wet weight. Number of birds per km^2 estimated by dividing global bird population (2-4 x 10^{11}) by land area minus area of extreme desert, rock, sand, and ice (approx 125 million km^2). Estimates of abundance for Collembola represent minimum global totals. Whale estimates presume a sex ratio of 1:1. Mean Asian elephant weight estimated at 3 500 kg; range estimated at 500 000 km^2. African elephant weight estimated at 4 250kg.

Group	No. of Species	No. of Individuals		Biomass (kg)	
		Per km^2 of habitat	Globally	Per km^2 of habitat	Globally
Bacteria[55]	400 000	7.84 x 10^{21} – 1.18 x 10^{22}	4 x 10^{30} – 6 x 10^{30}	1.39 x 10^6 – 2.14 x 10^6	7.06 x 10^{14} – 1.09 x 10^5
Collembola[56]	6500	2.4 x 10^7	2 x 10^{15}	64	5 x 10^9
Termites[55,57]	2760	2.3 x 10^9	2.4 x 10^{17}	1400	1.44 x 10^{11}
Antarctic krill[58]	1	1.43 x 10^7	5 x 10^{14}	4.29 x 10^3	1.5 x 10^{11}
Birds[59]	9946	1600 – 3200	2 x 10^{11} – 4 x 10^{11}		
Elephants[60,61,64]	2	0.07 – 0.1	4.26 x 10^5 – 6.31 x 10^5	85 – 126	5.34 x 10^8 – 7.29 x 10^8
Great whales[62,63,64]	10	<0.01	2.83 x 10^6 – 3.6 x 10^6	52 – 65	1.89 x 10^{10} – 2.33 x 10^{10}
Domestic livestock[34] (excluding pets)	c 15				7.3 x 10^{11}
Humans	1		6 x 10^9		3.9 x 10^{11}
Humans plus livestock					**11.2 x 10^{11}**

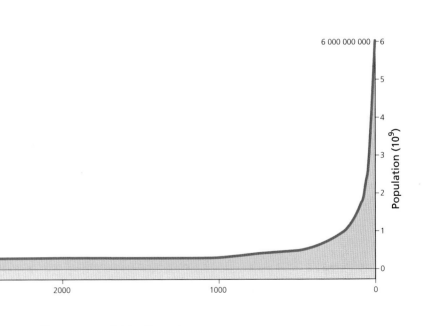

no animal of comparable size has a population remotely similar in size to that of humans.

Biomass provides a complementary measure, and is more indicative of the way in which global net primary production or NPP (see Chapter 1) is partitioned. If the standing crop of domestic livestock is added to the human biomass, amounting to around 11.2×10^{11} kg in total, then only the global biomass of bacteria as a whole is higher.

APPROPRIATION OF NET PRIMARY PRODUCTION

The human share of global resources

Agriculture can be viewed as simply a set of activities designed to secure a greater and more reliable share of the energy and materials in the biosphere for the benefit of the human species, ie. to divert an increased proportion of available energy toward production of human bodies. Foodstuffs specially grown by people require energy and nutrients that are not then available to wild species. In other words, the expansion of human populations is equivalent to human success in appropriating an increasing share of global net primary production.

Naturally occurring plants and animals are replaced by specially cultivated or bred varieties that can produce nutrients efficiently from available resources, and in a form that humans can conveniently use. The growth and persistence of these selected species are in effect subsidised by humans. In less developed agricultural systems, this subsidy may be very small, perhaps just the removal of competition for light or grazing; ie. plots are cleared or weeded and wild herbivores discouraged. Efficiency can be high but output is usually low. In western industrialised agriculture the subsidy is enormous. Competitors, pests and predators are removed from vast areas (herbicides, pesticides), fossil fuel is consumed to process, transport and apply any nutrients that limit production (nitrogenous fertilizer), and to store produce. Efficiency is high in some respects, eg. use of space and labour, but much lower if all hidden costs of fossil fuel use and waste impact are considered. Output can be very high.

Just how successful have humans been in diverting more than an equitable share of global production toward growing human bodies and their domestic plants and animals? Five thousand years ago, the amount of agricultural land in the world is believed to have been negligible. There is no direct evidence from the greater part of this period on the rate of expansion of agricultural land. Useful historical data relate to the past few hundred years, and this evidence suggests that about 265 million ha of land globally were devoted to crops in 1700. At present, arable and permanent cropland covers approximately 1510 million ha of land, with some 3400 million ha of additional land classed as permanent pasture (this figure includes rangeland and wooded land used for grazing). This represent a nearly six-fold increase in cropped land over the past three centuries (Table 4.3). All the cropland is used to produce domestic plant material, and much of the pasture is used to produce domestic herbivore biomass.

Most domestic herbivores are destined in large part to become human biomass, so in a sense can be counted as surrogate humans, and the rise in livestock numbers is one indication of human effectiveness in appropriating net primary

Table 4.3

Land converted to cropland

Source: 1700-1980 estimates from Richards[65], 1997 datum from FAO[34]

Year	1700	1850	1920	1950	1980	1997
Cropland area (million ha)	265	537	913	1170	1501	1510

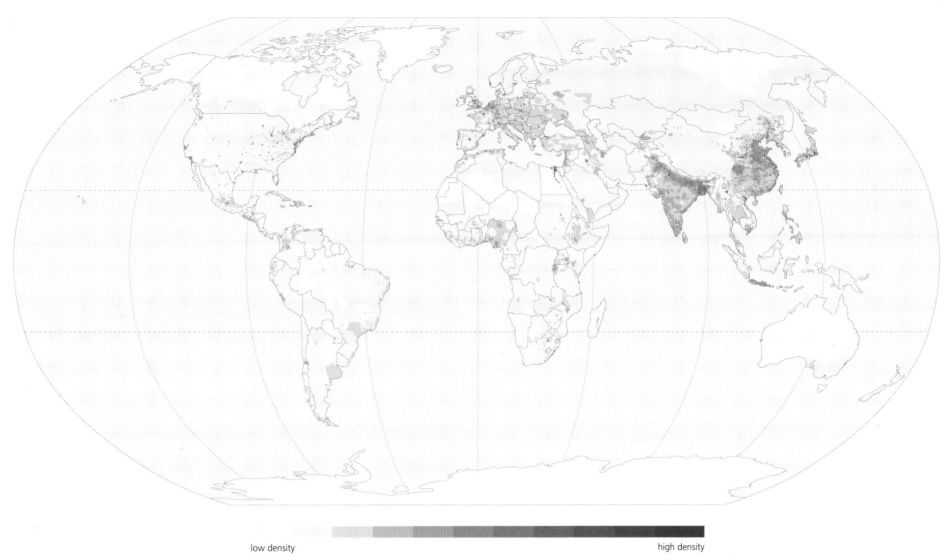

low density   high density

Map 4.2

Human population density

This map portrays the relative density of human population based on census data relating to administrative units of various sizes.

Source: demography project by Tobler *et al.*[87]; data available at http://www.ciesin.org/datasets/gpw/globldem.doc.html

Table 4.4

Comparison of large herbivore numbers and biomass in Mesolithic and Modern Britain

Source: after Yalden[88]

MESOLITHIC BRITAIN		n	kg	MODERN BRITAIN	n	kg
				Sheep	20 364 600	916 407 000
Wild boar	Sus scrofa	1 357 740	108 619 200	Pigs	853 000	127 950 000
Aurochs	Bos primigenius	99 250	39 700 000	Cattle	3 908 900	2 149 895 000
Red deer	Cervus elephus	1 472 870	147 287 000	Red deer	360 000	54 000 000
Roe deer	Capreolus capreolus	1 083 810	21 676 200	Roe deer	500 000	10 500 000
Moose	Alces alces	67 490	13 498 000			
				introduced deer	111 500	5 017 500
				feral sheep and goats	5 700	256 500
				domestic herbivore total	25 126 500	3 194 252 000
				wild herbivore total	977 200	69 774 000
herbivore total		**4 081 160**	**330 780 400**	**herbivore total**	**26 103 700**	**3 264 026 000**

production. This rise has been accompanied by a decline in wild herbivores. For example, the form believed ancestral to domestic cattle, the wild ox Bos primigenius, has been extinct through most of its former wide range in Eurasia and North Africa for about one thousand years, and was last recorded in the early 17th century, but there are now approximately 1318 million head of domestic cattle in the world[34]. Similarly, the American bison Bison bison, was reduced from perhaps 50 million head before European arrival on the continent to under 1000 at the end of the 19th century; although numbers have now recovered to over 200 000, there are now nearly 100 million head of cattle in the USA alone.

A recent study used contemporary data on the fauna of the Bialowieza forest in Poland (a forest remnant with populations of large wild herbivores and predators) as a basis to calculate the possible number and biomass of wild herbivores in heavily forested Mesolithic Britain. These estimates were compared with current herbivore populations in Britain. Indications are that there are now 40 times more domestic cattle than there were wild cattle (aurochs), and 20 times more domestic sheep than there were wild deer; the overall large herbivore biomass has increased by a factor of 10 (see Table 4.4). Such values are likely to apply to other heavily populated industrialised countries. Not only are domesticated mammals far more abundant than their wild relatives ever were, the latter are in many cases extinct or near extinction.

Human appropriation of NPP

Human appropriation of NPP may be defined as the difference between the NPP of potential vegetation and the amount of this production that remains after human harvesting of resources, and anthropogenic changes in land use and cover. On the basis of a number of well-founded estimates of land use and resource consumption, it has been calculated[30] that humans appropriate around 40% of global annual net production. This sum includes the amount directly consumed by humans and domestic stock, plus the amount indirectly used in anthropogenic systems (cropland, grazing land, shifting cultivation), plus the amount foregone because of human habitation, pollution, desertification and permanent changes in land use. The magnitude of the percentage figures of course depend on the value of estimated total primary production. These data are summarised in Table 4.5.

A subsequent study took account of production uses and did not assess biomass uses (such as fuelwood consumption, timber harvest and wastage), because the long term effect on production and wild species could not readily be assessed. This study suggested that between 20% and 30% of global terrestrial net primary production is appropriated by humans[31].

The estimated proportion of NPP that is appropriated by humans globally, whether 20% or 40%, is surprisingly large, but a figure of the same order of magnitude resulted from a recent study of the situation in Austria. This work[32] used very high resolution national survey data to estimate the amount of biomass currently produced in the country as compared with production by the potential natural vegetation. It was calculated

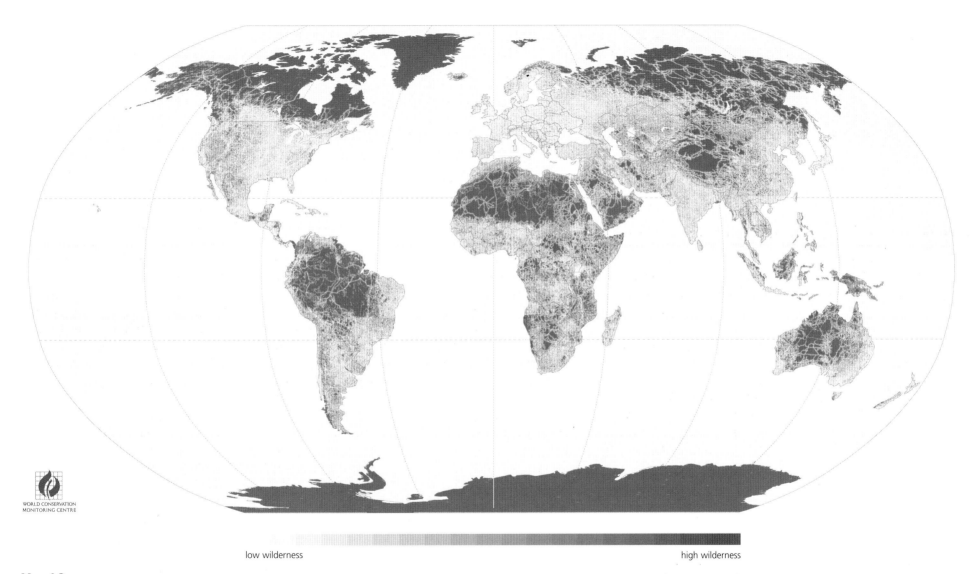

low wilderness   high wilderness

Map 4.3

Terrestrial wilderness

The wilderness value of any given point is essentially a measure of remoteness from human influence, assessed on the basis of distance from settlement, access and permanent manmade structures.

Source: GIS analysis by R. Lesslie (ANU), method developed for the Australian Heritage Commission.

WORLD CONSERVATION
MONITORING CENTRE

Table 4.5

Global human appropriation of net terrestrial primary production.

Note: values are given as in the original source, but see Chapter 1 for recent estimates of global production

Source: data extracted from Vitousek et al.[30]

	Production	% of total
Total terrestrial production (59% of global total (224 x 10⁹))	132 x 10⁹ t	
Amount used directly by humans and stock	5.2 x 10⁹ t	4%
Amount directly and indirectly	40.6 x 10⁹ t	30.7%
Amount foregone because of environmental change	17.5 x 10⁹ t	13.2%
Total terrestrial production appropriated by humans	58.1 x 10⁹ t	38.8%

that humans appropriate 51% of the potential available aboveground production.

If the proportion of NPP used or diverted by humans has increased over the past two million years to between 20% and 50% of the total, the amount of energy available to non-human biodiversity has therefore reduced correspondingly during the same relatively brief period. This amounts to a very radical change in the flow of energy and materials in natural ecosystems. It seems inconceivable that these changes will not have a significant and continuing impact on biodiversity, but the nature and size of these impacts are not at present clear.

One approach to assessing the effects of energy appropriation uses the empirical relationship between energy and species number[31]. As noted in Chapter 5, there is much evidence suggesting that at a range of spatial scales and within different taxonomic groups, the diversity of species present in an ecosystem tends to be positively correlated with the amount of energy available. Accepting the empirical evidence for this relationship, it has been argued that the number of species present will decline if the amount of energy available for use declines. At global level, a conservative estimate using this relationship predicts that 3% to 9% of terrestrial species will be extinct or endangered by the year 2000[31]. The evidence from the Austria study[33] is consistent with the species-energy relationship: the curve developed by Wright[31] predicts that with 41% of potential NPP at country level now being appropriated by humans, 5-13% of species should have been extirpated from the country, in fact, 8% of birds and 7-14% of reptiles have been lost.

Current foodstuffs

Food plants exemplify the most fundamental values of biodiversity. The variety of species used is no doubt limited much more by cultural factors, such as tradition and palatability, than by nutrient content. Estimates suggest that some 7000 of the 270 000 described plant species have been collected or cultivated for consumption[36].

Perhaps more remarkable is the fact that very few, some 200 or so, have been domesticated, and a mere handful are crops of major economic importance at global level[35]. A dozen crop plants together provide about 75% of the world's calorie intake. These comprise: bananas/plantains, beans, cassava, maize, millet, potatoes, rice, sorghum, soybean, sugar cane, sweet potatoes and wheat[36]. There are far fewer animal than plant species contributing to the human diet, and although most consumption of wild species will go unreported, this must be insignificant at global level in comparison to products from just three domestic forms: pigs, cattle and chickens.

At country level, a much wider variety of plant species are important in that they make a significant contribution to nutrient supply[66]. Around 22 species and groups of species provide more than 5% of the per capita supply of either calories, protein or fat in at least 10 countries of the world. Notes on the origin, uses and genetic resources in these 22 crops are given in Table 4.9 at the end of this chapter. The following table (4.10) includes briefer information on the remaining 50 or so crops that are also nutritionally important but to a smaller number of countries.

At world level, cereals are the most important single class of food commodities, providing around 50% of daily calorie supply

Table 4.6

The top ten food commodities, ranked by percentage contribution to global food supply.
Source: FAO food balance sheets[34], 1997 data.

% calories/person/day		% protein/person/day		% fat/person/day	
Rice (milled equiv.)	21	Wheat	22	Pigmeat	14
Wheat	20	Rice (milled equiv.)	15	Soyabean Oil	10
Sugar (raw equiv.)	7	Milk (excl. butter)	9	Milk (excl. butter)	9
Maize	5	Pigmeat	5	Rape and Mustard Oil	6
Milk (excl. butter)	4	Bovine Meat	5	Sunflowerseed Oil	6
Pigmeat	4	Poultry Meat	5	Palm Oil	5
Soyabean Oil	2	Maize	5	Fats, Animals, Raw	5
Potatoes	2	Vegetables (other than		Bovine Meat	4
Vegetables (other than		tomatoes, onions)	3	Poultry Meat	4
tomatoes, onions)	2	Eggs	3	Butter, Ghee	4
Cassava	2	Pulses (other than			
		peas, beans)	2		
other	31	other	25	other	34
Global mean: 2745 cal/cap/day		Global mean: 73.4 grams/cap/day		Global mean: 70.1 grams/cap/day	

and 45% of protein. In contrast, meat provides around 16% of the protein, and fishery products only some 6%, or 15% of all animal protein.

If the sources of calories, protein and fat in the global human diet are assessed, in each case just two or three commodities stand out from a large number of commodities of lesser importance, many of them of very small importance globally. Rice and wheat together provide around 40% of the world supply of both calories and protein, while milk and pigmeat are key sources of fat, calories and protein (see Table 4.6).

Global food supply

National level data on reported food commodity supplies are collated by the FAO[34]. Given information on the food value of these commodities, and the size of the human population, it is possible to estimate the average national food supply per person.

Dietary food value can be broadly assessed in terms of energy or materials. The former is conventionally measured in calories

([kilo]calories/person/day); the latter in terms of weight of protein or fat (grams/person/day). These standard measures take no account of vitamins and minerals (micronutrients) that are required for maintenance of full health. The nutritional value of the human diet varies geographically and over time, and to a great extent, according to the state of development and purchasing power of the people concerned.

The human population of the world doubled between 1950 and 1990, reached around 6000 million in the late 1990s, and will continue to grow for decades, albeit at a slower rate because of decreased fertility. At global level, there has been enough food available in recent years to supply the entire human population with a very basic diet, largely vegetarian, providing 2350 kcal/day. In 1992 the average food supply was estimated at 2718 kcal/day (after losses during storage and cooking), comprising 2290 kcal from plants and 428 from livestock products[37]. Thus the global food supply was nominally sufficient for 15% more people than the actual population, and a similar small annual surplus has existed since the 1970s[38]. Taking these aggregated global data at face value, sufficient food has been produced annually during the past two decades to feed the world's human population.

Each year during the past two decades, between 850 million and 900 million people have been undernourished[38]. Given that there has been sufficient food available in the world overall, undernourishment must be an effect of unequal access to food. Unequal distribution is evident at macro and micro scales: some countries are more favourably endowed than others, and whether at national or village level, high status social groups secure better diets than others, compounded in some cases by gender differences. Poverty is a key cause of undernutrition, and often also an effect of it, forming a self-reinforcing cycle that is difficult to escape from without appropriate outside intervention. Put in different terms "food insecurity is a problem of lack of access resulting from either inadequate purchasing

Table 4.7

Proportion of the world human population in each diet class, and predicted increases needed by 2050.

Note: Population calculation based on 1997 data, from FAO website. Excludes Japan and Malaysia, in anomalous position using six part classification. Total population in 117 countries in sample: 5 462 493 000. The penultimate row in fact shows the percent each class forms of the total population in the sample countries; this is assumed to be an acceptable surrogate for the total world population.

Source: FAO[37]

diet class	class 1: rice	class 2: maize	class 3: wheat	class 4: milk, meat, wheat	class 5: millet, sorghum	class 6: cassava, plantain, taro, yams
geographic distribution	Asia	Central and S America (north)	N Africa, West Asia, S America (south)	Europe, N America, Australia	Sahel, Namibia	Central Africa, Madagascar
human population (x 1000)	2 920 923	514 911	664 507	942 924	61 867	357 361
% of world population	54	9	12	17	1	7
increase needed by 2050	x 2.37	x 1.96	x 2.84	x 1.13	x 4.82	x 7.17

power or an inadequate endowment with the productive resources that are needed for subsistence"[37].

Regional variation

The 'global average diet' is a simplifying abstraction that ignores an enormous amount of regional, national and local variation in food sources and in supply. Nor does it take account of micronutrient availability (vitamins, minerals, trace elements), ie. substances that do not directly contribute to energy or protein intake but which are nevertheless essential elements in a healthy diet.

The human diet can be assessed in several ways. The FAO devised a simple classification to serve as the basis for an analysis of food requirements in relation to population growth[37]. In this scheme, diet is assessed in terms of calorie sources as recorded at national level in the FAO food balance database, and countries with similar diet structures are clustered together in one of six diet classes (Table 4.7, Map 4.4). Each class is named after the food product that best characterises the diet (although this is not necessarily the principal calorie source). Although the method used to produce the classification is not made explicit, and it has generated some anomalies (FAO note

the uncertain placement of Japan and Malaysia, and the absence of Madagascar from the rice class also appears anomalous) it does provide a useful sub-global overview of major diet patterns.

The countries in each food class tend to share broad demographic features. Rice countries (class 1) have very high population densities, higher than average mortality rates and little diet diversification. Maize countries (class 2) generally have population densities near the world average and low mortality, especially infant mortality. Wheat countries on average have low population density, but this masks serious land and water shortage in many. Class 4 countries include the world's most highly developed nations, characterised by fertility, mortality and population growth rates well below global average.

Millet countries (class 5) tend to have high population growth, high fertility and low life expectancy; the diet provides only a marginal surplus of energy supplies over requirements. Diet class 6 countries hold the human populations most at risk from food insecurity. The diet is characterised by roots and tubers, and on average does not provide basic energy

Map 4.4

FAO world diet classes

This map is based on a six-part classification of country dietary patterns, in particular on calorie sources. Each class is named after the foodstuff that best characterises the diet (not necessarily the main calorie source). The classification of a few countries (eg. Japan, Malaysia) appears anomalous.

Source: analysis and data by FAO[37].

| no data | class 1 rice | class 2 maize | class 3 wheat | class 4 milk,meat, wheat | class 5 millet, sourghum | class 6 cassava,plantain, taro, yams |

WORLD CONSERVATION
MONITORING CENTRE

requirements. These populations show high fertility, high mortality, and rapid growth in numbers. Poverty is high, infrastructure is weak, but there are significant reserves of under-exploited arable land.

Meeting future needs

At global level, human numbers and overall food supply have both increased during the present century, and the adequacy of food supplies at this level is determined by the relation between these parameters. Global food supply has doubled since the 1940s, and there is currently just sufficient food to meet the basic needs of the world human population. Increased food supply in the last two decades is attributed mainly to increased productivity (69%), and secondarily to increase in production area (31%).

The predicted severe deterioration in the already poor food supply situation in sub-Saharan Africa, where most of the people in diet classes 5-6 live, is in large part due to the delay the region has experienced in moving through the demographic transition shown by other countries. This involves a lowering of fertility in compensation for reduced mortality, and a consequent reshaping of the age structure of the population, with increasing numbers in mature age classes. The average rate of population growth is likely to remain higher than the world average in sub-Saharan Africa. It has already been pointed out[37] that meeting the anticipated needs in this region would demand a complete change in the current scale of development.

However, numbers are set to rise further until around 2050. In order to meet the per capita energy needs of the human population at that time, it is predicted that the food supply could remain little changed in Europe and North America, but would need to approximately double in Central and South America, and Asia, and to increase fivefold in Africa. In terms of diet class, class 4 could remain little changed, classes 1, 2 and 3 would need to double, and classes 5 and 6 would need to increase fivefold and sevenfold, respectively.

A global increase in crop yields appears to be technically feasible[37]. A large part of such increase is likely to arise from use of genetically improved cultivars, thus continuing recent trends in which, for example, modern varieties of rice and wheat have increased from 20-30% of total seeds to more than 70%[37]. However, some staple crops have been relatively little improved; these include the millets, sorghum, roots and tubers typical of diet classes 5 and 6, and so of key importance to food security in large areas of the world. In fact, it is estimated that over half the world's farmers, approximately 1400 million people, support a quarter of the world's population without the use of modern high-yielding crop varieties[39]. Most development in transgenic crops, as with other modern crops, are targeted towards improvements in economically-valuable traits and production in highly intensive systems[40]. As a further challenge to the transfer of plant breeding technology and its products to developing countries, most current research is carried out and patented by private corporations.

In some cases, food security may be better ensured by diversification rather than increasing yields. Although contemporary agricultural methods help to ensure high productivity of individual crops, they also employ physical space and limiting resources such as nitrogen less efficiently, and are associated with a lower total primary productivity[41]. The limitations of agriculture based on uniform varieties and a high input of fertilizers and pesticides become more acute for farmers who rely on poor resources or marginal environments. By growing diverse and locally adapted crops farmers can bring about greater security in food production and more efficient use of limiting resources. Many traditional agricultural systems manage to varying degrees a high diversity of both cultivated and wild food species (Table 4.8), and in a sense form an interface between wild plants and domesticated crops. Strong selection pressures exerted by humans over several millennia

Table 4.8

Examples of diversity in agricultural systems

Source: multiple sources[72-85]

Country	People	Food production and diversity
Brazil	Kayapo	Over 45 tree species planted for food or to attract game; 86 varieties of food plants planted
Ecuador	Siona-Secoya	Major staples: 15 vars of manioc, 15 vars of plantain, 9 vars of maize; pre 1978 Traditional gardens yield 12 300kg of food or 8.8 million calories. 72% calories & 14.8% protein, 22.2% fats, 90.9% carbohydrates; post 1978 horticulture provides 67.8% calories, 10.2%, purchased
Indonesia	Java	500 species in home gardens in a single village. Ability to support 1000 people per sq.km.
Indonesia	West Sumatra	6 main tree crops, many vegetables and fruit. 53 species of wild plants also protected and harvested
Kenya	Bungoma	100 species of fruit and vegetables; 47% households collect wild plants, 49% maintain wild plants in gardens
Kenya	Chagga	Over 100 species produced in gardens
Mexico	Huastec	Over 300 useful species found in managed forest plots called te'lom, 81 food species
Mexico	Migrant rural community in south east	338 species of plants and animals in home gardens, including 62 species of wild plants
Papua New Guinea	Gidra	Approximately 54% calories and 82% of protein come from non-purchased sources (wild, sago, coconut gardens)
Peru	Bora	22 varieties of manioc and 37 tree species are planted; 118 useful species found in fallow fields
Peru	Santa Rosa	168 species identified in 21 home gardens
Philippines	Hanunoo	System of intercropping with 40 crops in a single field. Over 1500 plants considered useful, of which 430 may be grown in swiddens
Sierra Leone	Gola forest	Of food items 14% are hunted, 25% are from fallow land, 8% are from plantations, 19% are from farm, swamp or garden, 21% are from streams and rivers, 13% is bought or given
Thailand	Lua	110 varieties of food plants and 27 wild food plants are found in swidden fallows

and wide geographical areas have resulted thousands of varieties within most crop species.

High agricultural diversity can not only provide insurance against crop failure in difficult agricultural environments, but tends to have nutritional benefits. Transition to more uniform diets, with high intake of fats and sugar, has resulted in declining nutritional status among numerous indigenous groups[42-44]. Low dietary diversity is associated with micronutrient deficiency, a problem far more common than general protein-energy malnutrition, and particularly prevalent in children, pregnant women and breast-feeding mothers. Diverse cropping systems in backyard and home gardens, whether rural or urban, can lead to direct improvement in family nutrition and in some cases provide cash income. Even a very small mixed vegetable garden is capable of providing 10-20% of the recommended daily allowance of protein, 20% of iron, 20% calcium, 80% Vitamin A and 100% Vitamin C[45].

The success of new commercial varieties has resulted in rapid declines in all forms of diversity in agriculture. Reported losses of over 80% of varieties in species such as apple, maize, tomato, wheat and cabbage have occurred in developed and developing countries alike. Increasing concern over genetic erosion in the 1970s led to establishment of what is now the International Plant Genetic Resources Institute (IPGRI) and increased efforts to collect germplasm for ex situ collections. The main incentive came from the recognition of the importance of plant genetic material in the development of new varieties. Between 1950 and 1992 the average number of landraces contributing to new wheat varieties from the International Centre for Maize and Wheat Improvement (CIMMYT) rose from 12 to 64. Genetic exchange, via artificial or natural means, is able to introduce valuable characteristics into commercial varieties, conferring for example, resistance to disease and pests, improved nutritional content and taste, and improved capacity for storage and transportation. About half of the increase in yields of wheat, rice and maize has been ascribed to genetic inputs.

For most major crops, breeders' collections are sufficiently large to provide an adequate source of additional genetic material. Material from landraces and conspecific wild populations (primary gene pool) are also frequently called upon. FAO have estimated that 30-40% of productivity gains overall have relied on genetic contributions from landraces[34]. The secondary gene pool, consisting of related species in the wild or in cultivation, has also provided important and economically valuable contributions to major crops. However, the difficulty of crossing different species using conventional methods has until now limited the use of this genetic reserve. Gene transfer technology has the potential to avoid some of the difficulties limiting conventional techniques, and brings the possibility of introducing into cultivars traits from an unlimited gene pool, including both close relatives and completely unrelated species. Such processes could perhaps provide new economic incentives to conserve agricultural biodiversity.

CURRENT USE OF WILD RESOURCES

As we have noted above, the vast bulk of human food supply on a global level is derived from farming. However, a very small number of people in the world still derive most of their requirements from wild sources. Ethnographers of these non-agricultural, non-industrialised societies have stressed the distinction between fishing and the hunting or gathering of terrestrial resources[46]. In a sample of 220 such societies, nearly 40% have a high dependence on fishing, around one third are highly dependent on gathering, and 28% on hunting of terrestrial resources. Fishing and gathering tend to be alternative activities both complementary to terrestrial hunting, and in contemporary conditions fishing tends to be more important where temperature is lower (especially in high northern latitudes) and the converse applies to gathering. In addition to food, humans in such communities still derive shelter, medicine, fuelwood, and aesthetic or spiritual fulfillment from wild species. Ethnographers have documented the immense variety of plant and animal species used by many such groups.

A far larger number of people worldwide, while principally agriculturalists or pastoralists, still make extensive use of wild resources. In parts of Africa, 'bushmeat' (meat from wild animals) may supply most of the protein intake; similarly some 80% of people in sub-Saharan Africa are believed to rely largely or wholly on traditional medicines derived almost exclusively from wild sources. Indeed, traditional medicine continues to be the source of health care for the majority of people living in developing countries, and is widely incorporated in primary health care systems. Wild plants in many areas are extremely important as famine foods when crops fail, or may provide important dietary supplements and use of fuelwood and charcoal from wild sources is almost universal in the developing world.

Even in developed societies a surprising number of commodities may be derived at least in part from wild sources. Foremost among these in economic terms are timber (hardwoods, some softwoods) and fishery products, while live animals and animal products such as furs, skins and ivory are also important, and an enormous number of species are harvested for food from natural or semi-domestic habitats worldwide. A few wild resources, mainly plants, provide substances of direct medicinal use, or indirectly have provided natural molecular models for the industrial synthesis of medicinal substances.

Wood is one of the few commodities used and traded worldwide that is mainly harvested from wild sources (see forests, Chapter 7); it is also one of the economically most important commodities in international trade. The annual value of wood exports worldwide totals several billion US dollars, and such exports form a significant part of the earnings of many tropical developing countries. Most timber in world trade consists of softwood (conifers) from north temperate sources, and a significant proportion is from plantation forests. Hardwoods, including highly valued timbers such as mahogany and teak, are produced mainly from natural forests in the tropics

Fishes and other fishery products (see Chapters 6, 8) make up the second class of commodities derived mainly from wild sources by direct harvest, and which are of great economic importance in world trade. These resources are also of crucial importance to global food security. Annual landings of aquatic resources have increased nearly five-fold in the past 40 years; while capture fisheries have stagnated since the start of the 1990s, the total harvest now exceeds 100 million tonnes with the deficit being filled by aquaculture production. This can have production benefits, but in many parts of the world the coastal environment has been severely degraded by conversion to mariculture operations.

Before the advent of modern medicine, all biologically active compounds were derived from natural sources. interest in traditional alternatives to modern medicine is increasing in many developed countries. The vast majority of botanical material for medicinal use is still collected from the wild, eg. some 70-90% of the herbal medicine imported into the major European centre for this trade is estimated to be of wild origin[47]. Few medicinal species are cultivated[48] and many wild populations are now at risk from over exploitation. Of the estimated 10 000 to 20 000 plant species used medicinally, the pharmaceutical properties of some 5000 have been laboratory tested[49,50,51]. A number of studies have highlighted the importance to modern drugs of ingredients derived from natural sources. They are used either as a direct source of beneficial compounds, as a blueprint for their manufacture, or as means for their study. Amongst the top 150 most prescribed drugs in USA, 56% contain compounds which are attributable at some point in manufacture or design to animals (23%), plants (18%), bacteria (4%) and fungi (11%)[52.] This contribution may be translated into an economic value of at least $80 billion per annum in USA alone[54].

In addition to these direct-use benefits there are, of course, enormous other less tangible benefits to be derived from natural ecosystems and their components. These include 'ecosystem services' (essentially the continued maintenance of a functioning biosphere in a state fit for human habitation) and the values attached to the persistence, locally or globally, of natural landscapes and wildlife, values which increase as such landscapes and wildlife become scarcer. A recent global level review of the subject[53] produced a first-order approximation of some 33 trillion dollars (ie. US$ 33×10^{12}) per year as the total value attributable to natural ecosystems, this being nearly twice the global economy as expressed as the sum of gross national products (18 trillion dollars per year). Although many of the assumptions made in deriving this estimate are highly questionable the study has at least served to demonstrate in economic terms the ultimate dependence of humankind on the natural world.

Table 4.9

The 22 nutritionally most important food crops

CEREALS

Notes: These species and groups of species are those that, according to national food supply data maintained by the FAO[66,34], provide 5% or more of the per capita supply of calories, protein or fat in at least 10 countries.

Sources: list of crops from Prescott-Allen and Prescott-Allen[66], other information from Mabberley[67], Smartt and Simmonds[68], Smith *et al.*[69], Vaughan and Geissler[70]; conservation status from Walter and Gillet[71].

Millets

Echinochloa frumentacea, Eleusine coracana, Panicum miliaceum, Pennisetum glaucum, Setaria italica (Gramineae)

Main uses

Ec. frumentacea: quickest growing of all millets, available in 6 weeks. Used for human consumption in India and East Asia and for animal fodder in USA.

El. coracana: important staple in East and Central Africa and India. Wild cereal is harvested during times of famine. In Africa it is the preferred cereal for brewing beer. Seed heads may be stored for 10 years.

Pa. miliaceum: cereal cultivated for human consumption, mainly in northern China, Russia, Mongolia and Korea, and as animal feed elsewhere. In Europe it is grown mainly as bird seed. Millets are generally tolerant of poor soils, low rainfall and high temperatures and are quick maturing.

Pe. glaucum: most widely grown of the millets. The main cereal of the Sahel and northwest India. Heat and drought resistant. May contribute to incidence of goitre.

S. italica: a once important cereal that has declined in popularity, still grown on relatively large scale in India and China, mainly for home consumption. Used also animal fodder and bird seed. Early maturing and good storage

Production

29 055 086 mt.

Origins

Echinochloa: different strains are thought to have partially different origins. Approximately 35 spp. exist in the genus, distributed in warm areas.

Eleusine: eastern and southern Africa highlands. 9 spp. in the genus in Africa and South America.

Panicum: unknown in wild state. The closest relative *P. miliaceum var. ruderale* is native to Central China. At least 500 spp. in the genus in tropical to warm temperate areas.

Pennisetum: cultigen originated in West Africa from *P. violaceum*. Total of 130 spp. in the genus, found in tropical and warm areas.

Setaria: native to temperate Eurasia. Approximately 150 spp. in the genus in tropical and warm areas.

Related species

Echinochloa: *E. pyramidalis* (tropical & S. Africa and Madagascar) used as fodder and locally as flour; *E. turnerana* Channel Millet (Australia) is a promising forage and grain crop. Several other spp. are weeds.

Panicum: *P. hemiotum* (pifine grass, N America) and *P. texanum* (Colorado grass, N America) used as fodder; *P. maximum* (Guinea grass, Africa, naturalised America) a cultivated forage crop; *P. sonorum* (Mexico) minor grain; *P. sumatrense* (little millet, Malaysia) a minor grain.

Pennisetum: fodders, lawn-grasses, some grains. *P. hohenackeri* (moya grass, E Africa to India) is suggested for paper making; *P. clandestinum* (Kikuyu grass, tropical Africa) pasture grass, erosion control, lawns; *P. purpureum* (elephant or Napier grass, Africa) fodder and paper. *P. violaceum* (Africa) harvested during times of famine.

Setaria: *S. glauca* (Yellow foxtail) cattle fodder; *S. pallidifusca* is a cereal in Burkino Faso; *S. palmifolia* (India) shoots eaten in Java; *S. pumila* cultivated as cereal; *S. sphacelata* (S Africa) an important silage crop.

Genetic base

Echinochloa: 1 spp. listed threatened in 1997.

Eleusine: 5 races of cultivated finger millet recognised from Africa and India. Excellent prospects for improvement. Significant annual yield increases in India, mainly due to the incorporation of African germplasm.

Panicum: 16 spp. listed as threatened in 1997.

Pennisetum: *P. violaceum*, the wild progenitor, is an aggressive coloniser and may be found in large populations around villages in West Africa. The cultivated crop is relatively undeveloped. Open-pollinated cultivars are popular in Africa and India. 5 spp. listed threatened in 1997.

Setaria: largely a crop of traditional agriculture systems. 2 spp. listed threatened in 1997.

Breeding

El. coracana: wild spp. in Africa crosses with domesticated finger millet to produce fertile hybrids which can be obnoxious weeds.

Pe. glaucum: genetic exchange with related wild forms in the same geographical area is possible.

S. italica: hybridises easily with wild relative *S. italica* var. *viridis*, to produce fertile offspring.

Germplasm collections

90 500 general millet accessions, 45-60% of landraces and 2-10% of wild spp. represented.

Barley

Hordeum vulgare (Gramineae)

Main uses

Early maturing grain with high yield potential, can be grown where other crops fail: eg. above Arctic circle, at high altitude and in desert and saline areas. Most important as animal feed, also for brewing beer and human food.

Production

133 419 989 mt. Main producers in Europe, North Africa, Near East, Russia, China, India, Canada, USA

Origins

Southwestern Asia. Approximately 20 spp. in the genus, distributed in the North Temperate region.

Related species

H. *distichon* (2-rowed barley) is possibly H. *vulgare* x H. *spontaneum*.

Genetic base

More genetic diversity in wild barley exists in Israel than in entire domesticated crop. Landraces have been almost completely replaced by pure line cultivars and the change in genetic structure of barley populations has been profound. Important contribution of Ethiopian barleys highlights need to broaden genetic base. 2 spp. were listed as threatened in 1997.

Breeding

Fertile hybrids between wild and cultivated forms occur naturally where ranges overlap. Crosses possible with other spp. in the genus but not utilised in barley cultivars. Ethiopian barleys have been important as genetic source of disease resistance and improved nutritional value.

Rice

Oryza glaberrima, O. *sativa* (Gramineae)

Main uses

Highest world production of all grains. Primary source of calories and protein in humid and subhumid tropics. The grain is relatively low in protein; brown rice a source of some B vitamins. Rice bran is used in animal feeds and industrial processes. Can grow in flood prone areas.

Production

586 787 182 mt. Main producers are China, India, Indonesia, Bangladesh, Vietnam. Only 4% of world production is exported. Main exporters Thailand, Vietnam, Pakistan, USA.

Origins

2 cultigens appear to have been domesticated independently. The origin of O. *sativa* is uncertain, possibly derived in several centres from O. *rufipogon* (selected weed in *Colocasia* fields). Archaeological evidence suggests origin in China or southeast Asia. Centre of diversity of O. *glaberrima* is the swampy area of the Upper Niger. Approximately 18 spp. in the genus, distributed in the tropics.

Genetic base

Following agricultural intensification many populations of wild relatives have disappeared or intergraded with domesticated rice. Reduced genetic base has led also to repeated pest epidemics. Great genetic diversity exists in O. *sativa* cultivars; much less in O. *glaberrima*. Rapid spread of improved rice varieties has displaced tens of thousands of landraces, many now extinct. O. *glaberrima* is rapidly being replaced by O. *sativa*. Genetic erosion is reported in China, Philippines, Malaysia, Thailand and Kenya. 3 spp. were listed as threatened in 1997.

Breeding

O. *sativa* has formed numerous hybrids with wild spp., O. *nivara* and O. *rufipogon*. Genes improving tolerance to diseases or adverse conditions have been derived from African rices and wild relatives.

Rye

Secale cereale (Gramineae)

Main uses

Cereal crop used as animal feed and for human consumption. Eaten mainly as rye bread and crispbread. Higher in minerals and fibre than wheat bread. Previously more popular as a bread flour, now largely replaced by wheat. Still important in cooler parts of northern and central Europe and Russia, cultivated up to the Arctic circle and up to 4000m altitude. Tolerates poor soils. Also used in brewing industry and young plants produce animal fodder.

Production

21 424 560 mt. Main producers are Russia and Europe.

Origins

Probably originated from weedy types in eastern Turkey and Armenia. *S. montanum* is probable ancestor. Total of 3 spp. in the genus in Eurasia.

Genetic base

A number of weed ryes, found associated with agriculture throughout the Near East, are now considered to be subspecies of *S. cereale*. A complex of subspecies of *S. montanum* extends from Morocco east to Iraq. 5 spp. were listed threatened in 1997.

Sorghum

Sorghum bicolor (Gramineae)

Main uses

Staple cereal in semi-arid tropics. Mostly grown in developing countries, especially for domestic consumption by small farmers in Africa and India. Used in brewing beer and as animal fodder. Grain stores well.

Production

65 810 360 mt. Main producers are USA, India, China, Nigeria, Sudan.

Origins

Developed primarily from the wild *S. arundinaceum* in Africa. Total of 24 spp. in the genus in warm areas of the Old World and Mexico.

Related species

Backcrosses with *S. arundinaceum* gave *S. drummondii* cultivated for forage; *S. halepense* (Mediterranean), is a widely naturalised fodder plant, often weedy.

Genetic base

Wide variation in landraces. Genetic erosion reported in Sudan. Modern varieties have not been widely popular for use as human food. 2 spp. listed threatened in 1997.

Breeding

Wild relatives may be important source for disease resistance.

Germplasm collections

168 500 accessions. 21% in the International Crops Research Institute for the Semi-Arid Tropics. 80% of landraces, 10% wild spp. represented.

Wheat

Triticum aestivum, T. turgidum (Gramineae)

Main uses

Most widely cultivated crop. Grain is gluten rich and highly valued for bread making. 90% of wheat grown is *T. aestivum*. Wheat germ oil is highly unsaturated and high in vitamin E. Durum wheat, *T. turgidum*, has higher protein content. Used for making pasta. High nutritive value, easy processing, transport and storage

Production in 1999

578 336 671 mt. Main producers are China, India, USA, France and Russia

Origins

Mediterranean and Near East. Origin is complex and not fully understood, probably involving *Aegilops* spp. Total of 4 spp. in the genus, distributed from the Mediterranean to Iran.

Genetic base

Large variation in the crop, around 25 000 different cultivars. However, large areas are planted with genetically uniform crops and the inflow of landrace material into breeding programmes is low (8%). Genetic erosion is reported in China, Uruguay, Chile and Turkey.

Germplasm collections

Approximately 850 000 accessions. Largest collection (13%) in Centre for Maize and Wheat Improvement. 95% of landraces and 60% of wild spp. collected.

Maize

Zea mays (Gramineae)

Main uses

Mostly grown for human consumption in parts of Africa and Latin America, elsewhere mainly for animal fodder. Starch may be extracted and used in food processing. The germ oil is important. Also used in the brewing industry.

Production

599 708 424 mt (maize), 8 425 132 (green corn). Main producers are USA, China, Brazil, Mexico, France.

Origins

Probably derived from teosinte, *Zea mays* ssp. *mexicana*. Total of 4 spp. in the genus, confined to Central America.

Genetic base

Most of the world's maize crop is derived from a few inbred lines. Landraces represent 40% of the crop grown in developing countries. Genetic erosion is reported in Mexico, Costa Rica, Chile, Malaysia, Philippines, Thailand. *Z. perennis*, was presumed extinct in the wild until its rediscovery in 1977. *Z. diploperennis* was recently discovered and is now protected in the Sierra de Manantlan Biosphere Reserve, Mexico. 3 spp. were listed as threatened in 1997.

Breeding

Teosinte crosses readily with maize to produce fertile offspring, *Tripsacum* crosses with less success. Neither has been widely used in breeding programmes.

Germplasm collections

277 000 accessions, the largest existing at Indian Agricultural Research Institute. 95% of landraces and 15% wild spp. are represented.

Yams

Dioscorea spp. D. *alata*, D. *bulbifera*, D. *cayenensis*, D. *dumetorum*, D. *esculenta*, D. *rotundata*, D. *trifida* (Dioscoreaceae)

Main uses

Edible stem tuber, 28% starch and limited vitamin C. Important staple in the humid and subhumid tropics. Also major ingredient in oral contraceptives. Religious and cultural role. Good storage properties.

Production

36 032 584 mt. West Africa produces 90% of world production, Nigeria alone produces 70%.

Origins

3 main independent centres of diversity or domestication in Asia, Africa and America. Approximately 850 spp. exist in the genus, distributed in tropical and warm regions.

Genetic base

Predominantly a subsistence crop. Apparently little genetic erosion. Large genetic variability in wild edible forest yams. 68 spp. were listed as threatened in 1997.

Breeding

New World and Old World spp. show strong genetic isolating barriers and crosses between them are not successful.

Cassava

Manihot esculenta (Euphorbiaceae)

Main uses

Edible tuber containing 35% starch and vitamin C. Cultivated in almost all tropical and subtropical countries, mainly by smallholders. One of the most efficient crops for biomass production. A good famine reserve, able to withstand harsh conditions. Also animal feed.

Production

165 469 465 mt. Main producers are Brazil, Nigeria, D.R. Congo, Thailand, Indonesia.

Origins

Unknown in the wild state. Total of 98 spp. exist in the genus, occurring between southwest USA and Argentina. Most diversity occurs in northeast Brazil and Paraguay and in west and south Mexico.

Related species

M. *glaziovii* is the source of Ceara or Manicoba rubber and oilseeds.

Genetic base

Estimated 7000 landraces. Local preferences in flavour, root texture and growth habit vary greatly; many farmers retain traditional cultivars despite improvements in new cultivars. Genetic erosion reportedly a risk in South and Central America, Thailand and China. 65 spp. were listed as threatened in 1997.

Breeding

Variability of cultivated forms has probably increased through crosses with wild forms. M. *glaziovii* and M. *melanobasis* have contributed to improvement of cultivated form. High diversity in germplasm provides good improvement potential. Interspecific crossing with wild relatives may be employed further to broaden tolerance of different conditions.

Germplasm collections

28 000 accessions, mostly in international centres of research. 35% of landraces and 5% wild spp. collected were listed as threatened in 1997.

Potato

Solanum tuberosum (Solanaceae)

Main uses

One of the most important world crops. Cultivated in 150 countries, mainly for local consumption. Little international trade. Tubers are cooked or processed into a range of products. Starch, alcohol, glucose and dextrin are also major products. Tubers also make animal feed. Potatoes are 80% water, 18% carbohydrates, range of minerals and a good source of vitamin C.

Production

289 780 417 mt. Main producers are Russia, China, Poland, Germany, India.

Origins

Maximum diversity in cultivated and wild spp. on the high plateau of Bolivia and Peru. A number of ancestral spp. involved. The gene pool consists of S. *tuberosum* ssp. *andigenum* and *tuberosum*, S. *stenotomum*, S. *ajanhuiri*, S. *goniocalyx*, S. *x chauca*, S. *x juzepczukii*, S. *x curtilobum*, S. *phureja*. Total of 1700 spp. in the genus, distributed worldwide.

Related species

S. *melongena* (India) (eggplant). S. *centrale* (arid Australia) and S. *muricatum* (pepino) (Andes) have edible fruit; S. *quitoense* (naranjillo) (Andes) is used for fruit juice; S. *melanocerasum* (?cultigen) (cultivated tropical W Africa) fruit; S. *hyporhodium* (upper Amazon); S. *americanum* (yerba mora).

Genetic base

Between 3 000-5 000 varieties of potato recognised by farmers in the Andes. Genetic erosion is reported in centres of origin, including Chile and Bolivia. In Peru, of the 90 wild potato spp. 35 are now extinct in the wild. Wild spp. and ancient cultivars largely replaced by modern varieties. Attempts to broaden the narrow genetic base have been slow. 125 spp. were listed as threatened in 1997.

Breeding

Much introgression from wild relatives has been attempted, improving disease resistance and other traits.

Germplasm collections

31 000 accessions worldwide. 20% are held by the Centro Internacional de la Papa, Lima, Peru. 95% of landraces and 40% of wild spp. are collected.

Cabbage

Brassica oleracea (Cruciferae)

Main uses

Mainly temperate vegetable crop, but grown worldwide. Large number of edible and ornamental varieties, including cauliflower, calabrese and kohlrabi. Important component of human nutrition throughout the world, a good source of fibre, vitamins E, B, and C, also Vitamin A in the greener parts.

Production

47 573 516 mt (cabbage), 13 690 094 mt (cauliflower). Main production in Europe and Russia.

Origins

The wild cabbage is native to Europe; development of cultivars took place in the Mediterranean region. Total of 35 spp. exist in the genus, distributed in Eurasia.

Related species

Wide range of crops (variously leaves, buds, florets, stems and roots eaten); also used for oil production. B. *campestris* and B. *napus* (rapeseed), B.*carinata* (Texsel greens) (NE Africa); B. *hirta* (white and yellow mustard) (Mediterranean); B. *juncea* (Indian mustard) (Eurasia); B. *juncea var. crispifolia* (Chinese mustard).

Genetic base

Outbreeding nature. Large amounts of genetic variation in most crops, where not highly selected. Continuing emphasis on uniformity in recent decade and controls on release of new cultivars have led to significant reduction in genetic variation in commercial cultivars. Wild relatives in Mediterranean are threatened, 14 spp. listed as threatened in 1997.

Germplasm collections

Efforts made to ensure different crops are represented, including obsolete and locally popular varieties. Cultivars from southern Europe are less well collected.

Beans

Lablab purpureus, Phaseolus lunatus, Phaseolus vulgaris, Vigna unguiculata (Leguminosae)

Main uses

L. purpureus: young pods and young and mature seeds of lablab are eaten; pulse contains 25% protein, little fat and 60% carbohydrate.

P. lunatus: dried or immature seeds of lima bean are used as pulses; seeds contain 20% protein, 1.3% fat, 60% carbohydrate; flour also obtained from seed.

P. vulgaris: most widely cultivated of all beans; in temperate areas grown mainly for the pod, which contains 2% protein, 3% carbohydrate with vitamins A, B, C and E; seeds have 22% protein, 50% carbohydrate, 1.6% fat and vitamins B and E.

V. unguiculata: Cowpea is a nutritionally important minor crop in subsistence agriculture in Africa; dry and green seeds, green pods and leaves are eaten; highly resistant to drought.

Production

19 366 815 mt (dry *Phaseolus* beans), 4 294 184 mt (green *Phaseolus* and *Vigna* beans), 1 539 567 mt (string *Phaseolus* and *Vigna* beans), 3 033 057 mt (*Vigna*), 3 483 526 mt (*Lablab* plus others). Main producers of L. *purpureus*: India, South East Asia, Egypt, Sudan; P. *lunatus*: USA; P. *vulgaris*: Brazil.

Origins

Lablab: African or Asian origin; only 1 species in the genus (previously *Dolichos lablab*).

Phaseolus: it is thought that separate domestications occurred in Central and South America from conspecific races; total of 36 spp. in the genus, found in tropical and warm America.

Vigna: centre of diversity of wild relatives in southern Africa; greatest diversity of cultivated form exist in W. Africa; subspecies *dekindtiana* is probable progenitor; total of 150 spp. in the genus, mainly in the Old World tropics.

Related species

Phaseolus: 5 cultigens exist in the genus; apart from P. *lunatus* and P. *vulgaris*, there is P. *acutifolius* (tepary bean, America); P. *coccineus* (scarlet runner, C America) and P. *polyanthus* (year bean, C. America); various other spp. are important pulse crops, previously listed as *Vigna* spp.

Vigna: other spp. are used for forage & green manure etc.; other pulses include: V. *aconitifolia* (moth bean, S. Asia; V. *angularis* (Aduki bean, Asia);V. *mungo* (urad, tropical Asia); V. *radiata* (mung bean, ?Indonesia); (Bambara groundnut, W Africa); V. *umbellata* (rice bean, S Asia); V. *unguiculata* (cowpea, Old World); V. *vexillata* (tropical Old World) roots edible.

Genetic base

Lablab: mainly grown in small plots and home gardens; larger areas under cultivation in Australia; no threat of genetic erosion.

Phaseolus: much dry bean cultivation in USA depends on very small germplasm base; improved varieties also widely adopted by smallholder farmers; relatively wide genetic base provided by landrace groups, if conserved; most wild relatives widespread but populations of several taxa being lost to overgrazing in southwest USA and northern Mexico. 2 spp. listed as threatened in 1997.

Vigna: breeding relies on narrow genetic base and hybridisation with other Vigna spp. is important; more variability in wild relatives in the primary gene pool than in cultivated cowpea. 4 spp. were listed threatened in 1997.

Breeding

Phaseolus: several wild relatives are fully or partly compatible; populations of wild lima bean with larger seeds recently discovered in northwest Peru and Ecuador.

Vigna: cowpea crosses successfully with wild subspecies of V. *unguiculata*.

Germplasm collections

Lablab: 11 500 accessions in Africa and Caribbean.

Phaseolus: 268 500 accessions of *Phaseolus* spp. in total. 15% are held by Centro Internacional de Agricultura Tropical, Cali, Colombia. On average 50% diversity in the genus is represented.

OIL CROPS

Groundnut
Arachis hypogaea (Leguminosae)

Main uses

The edible nut contains 50-55% oil, 30% protein, good sources of essential minerals, E and B vitamins. Cultivated for the nut or for oil in many tropical and subtropical countries. Seed residue useful as animal feed. Nutshells are used as fuel and in industry. Stems and leaves used as forage.

Production

32 219 427 mt. Main producers are India, China, USA, Argentina, Brazil, Nigeria, Indonesia, Myanmar, Mexico, Australia.

Origins

Mato Grosso in Brazil is the primary centre of origin and diversity for the genus. The cultivated groundnut is thought to have originated in southern Bolivia and northwest Argentina. Total of 22 spp. in the genus, all from South America.

Genetic base

Cultivated as a marginal crop with relatively little selection pressure. Many varieties exist worldwide with broad adaptability.

Breeding

A. *monticola* freely crosses with A. *hypogaea*. Wild *Arachis* material confers resistance on domestic form.

Germplasm collections

13 000 accessions at the International Crop Research Institute for the Semi-Arid Tropics.

Coconut
Cocos nucifera (Palmae)

Main uses

The endosperm of the nut contains 65% saturated oil, used in manufacture of margarine, soap, cosmetics and confectionery. Also eaten fresh, desiccated or as a coconut milk. Residue is high-protein animal feed. There are many more uses: source of naturally sterile water, fibre, wood, thatch. Mainly a smallholders' crop.

Production

48 507 022 mt. Main producers are the Philippines, Indonesia, India, Sri Lanka, Malaysia, Mexico, Pacific Islands.

Origins

Possibly originated in Melanesian area of Pacific. Wild types predominate on the African and Indian coasts of the Indian Ocean, and scattered in southeast Asia and the Pacific. Single species in the genus.

Genetic base

Tendency to plant uniform, improved hybrids is reducing genetic variation particularly in domesticated types.

Breeding

Wild and domestic coconuts are fully compatible. Hybridisation with wild types has increased genetic diversity of cultivated crops.

Oil Palm

Elaeis guineensis (Palmae)

Main uses

The mesocarp on the fruit yields oil for human consumption. Unrefined oil is high in vitamin A. Oil may also be extracted from the kernel. An export crop and important for local consumption. Very high yielding.

Production

94 506 349 mt. Malaysia supplies 70% of world exports.

Origins

West Africa, originally a species of the transition zone between savanna and rain forest. Only 2 spp. exist in the genus.

Related species

E. *oleifera* (Tropical America) is less important as an oil crop than E. *guineensis*.

Genetic base

Populations in Africa are semi-wild. They are being thinned to make way for other crops. Plantations in Malaysia were based on material from only four specimens. New material is being introduced to broaden the genetic base.

Breeding

Fertile offspring produced with E. *oleifera*.

Soybean

Glycine max (Leguminosae)

Main uses

The most important oil crop and grain legume in terms of production and international trade. An important basis of Asian cuisine, developed into various forms of food from soy sauce to tofu. Immature green beans and sprouts also eaten. Seeds contain 18-23% oil and 39-45% protein. Oil is used in various forms. Most of the meal is used as high-protein animal feed.

Production

157 743 654 mt. Main producers are USA, Brazil, China, Argentina, India.

Origins

A cultigen, not known in the wild. Soybean is thought to have arisen as a domesticate in the eastern half of northern China ca 3000 years ago probably from G. *soja*. Total of 18 spp. exist in the genus, distributed from Asia to Australia.

Genetic base

The genetic base of varieties is narrow worldwide. Conservation of traditional landraces is urgently needed. 2 spp. listed as threatened in 1997.

Breeding

Wild spp. are increasingly used for improvement and there is good potential for further valuable characteristics to be found in wild *Glycine* spp. G. *soja* easily crosses with soybean.

Germplasm collections

174 500 accessions. 9% in Institute of Crop Germplasm resources CAAS, Beijing China. 60% of landraces and 30% wild spp. are represented.

Cotton seed

Gossypium barbadense, G. *hirsutum* (Malvaceae)

Main uses

Cotton is the second most valuable oil crop, as well as being the most important textile fibre. Crop development is concentrated on fibre production because value is three or four times greater. New World cottons took over from Old World forms after the European exploration of the Americas.

Production

55 572 852 mt. Main producers of G. *barbadense* are Russia, Egypt, Sudan, India, USA, China

Origins

Unique in that 4 spp. were domesticated independently for the same use as a fibre and oil crop. In Africa and India: G. *arboreum* and G. *herbaceum*; in Central and South America: G. *hirsutum* and G. *barbadense*. Total of 39 spp. in the genus, found in warm temperate to tropical zones.

Related species

G. *arboreum* is still important in India and Pakistan. G. *herbaceum* is only grown on a small scale in Africa and Asia.

Genetic base

Modern cultivars of G. *hirsutum* are responsible for over 90% of world production. New Gossypium spp. possibly occur in Arabia and Africa. Wild forms of G. *herbaceum*, G. *hirsutum* and G. *barbadense* are known. Past breeding involved much introduction of genetic material from different geographic regions, but a severe narrowing of the genetic base has occurred in the production of modern G. *hirsutum* varieties. Large amounts of fertilizers and pesticides required in modern cotton production. 8 spp. listed as threatened in 1997.

Breeding

At least 6 related spp. have contributed genes of importance to the cultivated crop. Material from wild gene pool used in genetic engineering. G. *herbaceum* and G. *arboreum* are able to interbreed, although later generations have a high probability of failing reproductively.

Sunflower seed

Helianthus annuus (Compositae)

Main uses

Seeds contain 27-40% polyunsaturated oil and 13-20% protein. Oils and margarines used for human consumption, also industrial uses, and waste products useful in animal feed. Pollinating bees frequently used for honey production.

Production

29 947 687 mt. Main producers are Russia, USA, Argentina.

Origins

Probably originated in southwest North America. Total of 50 spp. exist in the genus, distributed in North America.

Related species

Also ornamental; H. *tuberosus* (Jerusalem artichoke) is also eaten. H. *petiolaris* used for hybridisation

Genetic base

Increased yields in hybrids led to increased interest and production in 1960s. Large gene pool exists in wild and weed sunflowers in North America, although habitat loss is resulting in population declines. 16 spp. listed threatened in 1997

Breeding

Resistance to several diseases was secured through hybridisation with H. *tuberosus*.

Olive

Olea europaea (Oleaceae)

Main uses

Fruit with 40% oil content. Highly superior oil for cooking, margarines, dressing, also used in cosmetics and pharmaceutical industry. Fruit eaten pickled. Despite competition with more modern oil-producing crops, olive oil still command premium price. Recent rise in popularity and recognition of nutritional value.

Production

14 074 198 mt. Main producers are Spain, Italy, Greece, Turkey, Tunisia

Origins

Olive is a cultigen, evolved in eastern Mediterranean. O. *europaea* ssp. *oleaster* recognised as progenitor. Total of 30 spp. in the genus, in tropical and warm temperate parts of Old World.

Related species

Related species provide good timber.

Genetic base

A long-lived tree. The turnover of clones should be slow. Hundreds of distinct cultivars, found in different geographic groups. Olive production still relies on traditional cultivars. Few new varieties have been released. Decline in area under cultivation. Marginal groves have been abandoned with serious consequences for Mediterranean wildlife. Wild populations outside area of cultivation under pressure from cutting and land clearance; 2 spp. were listed as threatened in 1998.

Breeding

Closely related to wild subspecies in the Mediterranean, Africa, Arabia, Iran and Afghanistan.

Sugarcane

Saccarhum officinarum (Gramineae)

Main uses

Major source of calories worldwide. Cultivated in about 70 countries, mainly in tropics. Requires good rainfall and rich soil for successful growth. Stems are easily transported.

Production

1 252 904 783 mt. Main producers are Brazil, India, China, Thailand, Pakistan.

Origins

A cultigen with origin and centre of diversity in New Guinea. Between 35 and 40 spp. in the genus, distributed in tropical and warm zones.

Related species

Other cultivated sugar canes include S. *barberi*, S.*edule* and S. *sinense*. S.*robustum* and S. *spontaneum* are wild sugar canes.

Genetic base

Risk of genetic erosion reported in Assam and suspected in Indonesia, PNG and Thailand, where monocrop plantations have taken over from indigenous spp. Modern hybrids have a narrow genetic base. Plantations are prone to severe pest and disease epidemics. Attempts to incorporate more genetic diversity is slowly having effect. Only 10% wild germplasm used in breeding. S. *robustum* exhibits the most genetic diversity, but has had little application in breeding.

Breeding

Commercial varieties are derived from interspecific crosses with other wild and cultivated sugarcane spp.

Germplasm collections

19 000 accessions in total, nearly a quarter of them in Centro Nacional de Pesquisa de Recursos Genéticos e Biotecnologia, Brasilia, Brazil. 70% of landraces, 5% of wild spp. represented.

Banana, Plantain

Musa spp. (Musaceae)

Main uses

One of the most popular dessert fruits in industrial nations; a major source of calories and export earnings in developing countries. Bananas and plantains are high in carbohydrates and potassium, bananas are a good source of vitamins C and B6, and plantains contain high levels of vitamin A. Numerous other uses.

Production

56 404 895 (banana); 29 453 513 (plantain) mt. Main producers are India, Brazil, Ecuador, Philippines and China for the banana; Uganda, Colombia, Rwanda, D.R. Congo and Nigeria for the plantain.

Origins

Bananas evolved in South East Asia from M. *acuminata* or combinations of M. *acuminata* and M. *balbisiana*. Plantains probably originated in southern India. Primary areas of diversity exist in South East Asia. Secondary areas also occur in tropical Africa, Indian Ocean Islands and the Pacific. Fe'i bananas (2n), thought to be derived from M. *maclayi* and possibly other related spp. Greatest diversity of fe'i bananas is on Tahiti. Total of 35 spp. in the genus, distributed throughout the tropics.

Related species

Fe'i bananas are significant source of food in New Guinea and Pacific. M. *textilis* recent domesticate in Philippines used for Manila hemp. Related E*nsete ventricosum* cultivated in Ethiopia for starchy pseudostem. M. *balbisiana* produces edible fruit and contributed to present day cultivars.

Genetic base

About 500 genetically distinct cultivars. 90% of global banana production is from smallholdings. International trade in bananas relies on very few cultivars, based on the Cavendish type. Dangerously narrow genetic base and extremely susceptible to diseases. Increased disease resistance is extremely important given the economic importance of the export crop. The number of Fe'i banana cultivars has declined severely as a result of human demographic changes in the Pacific and spread of pests. Banana is an aggressive weed. Wild populations of Musa benefit from forest clearance if succession is allowed to take place. 3 spp. listed as threatened in 1997.

Breeding

An extensive contact zone between cultivated and weedy types exists in several areas, eg. Sri Lanka. Much introgression is believed to have enriched the gene pool of cultivated types. M. *balbisiana* has valuable traits. Several other wild relatives have useful characteristics. Germplasm collections have been poorly used; better selections could be made to suit subsistence farmers.

Germplasm collections

Edible bananas, being seedless, are not storable. Seeds from wild spp. may be stored. Field gene banks hold collections. 10 500 accessions in total. The International Network for the Improvement of Bananas and Plantains hold 10%. Most of the diversity of wild and cultivated bananas thought to be covered.

Cocoa

Theobroma cacao (Sterculiaceae)

Main uses

Seeds are fermented and roasted to produce cocoa powder and chocolate. Waste goes to produce animal feed, mulch or fertilizer. Cocoa is a nutritional beverage; the powder is 25% saturated fat, 16% protein and 12% carbohydrates.

Production

2 988 369 mt. Main producers are West African countries, Brazil, Malaysia

Origins

Upper Amazon basin. Centre of cultivation in Central America. Total of 20 spp. in the genus, confined to tropical America.

Related species

All the following are cultivated: T. *grandiflorum* (cupuaçu, Amazonia); T. *speciosum* (cacaui, Central & S America); T. *subincanum* (S America); T. *obovatum* (Amazon); T. *angustifolium* (C America); T. *bicolor* (Central & S America); T. *glaucum* (Amazonia).

Genetic base

Undoubted genetic erosion has occurred in recent years. Currently cacao plantations are established by seed with varying degrees of genetic heterogeneity. Production in West Africa is based on a particularly narrow gene pool. Originally 3 main cultivated types. Criollo yields the most superior chocolate but has been largely replaced because of low yields. Forastero dominates world production. Wild cacao is highly variable, especially in its core area. Dramatic increase in plantations of coca and pulp-producing spp. in various parts of the Amazon, agricultural expansion and movement of human populations have caused severe losses to the wild gene pool.

Breeding

Little use of or research into wild genetic reserves because they are relatively hard to cross.

Germplasm collections

Seeds do not remain viable for long. 4 000 to 5 000 accessions kept in field gene banks. International Cocoa Genebank in Trinidad has the most comprehensive collection. Close relatives are poorly represented. Vegetative germplasm is collected.

Table 4.10

Food crops of secondary or local importance.

Notes: These species and groups of species are those that, according to national food supply data maintained by the FAO[66,34], provide a significant amount of the per capita supply of calories, protein or fat, but on the criteria followed here are not of equal importance to the crops in Table 4.9 (ie. they provide below 5% of the total per cap. supply and/or do so in fewer than 10 countries.

Sources: list of crops from Prescott-Allen and Prescott-Allen[66], other information from Mabberley[67], Smartt and Simmonds[68], Smith *et al.*[69], Vaughan and Geissler[70]; conservation status from Walter and Gillet[71].

CEREALS AND PSEUDO-CEREALS

Oats *Avena sativa* (Graminae)

Origin: W and N Europe from weed oat components of wheat and barley crops.
One of the major temperate cereals, although currently declining in production and generally regarded as a secondary crop. Mostly used for animal feed. Oat kernel is higher in high quality protein and fat than any other cereal. Oat bran is a good dietary fibre. A. *byzantina* also cultivated. Genetic erosion from intensive breeding has resulted in effort to conserve landraces and early varieties. Crosses between A. *sativa* and A. *byzantina* have led to numerous cultivars. Fertile hybrids obtained from crosses between cultivated oats and weed species. Some success in incorporating desirable genes from more distant relatives.

Quinoa *Chenopodium quinoa* (Chenopodiaceae)

Origin: High Andes
An important and sacred pseudo-cereal in Inca times. Remains a staple in large parts of South America. Nutrient composition is superior to other cereals, being high in lysine and other essential amino acids, calcium, phosphorus, iron and vitamin E. Can grow in marginal conditions. Greatest diversity of genotypes in the highlands of southern Peru and Bolivia. Cultivation declined with Spanish Conquest until 1970s when grown as a sole crop only in parts of Peru and Bolivian-Peruvian Altiplano. Agricultural and nutritional benefits have now been recognised and acreage has increased significantly. Improvement so far has been based on inbred populations and pure lines. Considerable potential for improvement, both in the crop and its use.

Fonio *Digitaria exilis* (Gramineae)

Origin: West Africa, thought to be a cultigen
Popular cereal in parts of West Africa. Adapted to marginal agricultural land. Several species are harvested as cereals during times of famine

ROOTS AND TUBERS

Taro *Colocasia esculenta* (Araceae)

Origin: India
Edible tuber, 25% starch, low protein, good vitamin C source. Probably cultivated before rice. Widely cultivated in China and staple in many Pacific Islands. Also used in food and beverage industries, and pasta products. Young leaves eaten as spinach. Tolerates high temperatures and poor soils. More than 1,000 cultivars have arisen through subsistence farming. Lack of interest and germplasm exchange at a more commercial level. Serious danger of genetic erosion

Carrot *Daucus carota* (Umbelliferae)

Origin: Afghanistan
Root crop, grown worldwide, eaten raw, cooked or processed. The best plant source of provitamin A, low in other nutrients. Numerous wild and cultivated subspecies. Open pollinated crops almost entirely replaced by hybrids in USA, Japan and Europe. Environmental health concerns over level of pesticide has led to interest in genetic source of pest resistance. D. *capillifolius* has passed some pest resistance to cultivated crop.

Sweet Potato *Ipomoea batatas* (Convolvulaceae)

Origin: not known in the wild. Greatest species diversity occurs between Yucatan and the mouth of the Orinoco. Major variation is found in Guatemala, Colombia, Ecuador and Peru.
The tuberous root is an important staple in the tropics. Able to grow in high temperatures with low water and fertilizer input. Good source of fibre, energy and vitamins A & C. Also industrial source of starch and ethanol. Although acreage has declined, increases are likely as a crop able to respond to population growth in marginal areas. China accounts for 80% of production. Until recently, material used in breeding programmes represented a fraction of existing diversity. Genetic base is now broadened but requires further increase. Little work has been done on cultivar improvement in areas of highest production (ie. where sweet potato is a staple). Countries where breeding programmes exist have replaced native cultivars with improved varieties. Sweet potato is thought to have more potential for yield improvement than any other major crop in Asia.

Tannia *Xanthosoma sagittifolium* (Araceae)

Origin: New World
Similar use and nutritional composition as taro, but starch is more difficult to digest. Used in preparation of fufu in West Africa.

BEANS AND OTHER LEGUMES

Pigeonpea *Cajanus cajan* (Leguminosae)

Origin: cultigen. India
One of the major pulse crops of the tropics. Mature seeds contain 20% protein, 60% carbohydrate and little fat. Important in small scale farming in mainly semi-arid regions. A multipurpose species with good potential in agroforestry systems and on marginal lands. India contributes more than 90% of world production. Domestication has not altered the species as much as other crops. C. *cajanifolius* is closest relative – 12 spp. may be crossed with pigeon pea.

Chickpea *Cicer arietinum* (Leguminosae)

Origin: southeast Turkey. C. *reticulatum* is probably the progenitor.
One of the most important pulse crops. The seeds contain less protein (17% or more) but more fat (5%) than other pulses. Grown over large area from Southeast Asia to Mediterranean. Only 2-4% of world production is exported. Recently discovered C. *reticulatum* is confined to 10 populations in Turkey. 2 main cultivars have emerged. Traditional landraces have been selected to suit local ecological conditions. Commercial breeding is a recent phenomenon. C. *reticulatum* and C. *echinospermum* are compatible with the chickpea.

Lentil *Lens culinaris* (Leguminosae)

Origin: Near East. Wild progenitor L. *orientalis*
Seeds contain 25% protein, 56% carbohydrate and 1% fat. Young pods also eaten. Seeds are commercial source of starch for textiles and printing. Residues used as animal feed. Unique assemblages of landraces in different geographic regions. The crop has been altered little by modern breeding. Much variation in the crop unexploited.

Lupin *Lupinus mutabilis* (Leguminosae)

Origin: Andes. Other Lupins originated in 2 main centres of genetic diversity in Mediterranean and in Americas. A relatively minor pulse crop, obtained from several *Lupinus* spp. Seed contains 44% protein, 17% oil. Seed is human food in subsistence agriculture. Also used as coffee substitute and high protein animal feed. Seed flour used as soya. Species also important in soil improvement. Related spp. have ornamental value, used as fodder, coffee substitute, green manure or to stabilise sand dunes. May act as substitute for soybean, where climate is unsuitable for soybean growth. Other *Lupinus* spp. potentially suitable for cultivation.

Pea *Pisum sativum* (Leguminosae)

Origin: wild progenitor is unknown. Possible centres of origin are Ethiopia, the Mediterranean and Central Asia. The second most important pulse. 90% production as dried peas. Seed coats are source of protein, used in bread or health foods. Russia and China produce 80% of the world production of dried peas; USA and UK are largest producers of green peas. Breeding relies on a fairly narrow genetic resource base and efforts to conserve genetic variability of the cultivated crop have been fairly limited.

Broad bean *Vicia faba* (Leguminosae)

Origin: cultigen, wild ancestor unknown, possibly from C. Asia.
A temperate pulse crop. Both immature seeds and dry mature seeds are eaten. The latter are also used as animal feed. Dried seed contains 25% protein, 1.5% fat, 49% carbohydrate; the immature bean has much less of these nutrients but more vitamin A and vitamin C. Also used as green manure.

OIL CROPS

Mustard seed *Brassica juncea* (Cruciferae)

Origin: Central Asia - Himalaya. Probably B. *nigra* x B. *campestris* and other *Brassica* spp.
The most important spice in the world in terms of quantity. 4 species contributing to mustard exist. B. *juncea* took over from B. *nigra* in 1950s as it allowed completed mechanisation of harvesting. Also valuable as oil crop and salad, vegetable and fodder. Long lived seed allows easy maintenance of large collections. Wild material is widely distributed.

Rapeseed *Brassica napus* (Cruciferae)

Origin: probably a hybrid of B. *oleracea* x B. *campestris*.
The seed is an important relatively recent source of oil, containing 40% unsaturated fat, with industrial and culinary applications. The root crop provides animal and human food (swede). Uncertain whether B. *napus* exists in wild form. Domestication was a relatively recent event. The crop is tolerant of inbreeding, and landraces have been replaced by improved cultivars since the nineteenth century. Swedes, of which there are only a few varieties, are the result of hybridisation with B. *campestris*. Various valuable contributions to oilseed rape also from B. *campestris* and B. *oleracea*.

Safflower seed *Carthamus tinctorius* (Compositae)

Origin: Turkestan, Turkey Iran. Iraq, to Israel and Jordon
Oilseed crop, produces 2 types of oil for margarine and also cooking oils. Ingredient of animal feeds. Dried flowers serve as substitute for saffron. Applications in cosmetic industry and as medicine.
Originally domesticated for use as dye plant. Much diversity developed as the species was cultivated over a wide area. Large-scale cultivation in few countries. No reported genetic erosion. Related species cross easily with cultivated crop and form natural hybrids.

Sesameseed *Sesamum orientale* (Pedaliaceae)

Origin: origin and ancestors unknown, possibly Ethiopia or peninsular India.
An ancient oilseed crop. Seeds contain 50% unsaturated oil and 20-25% protein and are used widely in bread and confectionery. Oil used in cooking, margarine, soaps and other industries. Residues are valuable animal feed. Interest in the crop is in decline as difficulty mechanizing harvesting and low seed yields compared to other oil crops. Good genetic diversity in related species. S. *malabaricum* produces fertile offspring with S. *indicum*.

Shea nut *Vitellaria paradoxa* (Sapotaceae)

Origin: D.R. Congo, Sudan, Uganda
The roasted kernels are used to make purified shea butter, rich in vitamin E, used in cooking and as an alternative to cocoa butter for chocolate manufacture. Also commercial use in manufacture of soap, cosmetics, candles. Various local uses. Fruit is eaten and is a good source of carbohydrates, iron and vitamins B. A monospecific genus. Threatened by overexploitation as a timber and source of fuel, also by land clearance. Stands may be conserved for their valuable seed, but no official protection exists. Mostly grown for local consumption

Artichoke *Cynara scolymus* (Composite)

Origin: Mediterranean, Canary Islands
Flowerheads and the receptacle are eaten. Small amount of vitamin C.

Lettuce *Lactuca sativa* (Compositae)

Origin: probably evolved in Asia Minor or Middle East from L. *serriola*.
Lettuce leaves are useful source of fibre, minerals (esp. potassium), vitamins A, E and C. May be grown year round. Stem is boiled as a vegetable in China. A highly variable crop, resulting probably from long history of selection. Increasing diversity of lettuce types consumed. Wild species, including L. *serriola*, L. *saligna and* L. *virosa* have been used in breeding programmes.

Spinach *Spinacia oleracea* (Chenopodiaceae)

Origin: S.W. Asia
Edible leaves contain range of minerals, vitamins A, E and C and the B vitamin range.

FRUIT

Pineapple *Ananas comosus* (Bromeliaceae)

Origin: obscure origin in South America, probably on fringes of Amazon.
Seedless fibre-rich fruit, source of vitamin C, A and B. Highly suited for canning and as a juice. Unique in that timing of harvest can be controlled by externally applied growth hormone. Over 65 countries grow pineapple for domestic consumption and export. No wild populations. Genetically variable species, but genetic base of commercial plantations very narrow. 70% of world production and 96% of cannery industry comes from one variety. Highest diversity of near relatives in Paraguay and Brazil. Poorly known, but A. *ananassoides* has contributed several characteristics to cultivated crop. A. *erectifolius* also considered for improvement programmes.

Papaya *Carica papaya* (Caricaceae)

Origin: obscure – probably hybrid of several *Carica* spp., arising in lowland tropical forest in Eastern Andes or Central America.
Easily digested fruits produced all year round. Good source of vitamin C, red fleshed fruits also rich in vitamin A. Papain extract is exported as a meat tenderizer, also used medicinally, to tan leather and in brewing beer. May be produced by biotechnology in future. Commercially produced in over 30 countries, mostly for domestic consumption. High diversity in Eastern Andes. At least 6 other spp. domesticated and 12 spp. are harvested for their fruit. Several commercial cultivars come from highly inbred hermaphrodite lines. Most production from backyard papaya trees, where local variation is high. Many wild species have desirable characteristics, useful in breeding. Hybridisation already carried out with 5 wild *Carica* spp. Highly susceptible to viral and fungal diseases – some resistance detected in wild relatives but conventional crossing impossible.

Lime, Lemon, Pomelo, Tangerine, Sweet Orange
Citrus aurantiifolia C. limon, C. maxima, C. x paradisi, C. reticulata, C. sinensis (Rutaceae)

Origin: Lime: cultivated hybrid with obscure origins, possibly a hybrid of C. *medica* with another sp.; Lemon: probably a hybrid of lime with C. *medica*; Pomelo: probably a native of the Malay peninsula; Grapefruit: probably hybrid between orange and pomelo, arising in the Caribbean; Tangerine: possibly Indo-China; Orange: probably introgressed hybrids of C. *maxima* and C. *reticulata*, perhaps originating in China.
Fruits contain nearly 90% water, potassium, vitamins A, B, E and high vitamin C. They are eaten fresh, used as a flavouring and in marmalade. Orange accounts for 70% of *Citrus* production. Various other spp. are cultivated. Wild populations located in northern India. Wild species threatened by forest clearance. In Southeast Asia wild groves are being replaced by oil palm and cacao plantations. C. *taiwanica* is critically endangered in Taiwan, mainly because of extensive habitat loss but also because of use as a rootstock for citrus plantations. Wide variation within the genus. Can be crossed with several genera. Economic *Citrus* spp. are highly interfertile.

Pumpkin, Squash, Gourd *Cucurbita moschata, C. maxima, C. argyrosperma, C. pepo, C. ficifolia* (Cucurbitae)

Origin: C. *moschata* is most like the wild species and was domesticated independently in Central & South America. Fruits, containing 90% water, small amounts of starch, sugars, protein, fat and vitamins A, B & C, are used as vegetables and as animal feed. Leaves and flowers may be cooked. Seeds eaten and sometimes processed for oil. Grown worldwide in temperate and tropical zones, commonly in home gardens and as subsistence crops as well as commercially. Long shelf life. Broad gene pool because of wide use of traditional or unimproved varieties in subsistence farming and home gardens. Many *Cucurbita* spp. have restricted geographical ranges. Disease resistance is found in wild relatives, with some transfer to cultivated species through interspecific crosses. Crosses between crop species and wild or feral relatives have occurred and genetic exchange takes place where their ranges overlap.

Strawberry *Fragaria x ananassa* (Rosaceae)

Origin: a hybrid between 2 American species, F. *chiloensis* and F. *virginiana*.
Soft fruit, 90% water, high vitamin C, eaten fresh or in jams and confectionery. Grown in most temperate and subtropical countries. All *Fragaria* spp. produce palatable fruit. Hundreds of cultivars with wide ecological adaptability. Considerable genetic diversity lost in cultivated strawberries in the last 100 years. Attempts are being made to extend the genetic base of the crop. Much unused genetic variation in wild species.

Fig *Ficus carica* (Moraceae)

Origin: eastern Mediterranean
Fruits contain 10% sugar when fresh and 50% when dried. Also substantial amounts of potassium, especially in the dried fruit. Most world trade as dried figs. Figs are widely distributed in tropical, subtropical and warm temperate areas throughout the world. *Ficus* spp. are also source of rubber, fibres, paper, medicines and ornamental plants. Fig is largely grown for domestic consumption using traditional cultivars, hundreds of which exist, with local clones occurring in distinct geographical groups in the Mediterranean basin. Closely related wild forms are distributed throughout the Mediterranean basin. Fig culture is in decline. Many old groves have been abandoned or cleared. A number of wild relatives are considered threatened. 27 *Ficus* spp. were listed as threatened in 1998. Reproductive isolation is dependent solely on the specificity of the wasp pollinator. Artificial crosses can be made between species.

Apple *Malus domestica* (Rosaceae)

Origin: an aggregate of over 1000 cultivars, of ancient and complex hybrid origin.

Apples, with pears, are the most important fruit crops of cooler temperate region. Fruit is eaten fresh or cooked, as a juice or brewed as cider. Potassium is the main mineral with small amounts of vitamin C. Breeders in the nineteenth century used wild species in breeding. Genetic diversity accumulated in North America was greater than in Europe because propagation was by seed rather than by grafting. Current trend depending on few varieties has caused rapid loss in genetic diversity and potential breeding material. Widespread elimination of wild stands is also taking place. 3 spp. are listed as threatened in 1998. Hybridisation with many wild species within the genus occurs readily.

Mango *Mangifera indica* (Anacardiaceae)

Origin: N.E. India

Fleshy edible fruit, a good source of vitamin A and of vitamin C. Thrives on infertile marginal soils. Important tree in Hindu mythology and religion. Kernel oil may be used in chocolate manufacture. Demand for the fruit and its juice is increasing worldwide. India accounts for two-thirds of production. Fruits of more than 12 wild spp. collected. Several are cultivated. The majority of fruit-bearing trees are more or less wild. Genetically highly heterogenous. Over 1000 cultivars exist, many in Borneo and Malay peninsula. Feral populations are distributed throughout the tropics. Of the 40 to 60 spp. in the genus, many are poorly known, severely threatened or possibly extinct. 35 spp. were listed as threatened in 1998. Logging, forest clearance and replacement with commercial species in Southeast Asia are largely responsible for population extinctions. Various species are suitable for cultivation given further selection. Many display valuable traits, such as tolerance of waterlogged soils and more regular fruiting.

Avocado *Persea americana* (Lauraceae)

Origin: Mexico to northwest Colombia

A highly nutritious fruit, containing 15-25% monounsaturated fat, vitamins C, B and E. The oil is used in cosmetics. Trees fruit year round. Importance has increased over recent decades and crop now grown in most tropical and subtropical countries. Most production is for domestic consumption. Other *Persea* spp. used for timber and fruit. 3 geographically distinct varieties which are able to interbreed. Commercially important cultivars have arisen mostly in private orchards by chance rather than result of germplasm manipulation. Increasing use of grafting and uniform varieties. Serious genetic erosion in traditional varieties. Diversity appears greater in traditional growing areas, where farmers still propagate by seed. Genetic exchange occurs between cultivated forms and wild populations. Wild populations of the avocado and its close relatives are small and becoming increasingly isolated. Deforestation poses a severe threat to their survival. 15 spp. were listed as threatened in 1998. A number of wild relatives show resistance or tolerance to disease, drought and frost.

Date *Phoenix dactylifera* (Palmae)

Origin: western India or Arabian Gulf

Edible fruit with sugar content of 30 to 80%, corresponding to soft and dry dates. Vitamins are relatively low in quantity. Eaten as an ingredient in a variety of foods or as a juice. A staple where produced. Good storage. One of the oldest cultivated tree crops. Current cultivars resulted from thousands of years of selection. Perhaps over 3000 cultivars exist, only 60 grown widely. All commercial cultivars are female. Wild populations of some related spp. are highly restricted in geographical range. All *Phoenix* spp. intercross freely.

Apricot, Cherry, Plum, Almond *Prunus armenica*, *P. avium*, *P. domestica*, *P. dulcis* (Rosaceae)

Origin: Apricot: west China; Cherry: west Asia; Plum: an ancient 6n cultigen with complex origin, possibly in SW Asia. North American plums may be native American spp. or hybrids with *P. salicina*; Almond: central to west Asia; Peach: west China, possibly a cultigen derived from *P. davidiana*.

Apricot, cherry, plum and peach are soft fruit with up to 10% sugar, good potassium and Vitamin A in the case of apricots, but low vitamin C. Consumed fresh, dried or as an ingredient in jams and confectionery. Almond is the most important tree nut crop. The kernel contains 40-60% unsaturated oil and 20% protein. Eaten as a dessert nut and in confectionery and marzipan. A major trading commodity. Many other *Prunus* spp. have edible fruit. Many cultivars and much genetic diversity. Plums are genetically central to the genus and harbour the most useful genetic material. Narrow breeding has led cherries to be more isolated from the rest of the genus. Increasing loss of genetic diversity. Developing countries are tending to replace indigenous types and wild stands with western varieties, eg. the switch from seed to vegetatively propagated almonds in Turkey. A number of wild relatives are confined to narrow ranges. 23 spp. are listed as threatened in 1998.

Pear *Pyrus communis* (Rosaceae)

Origin: Asia minor, the Caucasus, Central Asia and China. Cultivars have come from *P. bretschneiderii*, *P. pyrifolia*, *P. sinkiangensis* and *P. ussuriensis*. *P. nivalis* for perry production.

With apples, pears are the most important fruit crops of cooler temperate region. The fruit is eaten fresh or cooked, as a juice or brewed as perry. The fruit are a good source of dietary fibre, potassium and reasonable amounts of vitamin C. Currently about 20 spp. and 5000 recorded cultivars. Major loss of genetic diversity through concentration on few varieties. Several wild species in Turkey are under threat. 5 spp. were listed as threatened in 1998. Hybridisation with high proportion of wild species within the genus is possible, providing useful rootstocks and possibly disease resistance. Much use of wild species in breeding in the past. Evolution of new varieties will be seriously limited unless stands of wild species conserved.

Blackcurrants, Redcurrants *Ribes nigrum, R. rubrum* (Grossulariaceae)

Origin: Europe and northern Asia, with the blackcurrant extending to the Himalayas.
Fruits with high vitamin C content. A luxury crop, largely produced for processing into juice.
Many spp. with edible fruits, cultivated and wild. Wide use has been made of wild or near-wild relatives.

Grape *Vitis vinifera* (Vitaceae)

Origin: Eurasia
The fruit has high sugar content. 68% of grape production is for the manufacture of wine, 20% for dessert grapes, 11% for dried fruit; raisins, sultanas, currants, 1% for juice. Other commercial products include grapeseed oil and vine leaves. Various other spp. produce edible grape. One estimate suggests there are 10 000 cultivars of grape. Wild species still occurs in Middle Asia. Wild relatives are suffering genetic erosion in the USA. All known *Vitis* spp. produce fertile offspring.

FRUIT VEGETABLES

Melon seed/Watermelon *Citrullus lanatus* (Cucurbitae)

Origin: tropical and sub-tropical Africa, domestication took place in Mediterranean.
The flesh of the fruit is 90% water, also vitamin C and A. The seeds contain 40% unsaturated oil and 40% protein. Wild plants still harvested in Kalahari. *C. colocynthis* is fertile with the watermelon. An African watermelon with extraordinarily long storage life has been identified.

Melon, Cucumber *Cucumis melo, Cucumis sativus* (Cucurbitae)

Origin: wild melon populations appear to be distributed south of the Sahara to Transvaal in South Africa. Cucumber's wild or feral relative and possible progenitor, var. *hardwickii*, is native to the southern Himalayan foothills.
Melon is grown worldwide in temperate and tropical countries. 90% water some sugar and vitamin C. Pink or orange coloured fruit have a high percentage of vitamin A. Also grown for their fragrance or ornamental value. Cucumber produces edible fruits, containing 96% water, some vitamin C and reasonable amounts of Vitamin A. Also used in production of fragrances, cosmetics and medicines. Young leaves and shoots may be cooked. Also cultivated *C. anguria* (West Indian Gherkin) and *C. metuliferus* (African horned cucumber or jelly melon). Wild and feral populations of melon occur throughout Africa and southern Asia.
Cucumber produces fertile hybrids with its wild counterpart *C. sativus* var. *hardwickii*. No interspecific hybridisations have been used to improve crops.

Tomato *Lycopersicon esculentum* (Solanaceae)

Origin: cultigen, from Mexico.
There are few growing areas, from the tropics to the Arctic circle, where the tomato is absent. Fruit, containing potassium, vitamins A, B, C and E., eaten fresh, dried or cooked as a vegetable or processed in a wide range of food products. Disease is common threat. The wild relatives of the tomato have limited ranges. Wild gene pools are prone to erosion by habitat destruction. Tomato can be hybridized with all spp. in the genus and wild relatives have been used as source of numerous useful traits, including disease resistance.

Eggplant *Solanum melongena* (Solanaceae)

Origin: India; wild progenitor, *S. incanum*, occurs throughout Africa and Asia.
Fruit is eaten as a vegetable, contains over 90% water, large amount of potassium, some vitamin A, E, B, C. Highly productive and useful smallholder's crop. Various spp. cultivated and used as grafting stock.

BEVERAGE CROPS

Tea *Camellia sinensis* (Theaceae)

Origin: probably lower Tibetan mountains or Central Asia
Tender shoots are used to make tea. Important plantation and smallholder crop throughout the tropics and subtropics. Planted commercially in at least 30 countries. Increasing consumption in developing countries. High diversity of forms or species in East and Southeast Asia.
Many distinct forms, hybrids and species continue to be discovered. Recent trend to propagate the plant vegetatively, which has led to large areas being planted with one or few clones. No threat yet of genetic erosion in the crop. 11 *Camellia* spp. in China and Vietnam were recorded as threatened in 1998.

Coffee *Coffea arabica, C. canephora* (Rubiaceae)

Origin: Arabica coffee originated in montane forest in southwest Ethiopia and the neighbouring Boma Plateau in Sudan, possibly also in Marsabit forest in Kenya. Robusta grows wild in West and Central Africa.
Important sources of foreign currency in many developing countries. Roasted seeds used for beverage containing 1-2.5% caffeine, also niacin and potassium. A number of other *Coffea* spp. also cultivated as coffee or source of edible berries. Commercial arabica cultivars have very narrow genetic base, especially in Latin America and the Caribbean and highly susceptible to disease. Robusta outcrosses and has wider variation. Much production by smallholders, but 40% of coffee from Americas and Caribbean from intensive monocrop plantations. Recent recognition of importance of conserving species-rich shade coffee systems. Significant percentage of Ethiopian coffee from uniform commercial cultivars; 400 000ha remain of wild coffee accounting for half of Ethiopia's coffee production. Several populations of wild species increasingly restricted in distribution and fragmented. 9 spp. from mainland Africa were listed as threatened in 1998.

Mate *Ilex paraguariensis* (Aquifoliaceae)

Origin: South America
Tea is made from leaves, contains 2% caffeine. Little use in export market.

SPICES AND FLAVOURS

Onion, Garlic *Allium cepa, Allium sativum* (Alliaceae)

Origin: onion exists only in cultivation, may have come from Afghanistan, Iran, and former USSR area. Possible progenitors of garlic are A. *longicuspis* or wild A. *ampeloprasum*. The greatest number of *Allium* spp. are in North Africa and Eurasia.
Underground bulbs more important for their flavour and antimicrobial properties than nutritional value. Garlic contains large amounts of potassium and significant vitamin C. Important component in the diet of a wide range of cultures. Numerous medicinal functions. The allins contained in *Allium* spp, especially garlic, may protect against cancer and cardiovascular disease. 7 economically important cultivated *Allium* spp; many other species consumed on a lesser scale. Open pollinated populations still represent most of the production in tropical and subtropical countries. The habitat of some wild *Allium* spp. is severely threatened. Poor results from interspecific hybridisations. The genetic variability available in wild and cultivated relatives has not been extensively used in crop improvement.

Chili Pepper, Sweet Pepper *Capsicum annuum* (Solanaceae)

Origin: tropical America
Fruits of varying pungency are used either fresh as vegetable or dried or powdered as a spice. Fresh fruits contain large quantities of vitamin A, also vitamin C. Breeding generally has depended on pure lines. Wild peppers are still collected and sold locally. Some interfertility with other *Capsicum* spp. Wild spp. offer valuable new traits.

Cardamom *Elettaria cardamomum* (Zingiberaceae)

Origin: India
Seeds used as a spice. Essential oil used in perfume and as flavour for liqueurs. Wild populations exist in monsoon forests in South India and Sri Lanka. There are no essential differences between wild and cultivated forms. Collection from the wild contributes to the commercial trade. Wild populations are disappearing.

Pimento *Pimenta dioica* (Myrtaceae)

Origin: West Indies and Central America.
Plants grow in semi-wild state. Populations of wild relatives are confined to small areas of remaining dry forest and coastal habitat in the Caribbean, especially Cuba. They are poorly studied and are severely threatened by habitat loss.

Pepper *Piper nigrum* (Piperaceae)

Origin: Western Ghats, India.
The dried fruits, high in alkaloid content, represent one of the oldest spice crops. Several other *Piper* spp. important for local pepper production. Grown as a smallholder crop in tropical countries. Large scale planting is based on one clone and is dangerously vulnerable to disease. Wild pepper still grows in the Western Ghats.

SUGAR CROPS

Sugar Beet *Beta vulgaris* (Chenopodiaceae)

Origin: evolved from sea-beet (*B. vulgaris var. maritima*) in Europe and west Asia.
Swollen taproot provides nearly half the world production of sucrose. Forms of the same species include leaf beets and chards used as garden vegetables, and other beets with swollen taproots, e.g. beetroot and mangold for human consumption and animal feed. All forms within the species may be crossed. Wild relatives have already provided some disease resistance. The only source of resistance against the beet cyst nematode is detected in relatives from a different section in the genus.

Hazel and Filbert *Corylus avellana*, *C. maxima* (Betulaceae)

Origin: Europe and West Asia
Edible tree nut and ornamental. Kernels are used in confectionery, 18% protein and 68% oil. All *Corylus* spp. have edible nuts. *C. colurna* is also cultivated for nuts. Populations of *C. chinensis* in China have declined, largely because of overexploitation.

Walnut *Juglans regia* (Juglandaceae)

Origin: Balkans to China
Edible nut, containing vitamins E, C and B. Uses as dessert and in confectionery, oil also extracted. The kernel contains 15% protein and 70% unsaturated oil. Leaves make good fodder. The timber is highly valued. All *Juglans* spp. produce edible seeds, timber, ornamentals. No apparent threat of genetic erosion in the crop. Major cultivation in USA but enormous unexplored potential elsewhere. Wild walnut forest has declined and become fragmented throughout its native range. Of the 21 species in the genus, 7 were listed as threatened in 1997.

Pistachio *Pistacia vera* (Anacardiaceae)

Origin: Near East and Western Asia
Tree nut, low in sugar, >20% protein, 50% oil. Important food for nomads during migration in Iran and Afghanistan. Highly drought resistant. Trees used for ornamental and shade purposes, also as a source of resin, dye, turpentine, mastic and medicine. Cultivated in the Mediterranean and western Asia for 3 000-4 000 years. None of the related species has a value as a nut crop, although 7 spp. are used as rootstocks and also for pollination. Largely harvested from wild in Afghanistan and parts of Pakistan. Iran has had commercial plantations for hundreds of years. Wild species may have a role in future improvements. Many wild populations have been destroyed by forest clearance, over-cutting for charcoal and grazing. 3 spp. were listed as threatened in 1998.

Brazil nut *Bertholletia excelsa* (Lecythidaceae)

Origin: tropical South America

Kernel contains 17% protein, 65-70% monounsaturated oil. Largely an export crop. Also a staple for indigenous people and important ecological component of rainforest in the Amazon basin. Oil used for cooking or as fuel or animal feed. Valuable timber. Attempts to establish plantations have generally failed. Well managed plantations have the potential of producing yields far exceeding natural groves. Almost all nut production is from wild trees. Distribution and density of groves may have been largely influenced by indigenous groups in the past. Little information exists on genetic variation. Populations appear to tolerate different soil types. Sustainable system of harvesting in extractive reserves, but considerable habitat loss and illegal tree felling continues elsewhere. Development in the Tocantins valley where there is high concentration of brazil nut trees continues to cause population decline. Developments elsewhere also resulting in serious genetic losses. The species was listed as vulnerable in 1998. Protected populations are found in biological reserves, Indian and extractive reserves and corporate property.

References

1 Coope, G.R. 1995. Insect faunas in ice age environments: why so little extinction?. In, Lawton, J.H. and May, R.M. (eds). *Extinction rates*. Oxford University Press.

2 Adams, J. 1999. Sudden climate transitions during the Quarternary. *Progress in physical geography* 23 (1):1-36.

3 Van Couvering, J.A. (ed.) 1997. *The Pleistocene boundary and the beginning of the Quarternary*. Cambridge University Press. xxi, 296p.

4 Petit *et al.* 1999. Climate and atmospheric history of the past 420 000 years from the Vostok ice core, Antarctica. *Nature* 399:429-436.

5 Larick, R. and Ciochon, R.L. 1996. The African emergence and early Asian dispersal of the genus *Homo. American Scientist* November-December 1996.

6 Wood, B. and Brooks, A. 1999. We are what we ate. *Nature* 400: 219-220.

7 Dennell, R. 1997 The world's oldest spears. *Nature* 385: 767-768.

8 Thorn, A., Grün, R., Mortimer, G., Spooner, N.A., Simpson, J.J., McCulloch, M., Taylor, L. and Curnoe, D. 1999. Australia's oldest human remains: age of the lake Mungo skeleton. *Journal of human evolution* 36: 591-612.

9 Dillehay, T. D. 1997. *Monte Verde: A Late Pleistocene Settlement in Chile*, Volume 1: The *Archaeological Context and Interpretation*. Smithsonian Institution Press, Washington, D.C.

10 Guidon, N. and Delibrias, G. 1986. Carbon-14 dates point to man in the Americas 32 000 years ago. *Nature* 321: 769-771.

11 Haynes, C.V. 1984. Stratigraphy and Late Pleistocene extinction in the United States. In, Martin, P.S. and Klein, R.G. (eds). *Quaternary extinctions*. Pp. 345-353. The University of Arizona Press, Tucson, Arizona, USA.

12 Harcourt, C.S. and Sayer, J.A. 1995. *The conservation atlas of tropical forests: the Americas*. IUCN, Cambridge and Switzerland.

13 Pimm, L.S., Moulton, M.P and Justice, L.J. 1995. Bird extinctions in the central Pacific. In, Lawton, J.H. and May, R.M. (eds). *Extinction rates*. Pp. 75-87. Oxford University Press.

14 Dewar, R.E. 1984. Mammalian extinctions and stone age people in Africa. In, Partin, P.S. and Klein, R.G. (eds). *Quaternary extinctions*. Pp. 553-573. The University of Arizona Press, Tucson, Arizona, USA.

15 Semaw, S., Renne, P., Harris, J.W.K., Feibel, C.S., Bernor, R.L., Fesseka, N. *et al.* 1997. 2.5-million-year-old stone tools from Gona, Ethiopia. *Nature* 385:333-336.

16 Thieme, H. 1997. Lower Palaeolithic hunting spears from Germany. *Nature* 385: 807-810.

17 Morwood, M.J., O'Sullivan, P.B., Aziz, F. and Raza, A. 1998. Fission-track ages of stone tools and fossils on the east Indonesian island of Flores. *Nature* 392: 173-176.

18 Bird, M.I. and Cali, J.A. 1998. A million-year record of fire in sub-Saharan Africa. *Nature* 394: 767-769.

19 Gowlett, J.A.J., Harris, J.W.K., Walton, D. and Wood, B.A. 1981. Early archaeological sites, hominid remains and traces of fire from Chesowanja, Kenya. *Nature* 294: 125-129.

20 Isaac, G. 1981. Early hominids and fire at Chesowanji, Kenya. *Nature* 296: 870.

21 de Heinzelin, J. *et al.* 1999. Environment and behaviour of 2.5 million year old Bouri hominids. *Science* 23 April 1999: 625-629.

22 James, S.R. 1989. Hominid use of fire in the lower and middle Pleistocene – A review of the evidence. *Current Anthropology* 30(1): 1-26.

23 Hoopes, J.W. 1996. In search of nature. *Imagining the precolombian landscapes of ancient Central America*. Working paper for the nature and culture colloquium.

24 Miller, G.H. *et al.* 1999. Pleistocene extinction of *Genyornis newtoni*: human impact on Australian megafauna. *Science* 283: 205-208.

25 Martin, P.S. 1984. Prehistoric overkill: the global model. In, Martin, P. and Klein, R.G. (eds). *Quaternary extinctions*. Pp. 354-103. The University of Arizona Press, Tucson, Arizona, USA.

26 Clutton-Brock, J. 1995. Origin of the dog: domestication and early history. In, Serpell, J. (ed.), *The domestic dog: its evolution, behaviour and interactions with people*. CUP, Cambridge, UK. Pp. 8-20.

27 McEvedy, C. and Jones, R. 1978. *Atlas of world population history*. Penguin Books.

28 Roberts, N. 1998. *The Holocene. An environmental history*. 2nd edition. Blackwell Publishers.

29 Smil, V. 1999. How many billions to go? *Nature* 401:429

30 Vitousek, P.M., Ehrlich, P.R., Ehrlich, A.H. and Matson, P.A. 1986. Human appropriation of the products of photosynthesis. *BioScience* 36:368-373.

31 Wright, D.H. 1990. Human impacts on energy flow through natural ecosystems, and implications for species endangerment. *Ambio* 19(4):189-194.

32 Haberl, H. and Schandl, H. 1999. Indicators of sustainable land use: concepts for the analysis of society-nature interrelations and implications for sustainable development. *Environmental management and health* 10(3): 177-190.

33 Haberl, H. 1997. Human appropriation of net primary production as an environmental indicator: implications for sustainable development. *Ambio* 26(3):143-146.

34 FAO. FAOSTAT database. Available at FAO website, http://www.fao.org.

35 World Conservation Monitoring Centre. 1992. *Global biodiversity: status of the Earth's living resources*. Chapman and Hall, London xx+594pp

36 FAO. 1998. *The state of the world's plant genetic resources for food and agriculture*. Food and Agriculture Organization of the United Nations. Rome 1998.

37 FAO. 1996. *Food requirements and population growth. World food summit technical background document* TBD VI E4. Available at: http://www.fao.org.

38 Uvin, P. *The state of world hunger*. Hunger report. Gordon and Breach. Available http://www.brown.edu/departments/world_hunger_program/hungerweb/intro/ stateofworldhunger.pdf.

39 Wolf, E. C. 1986. *Beyond the Green Revolution: New Approaches for Third World Agriculture*. World Watch Paper 73, WorldWatch Institute, Washington D.C., October 1986

40 Herrera-Estrella, L. 1999. Transgenic plants for tropical regions: some considerations about their development and their transport to the small farmer. *Proc. Natl. Acad. Sci. USA* 96: 5978-5981

41 Tilman, D. 1999. Global environmental impacts of agricultural expansion: The need for sustainable and efficient practices. *Proc. Natl. Acad. Sci USA* 96: 5995-6000.

42 de Garine, I. and G.A. Harrison. (eds). 1988. *Coping with uncertainty in food supply*. Oxford: Clarendon. xiv,483p.

43 Fernandes, E.C.M. Okingati, A. and Maghembe, J. 1984. The Chagga homegardens: A multi-storied agroforestry cropping system on Mount Kilimanjaro (Northern Tanzania). *Agroforestry systems* 2: 73-86

44 Juma, C. 1989. *The gene hunters: biotechnology and the scramble for seeds*. London: Zed xiv,288p; 22cm

45 Marsh, R.R., and Talukeder, A. 1994. *Effects of the integration of home gardening on the production and consumption of target interaction and control groups: a case study from Bangladesh*. Conference paper.

46 Pálsson, G. 1988. Hunters and gatherers of the sea. In, Ingold, T., Riches, D., Woodburn, J. *Hunters and gatherers* 1, *history, evolution and social change*. Pp. 189-204. BERG, Oxford.

47 Lange, D. 1996. *Untersuchungen zum Heilpflanzenhandel in Deutschland*. Bonn, Germany; Bundesamt fur Naturschutz.

48 Kuipers, S.E. 1997. *Trade in medicinal plants*. In: FAO. Medicinal plants for forest conservation and health care. Non-Wood Forest Products Series No. 11. Food and Agriculture Organization of the United Nations.

49 Farnsworth, N.R. 1994. *Ethnopharmacology and drug development*. In, Prance, G.T., Chadwick, D. and Marsh, J. (eds). Ethnobotany and the search for new drugs. Pp.42-51. Ciba Foundation

50 Lewington, A. 1990. *Plants for people*. Natural History Museum (London) Publications

51 Phillipson, J.D. 1994. Natural products as drugs. *Transactions of the Royal Society of Tropical Medicine and Hygiene*. 88:S1-17-S1-19.

52 Grifo, F. *et al.* 1997. The origins of prescription drugs. In, Grifo, F. and Rosenthal, J. (eds). *Biodiversity and Human Health*. Pp. 131-163. Island Press, Washington DC.

53 Costanza, R. *et al.*, 1997. The value of the world's ecosystem services and natural capital. *Nature* 387: 253-260.

54 Artuso, A. 1997. Capturing the chemical value of biodiversity: Economic perspectives and policy prescriptions. In, Grifo, F. and Rosenthal J. (eds). *Biodiversity and Human Health*. Island Press, Washington DC.

55 Whitman, W.B., Coleman, D.C. and Wiebe, W.J. 1998. Prokaryotes: the unseen majority. *Proc. Natl. Acad. Sci.* USA 95: 6578-6583.

56 Rusek, J. 1998. Biodiversity of collembola and their functional role in the ecosystem. *Biodiversity and Conservation* 7: 1207-1219.

57 Zimmerman, P.R., Greenberg, J.P., Wandiga, S.O. and Crutzen, P.J. 1982. Termites: a potentially large source of atmospheric methane, carbon dioxide, and molecular hydrogen. *Science* 218: 563-565.

58 Nicol, S. *Time to Krill?*. Antarctica Online: The website of the Australian Antarctic Program http://www.antdiv.gov.au/science/bio/issues_krill/krill_features.html

59 Gaston, K. J. and Blackburn, T.M. 1997. How many birds are there? *Biodiversity and Conservation* 6: 615-625.

60 Said, M.Y., Chunge, R.N., Craig, G.C., Thouless, C.R., Barnes, R.F.W. and Dublin, H.T. 1995. *African Elephant Database 1995*. IUCN, Gland, Switzerland. 225 pp

61 Santiapillai, C. 1997. The Asian Elephant conservation: a global strategy. *Gajah. Journal of the Asian Elephant Specialist Group* 18:21-39.

62 Jefferson, T.A., Leatherwood, S. and Webber, M.A. 1993. *Marine mammals of the world*. FAO Species Identification Guide. UNEP: Rome.320 pp.

63 Kemf, E. and Philips, C. 1995. *Whales in the Wild*. WWF Species Status Report.

64 Nowak, R.M. 1991. *Walker's Mammals of the World* (5th ed). Volume II. The John Hopkins University Press.

65 Richards, J.F. 1990. Land transformation. In, Turner II, B.L., Clark, W.C., Kates, R.W., Richards, J.F., Mathews, J.T. and Meyer, W.B. (eds.). *The Earth as transformed by human action*. Pp. 163-178. Cambridge University Press with Clark University. Cambridge.

66 Prescott-Allen, R. and Prescott-Allen, C. 1990. How many plants feed the world? *Conservation Biology* 4(4):365-374.

67 Mabberley, D.J. 1997. *The plant-book. A portable dictionary of the vascular plants*. 2nd edition. Cambridge University Press. 858pp.

68 Smartt, J. and N.W. Simmonds. 1995. *Evolution of crop plants*. 2nd edition. Longman Scientific & Technical. 531pp.

69 Smith, N.J.H., Williams, J.T., Plucknett D.L., and Talbot, J.P., 1992. *Tropical forests and their crops*. Cornell University Press. 568pp.

70 Vaughan, J.G. and Geissler, C.A. 1997. *The new Oxford book of food plants*. Oxford University Press. 239pp.

71 Walter, K.S. and Gillett, H. J. (eds) 1998. 1997 IUCN *Red List of threatened plants*. Compiled by the World Conservation Monitoring Centre. IUCN – The World Conservation Union, Gland, Switzerland and Cambridge, UK. 1xiv-862pp.

72 Juma, C. 1989. *Biological diversity and innovation: conserving and utilizing genetic resources in Kenya*. African Centre for Technology Studies, Nairobi, Kenya

73 Alvarez-Buylla Roces, M.A., Lazos chavero, E. and Garcia-Burrios, J.R. 1989. Homegardens of a humid tropical region in South East Mexico: an example of an agroforestry cropping system in a recently established community. *Agroforestry Systems*. 8, 133-156

74 Alcorn, J.B. 1984. Development policy, forests and peasant farms: reflections on Haustec-managed forest contribution to commercial and resource conservation. *Economic Botany*. 38(4): 389-406.

75 Davies, A.G. and Richards, P. 1991. Rain forest in Mende life: resources and subsistence strategies in rural communities around the Gola North Forest Reserve (Sierra Leone). A report to the Economic and Social Committee on Overseas Research (ESCOR), UK Overseas Development Administration, UK

76 Fernandes, E.C.M., Oktingati, A. and Maghembe, J. 1984. The Chagga homegardens: A multi-storied agroforestry cropping system on Mt. Kilimanjaro (Northern Tanzania) *Agroforestry Systems*. 2: 73-86.

77 Michon, G., Mary, F. and Bompard, J. 1986. Multistoried agroforestry garden system in West Sumatra, Indonesia. *Agroforestry Systems*. 4(4): 315-338.

78 Kunstadter, P. Chapman, E.C. and Sabhasri, S. 1978. *Farmers in the forest: Economic development and marginal agriculture in Northern Thailand*. An East-West Center Book, The University Press of Hawaii, Honolulu.

79 Padoch, C. and de Jong, W. 1991. The house gardens of Santa Rosa: diversity and variability in an Amazonian agricultural system. *Economic Botany*. 45(2): 166-175

80 Denevan, W.M. and Treacy, J.M. 1987. Young managed fallows at Brillo Neuvo. In, Denevan, W.M. and Padoch, C. (eds). Swidden-fallow agroforestry in the Peruvian Amazon, *Advances in Economic Botany*. 5: 8-46. The New York Botanical Garden, New York.

81 Ohtsuka, R. 1993. Changing food and nutrition of the Gidra in lowland Papua New Guinea. In, Hladik, C.M. *et al.* (eds). Tropical forests, people and food. *Biocultural interactions and applications to development*. Man and the Biosphere Series. Volume 13.

82 Conklin, H.C. 1954. An ethnoecological approach to shifting agriculture. *Transactions of the New York Academy of Sciences*. 17: 133-142.

83 Posey, D.A. 1984. A preliminary report on diversified management of tropical forest by the Kayapo Indians of the Brazilian Amazon. *Advances in Economic Botany*.1:112-126

84 Vickers, W.T. 1993. Changing tropical forest resource management strategies among the Siona and Secoya Indians. In, Hladik, C.M. *et al.* (eds). Tropical forests, people and food. Biocultural interactions and applications to development. Man and the Biosphere Series. Volume 13.

85 Michon, G. 1983. Village forest gardens in West Java. In, Huxley P. (ed.). *Plant Research and Agroforestry*. Pp. 13-24. ICRAF, Nairobi.

86 FAO. 1996. *Assessment of feasible progress in food security*. World food summit technical background document TBD V3 E14. Available at: http://www.fao.org.

87 Tobler, W., Deichmann, U., Gottsegan, J. and Maloy, K. 1995. *The global demography project*. National Centre for Geographic Information and Analysis, University of California, Santa Barbara. Technical Report TR-95-6.

88 Yalden, D.W. 1996. Historical dichotomies in the exploitation of mammals. In, Taylor, V.J. and Dunstone, N. *The exploitation of mammal populations*. Pp. 16-27. Chapman and Hall, London.

Suggested introductory source

Lewin, R. 1999. *Human evolution. An illustrated textbook*. 4th edition. Blackwell Scientific Inc., Mass.

Smith, B.D. 1998. *The emergence of agriculture*. Scientific American Library.

Fischer-Kowalski, M. and Harberl, H. 1993. Metabolism and colonisation. Modes of production and the physical exchange between societies and nature. *Innovation* 6(4):415-442.

5 PATTERNS AND TRENDS IN GLOBAL BIODIVERSITY

Variation in global biodiversity exists because historical and ecological factors dictate that there are more species, and a greater variety of lineages, in some places than in others. The most striking overall pattern in global biodiversity is that species richness typically increases from higher latitudes toward the tropics, in parallel with variation in net primary production.

Any area contributes to global biodiversity by the number of species present and shared with other areas, and those present but found nowhere else (endemics). An area rich in either or both kinds of species has high biodiversity value. To enable cost-effective conservation measures, much effort has been devoted to identifying areas of high value for particular groups, and, more importantly but less conclusively, areas of congruence in richness, endemism, or risk among species in a range of different groups.

Natural global patterns in biodiversity have been radically altered by human action. In mammals and birds 10-20% of living species are at significant risk of extinction, and evidence suggests that the rate of extinction in historical time has been significantly above normal. Recently developed global biodiversity indicators show some conservation success against a background of declining ecosystem health.

THE DISTRIBUTION OF BIOLOGICAL DIVERSITY

The single most important fact about biological diversity is that it is not evenly distributed over the planet[1,2,3]. This comes about quite simply because more species live in some places than in others. This means that adverse change in the environment will have greater effect on global biodiversity in some areas than in others, and for a variety of different reasons some areas may be more deserving of conservation action than others. The discovery and description of patterns in biodiversity is not merely of intrinsic biological interest but links directly to any attempt to manage rationally the way humans interact with the biosphere.

Measuring variations in biodiversity

Comparing the diversity of different parts of the world is problematic, not merely because of the paucity of data but also because of the way diversity changes with scale. A wide range of observations has demonstrated that, as a general rule, the number of species recorded in an area increases with the size of the area, and that this increase tends to follow a predictable pattern known as the Arrhenius relationship, whereby:

$$\log S = c + z \log A$$

where S = number of species, A = area and c and z are constants (see Figure 5.1). The slope of the relationship (z in the equation) varies considerably between surveys, although is generally between 0.15 and 0.40, and some surveys do not fit the relationship at all. A common generalisation from this finding is that a ten-fold reduction in an area (ie. loss of 90% of habitat) will result in the loss of between 30% (with z = 0.15) and 60% (with z = 0.40) of the species originally present, or approximately half the species.

Because the slope of the Arrhenius relationship is not constant everywhere, the relative diversity of different sites will often depend on the scale at which diversity is measured. Thus 1m² of semi-natural European chalk grassland will contain many more plant species than 1m² of lowland Amazonian rain forest, whereas for any area larger than a few square metres this will be reversed. This is because as an area is sampled the number of species recorded increases with the size of the area, but this rate of increase varies from area to area.

The reason for this increase may be quite straightforward. When small areas are sampled they are likely to be relatively homogenous in terms of habitat type. At small scale, as sample area increases, so an increasing proportion of the species present in that habitat is likely to be included in the sample. Beyond a certain point, however, as larger areas are sampled so an increasing number of different habitats will be included in the sample area, each with new species which are likely to be included in the sample. The species/area relationship is thus probably a sampling effect at small scale and an effect of habitat heterogeneity at larger scales.

Figure 5.1

A typical species-area plot

Note: The data, consisting of species counts in a series of areas of different size, are plotted on logarithmic axes resulting in a straight-line graph, the slope of which (z) indicates the rate at which species number changes with changing area.

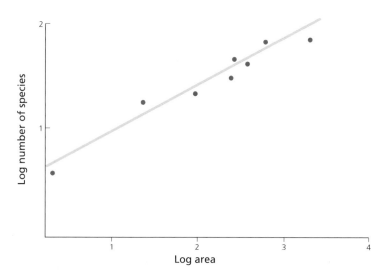

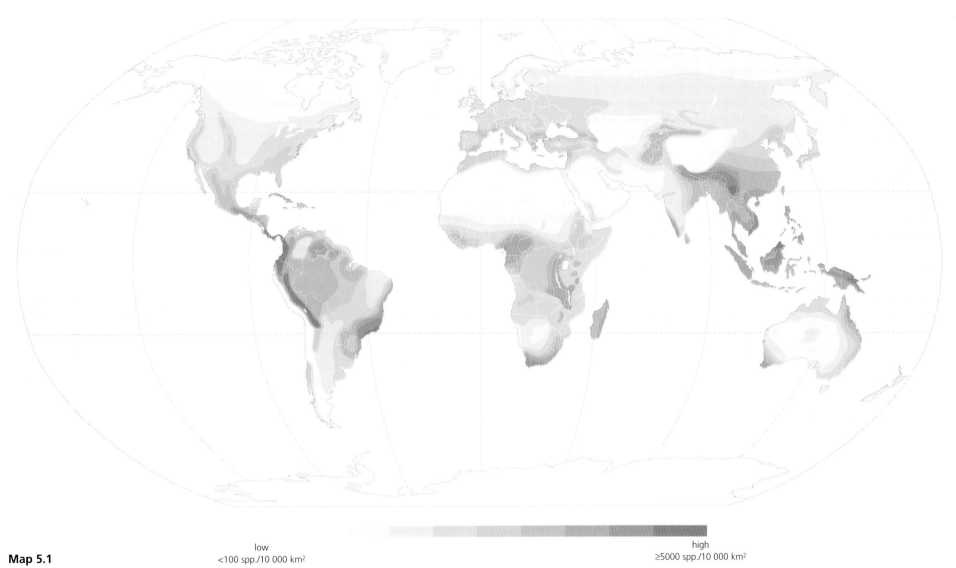

Map 5.1

Diversity of vascular plants

low
<100 spp./10 000 km²

high
≥5000 spp./10 000 km²

This map shows the species richness of vascular plants, plotted as a world density surface. It is based on some 1400 literature records from different geographic units, with richness values as mapped calculated on a standard area of 10 000 km² using a single species-area curve.

Source: Data and analysis © Wilhelm Barthlott (Botanic Institute and Botanic Gardens, University of Bonn). Reproduced by permission, with modification. For further details see[41,51,52].

Ecologists attempt to take account of this by recognising different kinds of biological diversity. The diversity within a site or habitat is often referred to as alpha (α) diversity while the differences between habitats are referred to as beta (β) diversity. Thus an area with a wide range of dissimilar habitats will have a high β-diversity, even if each of its constituent habitats may have low α-diversity. Differences in site diversity over large areas, such as continents, are sometimes referred to as gamma (γ) diversity.

Measures of diversity can refer simply to species richness but can also be more sophisticated statistical measures that take into account the relative abundance of different species in a given place. A variety of different measures of this kind have been developed (of which H', the Shannon-Wiener function, is frequently used). Under most of these measures, an area in which all species are of similar abundance would generally be given a higher diversity measure than an area with the same number of species, a few of which were very abundant and the remainder rare. Deriving these statistical measures requires intensive sampling and is of limited application at any other than the smallest scales. For large scale comparisons, measures of species richness are by far the most useful.

Present global patterns in species richness

Analysis of global variation in biodiversity is usually focused on species richness, or some other value that can stand as a reasonable surrogate for this, eg. generic or family richness[3,4,5] because this is the only indicator of diversity for which anything approaching adequate data are available.

Global variation in biodiversity is essentially the sum of individual species' distributions. The available distribution information is geographically very incomplete, and relates to only a small fraction of the 1.7 million known species. Geographically, the waters and land surface of western Europe have been more thoroughly sampled than elsewhere, while large areas in the tropics, particularly of South America and Central Africa are poorly known. Taxonomically, the larger mammals, birds, fishes of commercial interest, vascular plants and a few invertebrate groups, such as lepidoptera and odonata, are better known than other groups of species.

Nevertheless a wealth of empirical observations indicates that species richness in eukaryotes tends to vary geographically according to a series of fairly well defined rules, ie. for terrestrial environments:

- warmer areas hold more species than colder ones;
- wetter areas hold more species than drier ones;
- areas with varied topography and climate hold more species than uniform ones;
- less seasonal areas hold more species than highly seasonal ones;
- areas at lower elevation hold more species than high elevation areas.

The single most obvious pattern in the global distribution of species is that overall species richness increases as latitude decreases toward the equator. At its simplest this means that there are more species in total and per unit area in temperate regions than in polar regions, and far more again in the tropics than in temperate regions. This applies as an overall general rule, and within most individual higher taxa (at order level or higher), and within most equivalent habitats. The pattern can be seen in Maps 5.1 and 5.2, which are density surfaces representing the global diversity of flowering plant species and freshwater fish families, respectively. Several further examples and wider discussion are available[3], including at the Natural History Museum (London) website[4].

The overall pattern conceals a large number of minor trends where species richness in particular taxonomic groups or in particular habitats may show no significant latitudinal variation, or may actually decrease with decreasing latitude (eg. pinnipeds and seabirds, see Chapter 6; crayfish, see Chapter 8).

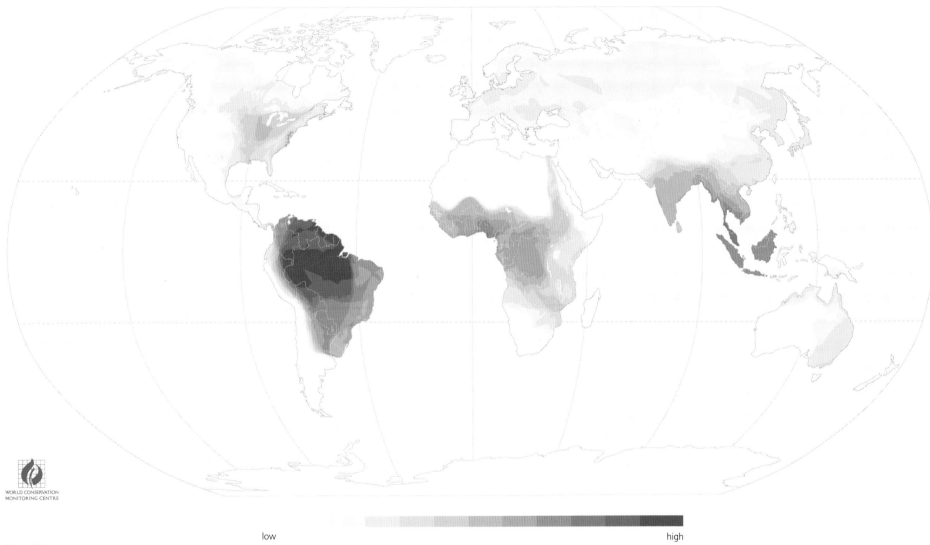

low   high

Map 5.2

Diversity of freshwater fishes

This map shows the family richness of fishes in inland waters, plotted as a world density surface. It is based on generalised range maps of 142 families, and colour depth represent number of families, from 2 to 49, potentially present at any point. The family level taxonomy of fishes has been relatively fluid, but the broad pattern of global diversity appears quite consistent.

Source: produced by WCMC using range maps digitised from Berra[40].

In some circumstances, and perhaps generally in terrestrial systems, geomorphological heterogeneity, particularly topography, plays a significant part in determining species number. This would be expected because, depending on the size and mobility of organisms, the chances of geographic isolation and speciation increase in topographically diverse landscapes. The rôle of topography has been demonstrated statistically at continent scale for North American mammals[12], and at both landscape and patch scale for vascular plants[13,14].

There is good evidence that moist tropical forests are in general the most species-rich environments on Earth. If current estimates of the number of as yet unknown species (see Chapter 2) in the tropical forest microfauna are accepted, these regions, extending over perhaps 7% of the world's surface, may hold more than 90% of the world's species. If tropical forest small insects are discounted, then coral reefs and (especially for flowering plants) areas of Mediterranean climate in South Africa and Western Australia, may be similarly rich in species.

Species and energy

Latitudinal variation in diversity on land is strongly correlated with, and may be largely explained by, variation in incident energy over the Earth's surface. The relationship between diversity and productivity, and related measures, has been the subject of long-standing debate in ecology, but recent studies have shown that at global or continental scale, organismal diversity, particularly as measured at higher taxonomic levels, is strongly correlated with available energy[6,42]. This kind of relationship has been demonstrated[7-10], for example, for flowering plants, for trees, lepidoptera, land birds, and land mammals in a range of countries and continents, and for fishes in river basins at a global level[47]. One simplistic explanation for this may be that higher energy availability leads to increased net primary production (NPP), and this broader resource base allows more species to coexist.

While the general relationship appears robust, the details are complex. Energy availability can be measured in several ways: as heat energy, as potential (PET) or actual evapotranspiration (AET), or as net primary production, and which is the best predictor of diversity has yet to be determined. Some measure of the simultaneous availability of water and radiant energy may provide the best general predictor of potential macro-scale species richness[6,42]. More complete explanation for richness variation would need also to consider the rôle of topography, history and edaphic factors.

Biogeography and endemism

While ecological factors determine which kinds of species, and almost certainly how many of them, can persist in a given area, history has already determined which actual lineages are present. A complete explanation for global variation in biodiversity must therefore involve both historical events and current ecological processes. The former are implicit in any explanation of the origin of diversity, the latter in explanations of its maintenance; these being two separate, although intimately linked, problems. On land, continental drift resulting from plate tectonics, climate change, mountain building or sea level change, and probably the evolutionary lability of different lineages, are among the important historical factors. Geographic features commonly restrict or prevent the further dispersal of species, eg. a large river can present a barrier to a terrestrial species, the sea is a barrier to non-flying island forms, and land is a barrier to freshwater species.

Barriers to dispersal explain why the fauna and flora of ecologically similar areas in geographically separated parts of the world tend to be composed largely of different individual species. They also underlie the phenomenon of endemism. An endemic species is one restricted to some given area, which may be a continent or country, or more significantly, a relatively small area, such as a mountain block, island or lake. Discrete areas of complex topography, particularly in the tropics, often have high endemism in a range of taxonomic groups, possibly because climate change has encouraged speciation by isolating different lineages at different times. For example, mountain blocks in the Eastern Arc mountains of Tanzania hold many bird, reptile, amphibian, and other endemics.

Assigning importance to areas

There are two principal approaches to evaluating the diversity of different areas. The first, and most obvious, makes reference only to intrinsic diversity, so that an area with higher numbers and kinds of species is deemed more important than one with lower diversity. The overall diversity of any given area will be a reflection both of the range of habitats it includes and the diversity of the component habitats. The greater the differences between the various component habitats in terms of species composition, then the greater the overall diversity will be. The second approach entails assessment of the contribution any given area makes to the overall diversity of a given geographic region, such as a country, continent, or to the world overall. From this perspective, some areas with lower intrinsic diversity may, if many of the species present occur there and nowhere else (ie. are endemic to the area defined), be more important than others with higher diversity. Such an area with relatively low species diversity may therefore still make an important contribution to the overall diversity of the larger region it is embedded in. Oceanic islands and continental montane regions are examples of geographical entities which typically have comparatively low species diversity but high rates of endemism. Assessing the relative importance of areas with high species diversity and low rates of endemism compared to areas with lower rates of diversity and high endemism remains an intractable problem unless other criteria are considered.

DIVERSITY AS A BASIS FOR CONSERVATION ACTION

Historically the impetus for much conservation activity has been the desire, seen by many as a moral imperative, to prevent where possible the extinction of individual species. Not only must the extinction of species be prevented, but their populations must be maintained at levels sufficient to minimise the risk of future extinction. In practice, the focus has tended to be on species that are large, charismatic and possibly also ecologically important or highly threatened, or both. The tiger *Panthera tigris*, Arabian oryx *Oryx leucoryx*, white rhinoceros *Ceratotherium simum*, and blue whale *Balaenoptera musculus* are familiar mammalian examples. In developed countries national legislation may be enacted in which avoidable impact on the species and its habitat are prohibited, and targets are set by which to gauge the success of action taken. Considerable success has been achieved with the species named above and the many others, mainly animals but some plants, subject to similar efforts. However, this approach could never be extended to cover more than an insignificant fraction of even the approximately 300 000 larger organisms (plants and vertebrates) in the biosphere, let alone the other several million that probably exist. In fact, its primary benefit appears to be that large organisms, terrestrial vertebrates in particular, generally require large areas of useful habitat, and if such areas can be managed to minimise risk, other species in the system may be safeguarded.

Area-based approaches essentially extend the emphasis on identifying, delimiting and conserving large areas of useful habitat. The approach is generally based on the premise that more is better, ie. an area with more biodiversity is more worth conserving than an area with less. At its simplest 'biodiversity' may be equated with species number overall or per unit area, but other selection criteria are possible. Higher value may be placed on an area with populations of threatened species, or one rich in endemic species (especially if there are endemics in several different groups), or in species of commercial or cultural importance, or which is particularly representative of an ecosystem (perhaps one that is widely degraded elsewhere).

These approaches converge on the more holistic current emphasis on ecosystem conservation. Many would argue that the primary goal of conservation action is the long-term maintenance of ecosystem processes at global scale and over foreseeable human generations. This is in effect the meaning of sustainability. But measuring the organisation, vigour and resilience of ecosystems, ie. *ecosystem health*, is fraught with practical difficulties. Much conservation activity therefore quite properly works from a strong form of the precautionary principle, and focuses on maintaining so far as possible the elements of ecosystems, ie. species and populations and their physical environment, in the hope that the system will thus be perpetuated

Setting priorities for conservation action

Management action aimed at biosphere conservation demands financial resources, but these are limited, while human numbers and impacts have been ever-increasing. This implies that choices must be made between possible actions, whether by design or default. Rational and informed decision-making should seek to increase the efficiency with which conservation funds are used. Several studies have been undertaken that aim to identify and sometimes to rank areas of high biodiversity value. While the results are usually of considerable biological interest, they may also have direct application as a basis for choice between alternative courses of action, on the grounds that greater benefit will accrue from ensuring the integrity of areas of higher biodiversity value as opposed to areas of lesser value. A considerable body of largely theoretical work has focused on measuring diversity in relation to priority setting[4].

One early area-based approach identified a dozen 'megadiversity' countries, which between them include a large proportion of global biodiversity in selected major groups[15]. Another delineated 18 endemic-rich botanical 'hotspots', which between them include around 20% of the world's known plant species in less than one percent of the land surface[16,17]. A strength of the former approach is that it focused at country level, and it is at this administrative level that most conservation action is undertaken; a weakness is that by evaluating species richness only it was not able to address uniqueness, and adjacent countries very rich in species are likely to have many species in common. A strength of the latter study is that it focused on areas rich in restricted-range endemic species and such areas by definition make a large contribution to global biodiversity; the areas identified were also believed to be subject to high rates of habitat modification. The hotspots approach has been further developed by Conservation International[18]. The term 'hotspot' is now often applied loosely (ie. without formal criteria for size and richness) to any area that has a concentration of diversity, whether by a measure of simple richness, or endemism, or number of threatened species.

The country-based approach has been extended in a study[19] using a database of estimates of richness and endemism in land vertebrates and vascular plants for all countries of the world (Table 5.5). Indices to overall diversity (weighting richness and endemism equally) and diversity adjusted for country area have been produced. Within the relatively wide margin of error associated with species inventory (see Chapter 2), these indices can yield a useful view of variation in diversity in geopolitical terms, or, if the data are treated as geographic samples, a view of general global variation in diversity (see Map 5.3).

Of studies based on biogeographic rather than geopolitical areas, the Centres of Plant Diversity project[20] remains one of the largest. It relied heavily on extensive consultation among botanists to identify several hundred important sites worldwide, defined semi-quantitatively on the basis of a general

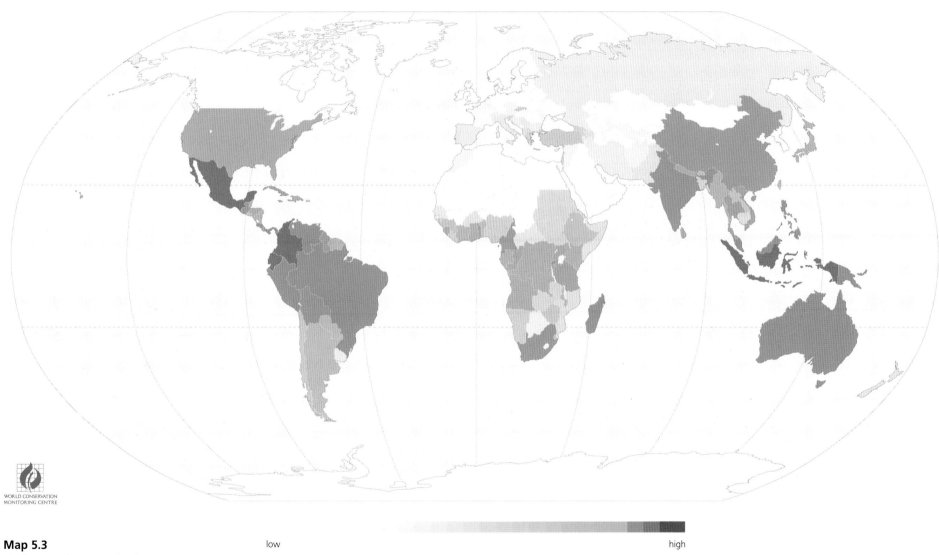

Map 5.3

Biodiversity at country level

low   high

This map illustrates relative biodiversity at country level, taking into account richness and endemism in the four terrestrial vertebrate classes and vascular plants, and adjusting according to country area. Countries at the high end of the scale have more diversity in these groups than would be expected on area alone. See data and AI column in Table 5.5. NB: to avoid ambiguity Alaska (USA) has for the purposes of this map been assigned to the same class as adjacent Canada rather than the conterminous USA.

Source: based on national biodiversity indices developed by WCMC[19].

combination of richness and endemism (Map 5.4). However, the most systematic and complete global level assessment to date has involved bird species. Birds are by far the best-known major group of organisms on the planet, with a relative wealth of distribution and population data available, and global analyses by BirdLife International have set a standard yet to be matched for other taxonomic groups. The distributions of all restricted-range bird species (defined as those in which the area encompassing all distribution records is less than 50 000 km²), amounting to 25% of all birds, have been mapped in digital format. The co-occurrence of restricted range species defines a set of 218 Endemic Bird Areas (EBAs)[21]; these are shown, ranked in three categories according to biodiversity importance, in Map 5.5. 'Importance' here takes account of the number of restricted-range species in the EBA and the number of EBAs in which present, taxonomic uniqueness, and EBA area.

There are a number of limitations to area-based methods for establishing priorities among possible conservation actions. First, there is no single unequivocal way of comparing value in different categories: how many vulnerable species is a single critically endangered species worth? Or how many endemic beetles is a single endemic bird worth? Second, information is always incomplete, in that it never covers all taxa at all sites of interest. Resolving the former requires more or less arbitrary assigning of value. Attempts to resolve the latter entail the search for indicators, that is groups of species or, sometimes, other variables that can act as surrogates for wider measures of biodiversity.

The search for biological indicators of this kind has generated a great deal of research and discussion. Findings to date have generally been equivocal, and depend in part on spatial scale. In general, though, it seems that at coarse scales, there may be quite good agreement between different taxa, so that, for example, many EBAs (which may be up to 600 000 km² in extent) also hold significant numbers of other restricted range species[21]. At this scale, then, birds may serve as indicators of high biodiversity value more generally. In other instances, and perhaps more generally at finer resolution, the relationship appears to break down. Studies in areas as disparate as North America, South Africa[46], Cameroon[45] and the British Isles[43,44] indicate that areas important for rare species often do not coincide in different groups (and these may be negatively correlated with areas of high species richness), and richness levels in any one group do not necessarily serve as an indicator of species richness in others.

Finally, it should be emphasised that priority-setting exercises remain very often only theoretical. This is because the opportunities to establish entirely new protected areas, or redesign existing protected area networks in the light of priorities identified are extremely limited. The designation and management of protected areas, as so much other conservation action, is generally driven and constrained far more by socio-political and economic considerations than it is by conservation biology.

GLOBAL TRENDS IN BIODIVERSITY

Natural patterns in global biodiversity are now obscured by changes brought about by the human species, a factor of overwhelming significance in the distribution and status of species and habitats. Domestic species, livestock and crops are nearly ubiquitous, along with introduced game animals and a host of accidental introductions, including many pernicious weeds and pests as well as unnoticed species.

The central issue is: are things getting better or worse, and how quickly? The question can be asked at all spatial scales, from local field level to the biosphere as a whole, but answered only incompletely. Many problematic issues arise. The task requires that change, and trends in change can be identified, and also, if progress is to be monitored, that a desired state can be defined. Change in biological diversity has principally been

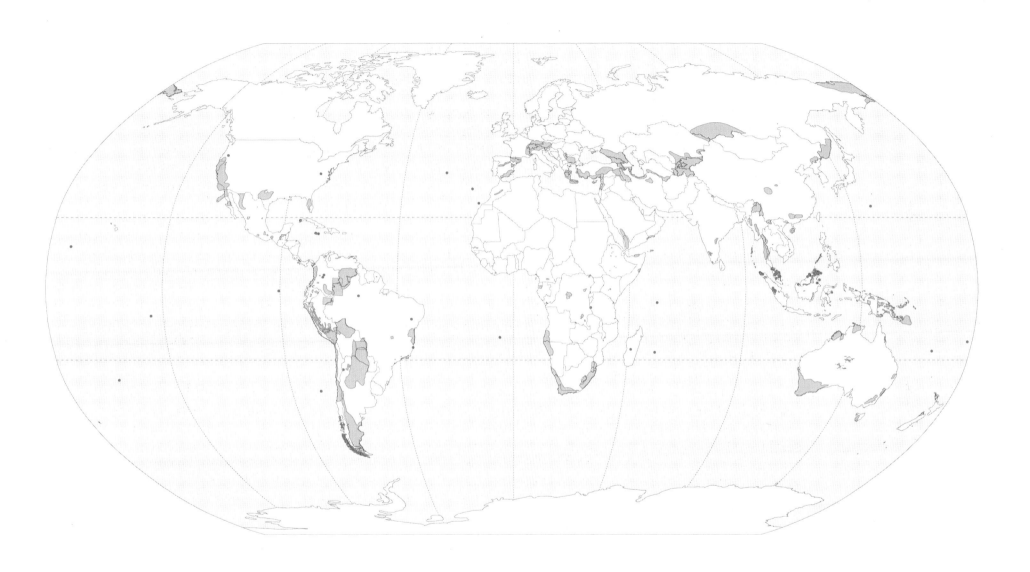

Map 5.4

Centres of Plant Diversity

This map shows the location of the sites and areas identified as important centres of plant diversity at regional and global level. See the three volume source cited below for further details and extended documentation.

Source: WWF and IUCN[20].

assessed in terms of declining populations and species, either individually, or collectively, when manifest as loss of habitats or ecosystems. In this context, there has always been special concern about extinction, because this is a threshold from which there is no turning back.

Tracking extinction

As discussed in Chapter 3, the extinction of species is natural and expected, and self-evidently there have always been species at risk of extinction. It seems very likely, however, that recent and current extinction rates are considerably higher than would be expected without the influence of humans.

For several reasons it is difficult to record contemporary extinction events with precision. The species involved may well be unknown. Even if they have been discovered and named, they may be too small to be noticed without special sampling procedures. The entire process of decline and eventual extinction may take place over many years or even centuries in the case of very long-lived organisms such as many trees. The near-terminal stages in the process of species extinction are unlikely to be observed. Where this has been possible, it is because the species has been destroyed principally by unusually extreme hunting pressure (eg. the passenger pigeon *Ectopistes migratorius*), or extreme ecological events (eg. extinction of many native land snails in French Polynesia and Hawaii following introduction of the predatory snail *Euglandina rosea*) and has been the subject of sufficient interest to be closely monitored.

In other cases, positive evidence of extinction is lacking. Typically, many years elapse before sightings of a species become sparse enough to generate concern, and many more years are likely to pass before negative evidence (ie. failure to find the species despite repeated searches) accumulates to the point where extinction is the most probable explanation. In other words, unless circumstances are exceptional, monitoring of recent extinction events has a resolution limit measured in years or decades. This is why it is not possible to state with precision how many species have gone extinct in a given month, year or even decade, nor to predict exactly how many species, let alone which ones, are going to become extinct this year or decade or century. This is demonstrated by the occasional rediscovery of species once feared extinct. The Fiji petrel *Pterodroma macgillivrayi* is a prime example; it was known by one specimen collected in 1855, and regarded as extinct, until a bird flew into a researcher's headlight one night in 1984. Jerdon's courser *Rhinoptilus bitorquatus* is another example, known by just two museum skins and last recorded in 1900, until rediscovered in 1986 in a patch of scrub forest in South India. Hidden survivors are even more likely in plants, which may have propagules that can remain viable but unseen for very long periods indeed.

Extinctions in the recent past are likely to be recorded with significant accuracy either where circumstances favour preservation of hard remains in good number (eg. in caves, pot-holes or kitchen middens) or where early naturalists recorded the fauna or flora with sufficient care that they set a firm baseline against which the composition of the modern biota may be assessed. The detailed record of bird extinction in Hawaii is a result of the former circumstance, and the record of mollusc extinction on many islands a result of the latter.

From the relatively sparse evidence that is available, it appears that amongst animals nearly 400 vertebrates, including around 88 mammals and more than 100 birds, and more than 320 invertebrates, have become extinct during the past 400 years (see numerical summary in Table 5.2 and list of extinct vertebrates in Table 5.4). Data for plants are much more equivocal, owing in part to the uncertain taxonomic status of many extinct plant populations. The WCMC Threatened Plants Database currently lists some 380 extinct plant taxa and a further 370 or so classified as extinct or endangered. These include a number of infraspecific taxa and a number that although believed extinct in the wild are extant, and sometimes abundant, in cultivation.

Because mammals and birds tend to be relatively well-recorded, and because they leave recognisable macroscopic skeletal remains, it is principally among these groups that known extinctions may be reasonably representative of actual extinctions. In each of these groups the known extinction rate over the past 400 years, based on data in Table 5.4, averages out at around 20-25 species per hundred years.

A crucial question then, is how this observed extinction rate compares with some hypothetical or expected background extinction rate. It is of course impossible to derive such a rate from observation of the modern world, as this has already been highly modified by human activity. The only reasonable comparison is thus with historical records, for which we must turn to the fossil record, discussed in Chapter 3. Although extinction rates have evidently been highly variable during the history of life on Earth, it seems that the average persistence time of species in the fossil record is around four million years. If 10 million species exist at any one time, on this basis the extinction rate would amount to around 2.5 species annually. However, it is unclear how representative the fossil record is of species as a whole. It is likely that many rare species or those with very restricted distribution never appear in it at all. These rare species may almost by definition be expected to be inherently more extinction-prone than the species that are recorded and may therefore be expected to have a lower persistence time. This would mean that average species duration was less – perhaps much less – than four million years and the actual extinction rate in geological time considerably higher than the rate observed in the fossil record.

Applying a mean persistence time of four million years to birds and mammals (and assuming some 10 000 species of the former and around 5000 of the latter), the 'background' extinction rate would be around one species every five hundred years and thousand years respectively so that current rates would be some 100 or 200 higher than background. Even if background rates in these groups were ten times this, the currently observed extinction rates would still be 10-20 times those expected. It seems therefore that even if a high background extinction rate is postulated, recent extinctions are still much higher than might be expected (the alternative explanation is that the background rate was higher still). This elevated rate is particularly noteworthy as the Holocene appears to have been a period of relative climatic and geological stability, in which extinction rates might have been expected to be low.

Most known extinctions have occurred on islands and most known or probable continental extinctions have been among freshwater organisms. Reasons for the former are probably twofold. First, island species do appear to be particularly extinction-prone, by virtue of their limited ranges and usually small population sizes, and also because they have often evolved in the absence of certain pressures (eg. terrestrial predators, grazing ungulates); if faced with these pressures (usually through human intervention) their populations may collapse completely. Second, it is much easier to arrive at some certainty that a given species is no longer present on an island of limited extent than that it has disappeared completely from a continental range, where the limits of its range were probably uncertain to start with. Many freshwater biota appear to have the characteristics of island organisms, in that they have limited and highly circumscribed ranges and they are similarly often sensitive to external pressures (eg. introduction of predatory fish species, complete habitat destruction through drainage or dam construction).

Clearly, in view of our very incomplete knowledge of the world's species, and the fact that only a minute proportion of living species are being actively monitored at any one time, it is extremely likely that more extinctions are occurring than are currently known. Indeed, most predictions of present and near-future extinctions suggest extremely high rates. Most are based on combining estimates of species richness in tropical forest with estimates of rate of loss of these forests, and predict species extinction on the basis of the general species-area relationship (which predicts a decline in species richness as area declines, see Figure 5.1). It is widely believed that the great majority of all terrestrial species occur in tropical forests, and most of these species will be undescribed arthropods (notably beetles). At present rates of forest loss, it has been predicted that between two and eight percent of forest species will become extinct, or committed to extinction, between 1990 and 2015[22]. Depending on whether higher or lower estimates of tropical species richness are used, extinction at this rate could entail loss of up to 100 000 species annually. As a cautionary note, it should be observed that very few extinctions to date have actually been recorded in continental tropical moist forests, although monitoring species in these habitats presents great difficulty.

Threatened species assessment

Extinct species are presumably drawn from a pool of species in decline or at risk, all of which face eventual extinction if negative trends in their populations are not reversed. Various national and other programmes have developed methods to assess the relative severity of risks faced by species, and to label species with an indicative category name. Conservation activities can then be prioritised on the basis of relative risk, taking account of other relevant factors, such as feasibility, cost and benefits, as appropriate. The revised system[23] developed by IUCN/SSC and collaborators in conjunction with its Red Data Book and Red List programme is now a virtual world standard. This has been designed to provide a more explicit and objective framework for assessment of extinction risk, and to be applicable to any taxonomic unit at or below the species level, and within any specified geographical or political area. To be categorised as threatened, any species has to meet one of five sets of criteria formulated to permit evaluation of all kinds of species, with a wide range of biological characteristics. The criteria are defined on population reduction, population size, geographic area and pattern of occurrence, and quantitative population analysis.

The size and connectedness of different populations of a species influence the likelihood of its survival. In general, small isolated populations will be more sensitive than larger connected ones to demographic factors (eg. random events affecting the survival and reproduction of individuals) or environmental factors (eg. hurricanes, spread of disease, changes in food availability). As noted above, islands tend to have a much higher proportion of their biota at risk than continental countries because they start with many fewer species, all or many of which face the risks associated with small range size (see Table 5.1 for an example using bird data). Biogeographic theory based around the species-area relationship, supported by much empirical evidence, predicts that each 'habitat island' created by fragmentation of a

Table 5.1.

Island diversity at risk: birds

Note: the first five rows are the continental countries (Indonesia and Philippines excluded) with most threatened bird species, the other five rows are the small island groups with most threatened bird species.

Source: threatened species numbers from Collar et al.[27], as in IUCN[24]; estimates of number of breeding birds from WCMC database.

	Number breeding bird species	Number globally threatened	% breeding species threatened
Continental countries			
Brazil	1492	103	7
China	1100	90	8
India	923	73	8
Colombia	1695	64	4
Peru	1538	64	4
Small islands			
French Polynesia	60	22	37
Solomon Islands	163	18	11
Mauritius (+ Rodrigues)	27	10	37
São Tomé & Príncipe	63	9	14
Saint Helena & dependencies	53	9	17

continuous habitat area will come at equilibrium to contain fewer species than previously. Human activities everywhere tend to promote fragmentation of natural and often species-rich habitats (eg. primary tropical forest or temperate meadow grassland), and the spread of highly-managed species-poor habitats (eg. teak plantations or cereal croplands). As a result, many species occur in just the kind of fragmented pattern that increases the risk of extinction.

Table 5.2.
Threatened and extinct species

	Number of species in group	Approximate proportion of group assessed	Threatened species	% of total in group threatened	Extinct species	% of total in group extinct
Vertebrates						
Mammals	4630	100%	1096	24	88	2
Birds	9946	100%	1107	11	107	1
Reptiles	7400	<15%	253	3	20	0.3
Amphibians	4950	<15%	124	3	5	0.1
Fishes	25 000	<10%	734	3	172	0.7
Invertebrates						
Insects	950 000	<0.01%	537	0.05	73	0.004
Molluscs	70 000	<5%	920	1	237	0.3
Crustaceans	40 000	<5%	407	1	10	0.03
others			27		4	
Plants	270 000	<20%	30 827	11	>400	0.2

Note: For animals, 'threatened species' are species regarded in 1996 as globally threatened with extinction (IUCN categories CR, EN, VU) as indicated in the Red List. These data are undergoing minor revision in late 1999. Other than mammals and birds only a small or extremely small proportion of the total species in any group have been assessed for threatened status; the proportions shown are coarse approximations. The extinct counts cover species known to have become globally extinct since 1600 AD as categorised in 1996, modified in mammals[37] and freshwater fishes[38] to incorporate recent revisionary work associated with CREO[39]. The latter changes include cases treated by CREO as fully resolved extinctions plus incompletely resolved extinctions (ie. some criteria not met). The extinct fish total here includes 102 Lake Victoria haplochromine cichlids whose extinction may be questionable[38]. Plants included are ferns, gymnosperms and flowering plants; the 1997 plant assessment[36] used the old IUCN/SSC system of threat categorisation, the categories Extinct/Endangered, Endangered, Vulnerable, Rare and Indeterminate for species-level taxa only are here summed as 'threatened'. 368 plant species were recorded as Extinct and another 348 as Extinct/Endangered.

Source: WCMC database, data in part compiled for IUCN[27], birds data from Collar *et al.*[24], plant data from Walter and Gillet[36].

Recent declines

Reduction in population numbers, or complete loss of a species from a site or an individual country (often termed extirpation) are far easier to observe than global species extinction, and appear liable to occur wherever humankind has modified the environment for its own ends. The conservation status of most species is not known in detail, and this certainly applies to the many million as yet undescribed species, but two large animal groups – the mammals and birds – have been comprehensively assessed and may be representative of the status of biodiversity in general. Approximately 24% (1096) of the world's mammals and 11% (1107) of the world's bird species are regarded as threatened on the basis of IUCN/SSC criteria (see Table 5.2). Proportions are much lower in other vertebrates, but none of these has been assessed fully. Empirical observations such as these give sufficient grounds for serious concern for biodiversity maintenance, regardless of any hypotheses that have been proposed regarding the future rate of extinction. Interestingly, the ratio of threatened mammals to threatened birds is very near 1:1, as is the ratio of recorded recently extinct mammals to birds, giving some indication that threatened species categories may be a reasonably reliable indicator of proneness to extinction, and also that mammals may as a group be somewhat more susceptible than birds to extinction.

Countless other species, although not yet globally threatened, now exist in reduced numbers and as fragmented populations, and many of these are threatened with extinction at national level. The significance of loss of diversity at gene level implied by loss of local populations of species is not clear, although it has been argued that loss of resilience in response to environmental change is inevitable.

At global level, most of the species assessed as threatened are terrestrial forms (Table 5.3), and among birds, for example, more than half (65%) of the threatened species occur in forest[24].

The preponderance of terrestrial species is because the great majority of mammals and birds are terrestrial, and little or nothing is known of the population status of most aquatic species in most groups. Where significant numbers of aquatic species have been assessed, eg. among crustaceans and molluscs, the proportion of threatened aquatic species rises markedly. Among fishes, the high number of freshwater species doubtless in part reflects the general lack of data on marine species, but to some extent indicates relative risk – many freshwater species being restricted to small and isolated habitat patches. Evidence of the vulnerable nature of freshwater habitats and the risk faced by many aquatic groups is accumulating. For example, in the United States, freshwater groups are considerably more threatened than terrestrial groups (specifically, nearly 70% of the mussels, 50% of the crayfish and 37% of the fishes)[25].

Numbers of globally threatened species have often been mapped at country level (eg. Figures 6 and 7 in[27]). Maps of this kind provide at best a broad overview of the occurrence of such species, shown as direct numbers or as a proportion of the total number in that group in the country. As with other country-level biodiversity analyses, this information could be used to help focus efforts to slow or reverse biodiversity decline. Data with improved spatial resolution (although plotted in a highly simplified way) are shown in Map 5.6, representing the individual ranges of all the mammals and birds assessed as Critically Endangered. BirdLife International has generated individual distribution maps of all threatened birds. Among other applications, these are being used to develop the first-ever world map of all the threatened species in an entire major group of organisms. A preliminary version of this, as a global density surface, is shown in Map 5.7. Being geo-referenced rather than country-based, and at relatively fine spatial resolution, this has considerable potential to focus bird conservation efforts effectively.

Driving forces

Humans are doing nothing fundamentally different from other living organisms in endeavouring to extract resources from the biosphere in order to maintain their structure and reproduce themselves, but with the benefits of society and technology they are extraordinarily successful at it. The continuing conversion of natural habitats to cropland[26] and other uses typically entails replacement of systems rich in biodiversity by monocultures or systems poor in biodiversity. Habitat modification or loss, although covering a range of different effects, is in general the most important factor acting to increase species' risk of extinction. Among globally threatened species[27], it is the principal factor affecting around three-quarters of threatened mammals, about half of the globally threatened birds[24], and the majority of threatened species in several other major groups

Table 5.3

Numbers of threatened animal species in major biomes

| Biome type | total | Animal group | | | | | | | | | | | | |
		Mammals	Birds	Reptiles	Amphibians	Fishes	Spiders	Crustaceans	Insects	Onychophora	Annelids	Molluscs	Enopla	Anthozoa
Marine	222	22	61	8	0	133	0	0	0	0	1	10	0	2
Inland water	1870	7	15	66	124	615	0	407	123	0	0	513	0	0
Terrestrial	3236	1067	1031	179	124	0	10	0	414	6	5	397	2	0

Note: Counts include globally threatened species (IUCN categories CR,EN,VU). Only mammals and birds have been comprehensively assessed. Amphibians and diadromous fishes (eg. sturgeon) are counted in more than one biome row, so data in columns are not necessarily additive. Counts of freshwater and terrestrial molluscs are highly provisional approximations only, largely based on family-level habitat preference where known. It is believed that virtually all the globally threatened plants listed by Walter and Gillet[36] are terrestrial. Inland water includes saline wetlands, cave waters, etc. as well as freshwaters.

Source: WCMC database, data in part compiled for IUCN[27], birds data from Collar et al.[24], plant data from Walter and Gillet[36].

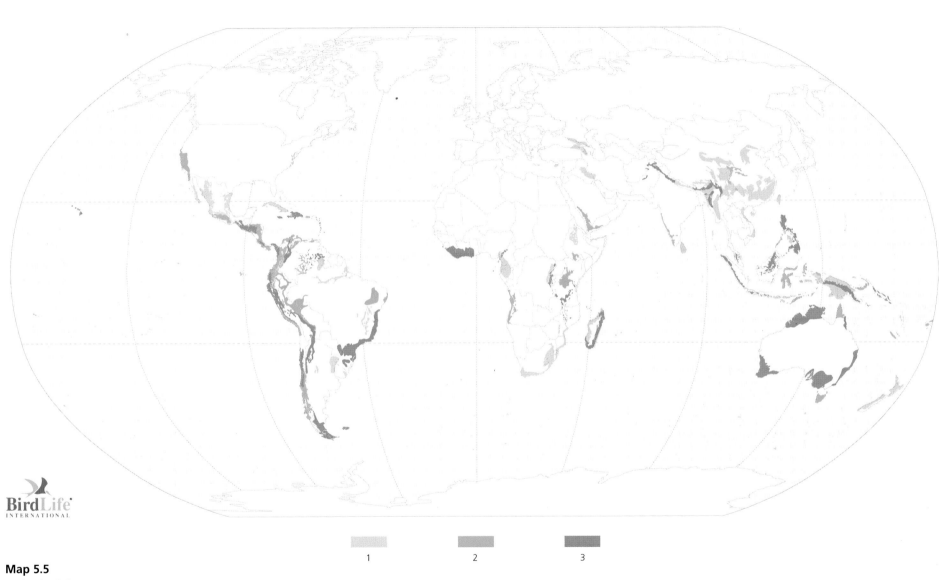

Map 5.5

Endemic Bird Areas

More than one quarter (2561) of the world's bird species, including 74% of the threatened birds, have a range restricted to less than 50 000km². Virtually all these occur within the 218 Endemic Bird Areas (EBAs) defined by BirdLife International[21]. The world's EBAs are shown on this map categorised 1, 2 or 3 according to increasing biodiversity importance (based on the number of restricted range species, whether shared between EBAs, taxonomic uniqueness, and EBA size).

Source: data and analysis provided by BirdLife International.

(most notably, in 95% of threatened bivalve molluscs). It is the main cause of loss in three-quarters of extinct freshwater fishes[38].

A second major source of biodiversity loss is the widespread introduction of species outside their natural range where they typically induce change at the community and ecosystem level. The effects of alien species are especially pronounced in closed systems such as lakes and islands. Introduced species, eg. rats and cats, are cited as a cause of extinction in nearly 40% of the approximately 200 species where cause could be attributed; the majority were island forms[1]. Seven endemic snails in French Polynesia have been extirpated following the late 1970s introduction of a carnivorous snail species (*Euglandina rosea*) intended to control another introduced species (the giant African snail *Achatina fulica*), itself an agricultural pest. Accidental introduction of the brown tree snake *Boiga irregularis* to Guam in 1968 led to decline of the entire avifauna, of which one species is now thought extinct, one extinct in the wild and one is Critically Endangered[24]. Introduction of the Nile Perch (*Lates niloticus*) to Lake Victoria has contributed to the decline or extinction of nearly 200 native and endemic cichlid fishes. Overall, introduction events probably number in the thousands globally[2], and although cases involving animals have been cited to illustrate the scope of the problem, introduced plants are in many places as pervasive and damaging.

High trade demand for certain species and products, whether for international markets consuming hardwoods, sea fish, live animals and plants and derivatives, or local markets consuming commodities such as bushmeat or turtle eggs, can readily push exploitation beyond the production capacity of the resource. International agreements such as the numerous regional fishery management schemes and CITES (the Convention on International Trade in Endangered Species) provide a framework for relieving some of the pressure on selected important resources, but it is clear that managing resource use to ensure sustainability in the face of ever-increasing human needs is a challenge yet to be met.

Rapid environmental change, such as El Niño Southern Oscillation (ENSO) events, can have significant impacts on natural habitats. For example, in 1997-98 climate fluctuation associated with El Niño is implicated in the persistence and spread of fires in Brazil, Indonesia and elsewhere: an estimated one million hectares of savannah woodland burned in Brazil and a similar area of forests in Indonesia. The effect of events of this type will be multiplied many times wherever habitats are already fragmented and species depleted.

Biodiversity indicators

During the past decade, considerable effort has been devoted to assessment of change in the environment, often at national[48,50] or regional level, and to the development of indicators to represent environmental change[28-32,49]. An indicator is a value, perhaps an index derived from a set of observational data, that can be taken to point to, or simplify, or otherwise represent, some broader issue beyond the indicator value itself. The general model for such indicators is provided by economic performance tools, such as the retail price index, derived from the shop price of a set range of household goods, or the Dow Jones index, derived from the value of a set of stocks. Their central purpose is to represent real-world complexity in a simplified and readily understood numerical or graphic form.

Much progress has been made in some subject areas, particularly in cases where individual substances can be measured and monitored over time. For example, phosphates and nitrogen compounds are often limiting factors to production in natural ecosystems, and if waterways are artificially enriched with these nutrients as a result of industrial or agricultural activities, this *eutrophication* can lead to dramatic and undesirable environmental change (anoxic conditions, change in community structure, algal blooms). The amount of these nutrients released can quite readily be measured and the

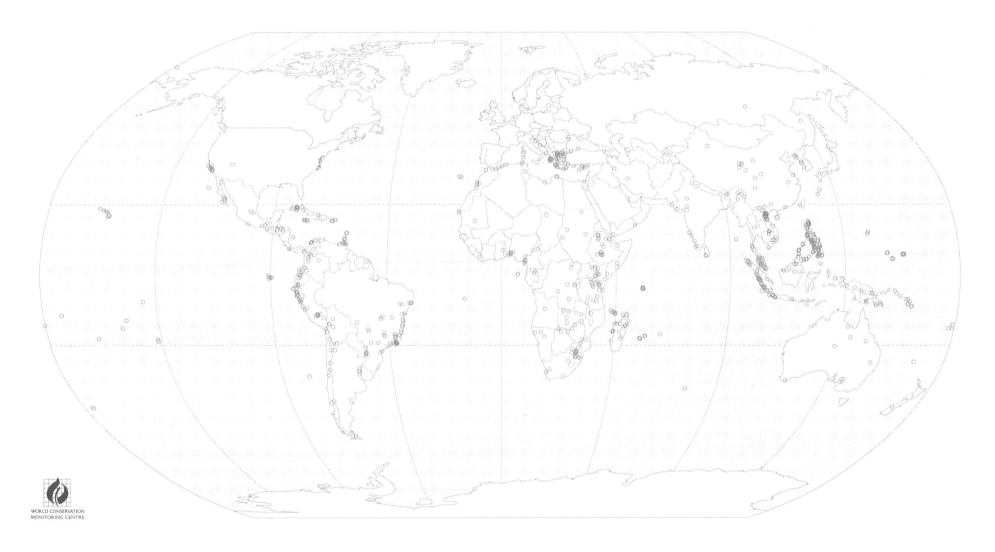

Map 5.6

Critically Endangered mammals and birds

This map shows the general distribution of almost all the 337 mammal and bird species categorised as Critically Endangered, the highest risk category, in the 1996 IUCN Red List[27]. Each circle represents one distribution record; some species are known from a single point locality, others are represented by a cluster of localities. On this map, a high density of symbols can represent many records of a single species, or single records of many separate species. Red and blue symbols represent mammals and birds, respectively.

Source: research and mapping by WCMC (supported by Rio Tinto).

values aggregated into a eutrophication index. Where feasible, a further function of indicators in the social and policy sphere is to show the effects of remedial policy and management measures, eg. falling levels of eutrophication risk in response to reduced nutrient release.

Existing economic indicators rely entirely on time-series of quantitative numerical data, as do the most effective environmental indicators, eg. those relating to time trends in measurable physico-chemical factors, such as temperature, carbon dioxide emissions, or acidification. One of the principal obstacles to the development of good biodiversity indicators is that time-series of numerical data, especially if applicable above local or national level, are scarce. For this reason, most existing biodiversity indicators are based on static status assessments, eg. the proportion of the mammal fauna that is threatened with extinction, the number of national endemic species, or amount of protected forest.

If the fundamental goal is biodiversity maintenance, trends in species extinction might be considered the ultimate indicator of the state of the biosphere, and trends in number of species appearing in lists of globally threatened with extinction might be a close second best indicator. In fact, although these data sets do provide a useful 'snapshot' guide to the state of the biosphere, problems of data availability and quality severely limit their use as indicators of global change. This is essentially because change from one period to another is attributable mainly to change in monitoring methods, or in species-level taxonomy, or in the amount of information available, rather than change in status of the species themselves. For example, 295 species categorised by BirdLife International as 'low risk' in their first world list of threatened birds[33] were upgraded to 'threatened' in the second version[24], but in only 10 cases (1% of the total of threatened birds) was this because of observed deterioration in status. This limitation may soon apply much less to birds, for which a third comprehensive and standardised

assessment is nearing completion, than to other groups, and applies even now much less at national level, where monitoring may be more standardised and loss of species from within national borders can be recorded with more accuracy than loss at global level.

Although much discussion has been focused on the design of potential biodiversity indicators, and some practical progress has been made at national level, few attempts have been made to design and implement a system that generates indicators of biodiversity change over time at global or continental level. The main exception appears to be the approach developed by WCMC for the WWF Living Planet Report[34,35].

This method is based on collating numerical time-series data on species or component sub-populations, allowing interpolation to fill gaps. It is designed to make use of the less-than-perfect data that are available, and so incorporates information on national or geographic populations, as well as full species, and uses some short and asynchronous time-series of population estimates. Provided that the data are suitable for use, no selection or weighting is imposed, and so the index includes non-threatened and threatened species, and widespread and local ones. The final index is derived by normalising the geometric mean change between years. The data set can be sorted geographically, taxonomically or by major biome type.

This system is limited ultimately by the number of populations for which quantitative size (or area) estimates are available. Globally, by far the greatest monitoring effort for any group of species is devoted to marine fishes of economic importance. By far the greatest volume of time-series data relate to stock estimates and catch levels (not used in the index) in the marine fish populations targeted by industrialised fisheries of developed countries. Birds come second to marine fishery stocks. The bird species that are surveyed regularly by networks of mainly amateur ornithologists in developed countries are by

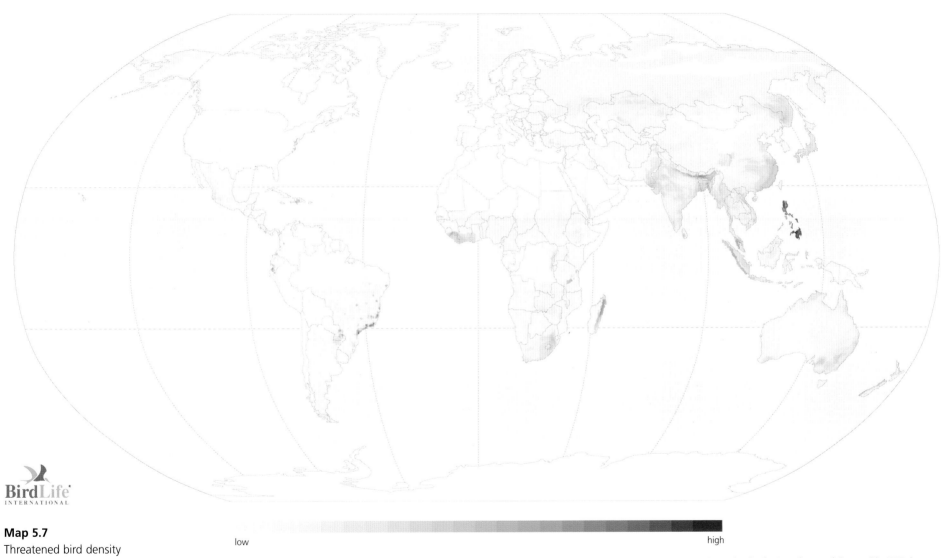

Map 5.7

Threatened bird density

low   high

BirdLife International has developed digital range maps of threatened bird species, and work is in progress on validating these data. This map is a preliminary result of this work, and shows the distribution of 1170 of the possibly 1200 that are likely to be included in the 2000 list of threatened birds. The data are plotted as a global density surface, representing the number of species potentially present in each location. The map is the first such treatment of any large group of threatened species, and shows at a glance where greatest numbers of species are at risk of extinction.

Source: distribution data created by BirdLife International; density analysis and plot by WCMC.

far the best known large terrestrial group. In recent years considerable attention has been devoted to the monitoring of amphibian numbers, against a background of rising concern for the widespread decline and extirpation of local amphibian populations.

The extent to which trends in species characteristic of a given habitat class can be taken as indicative of prevailing trends in that habitat, and whether this might be a property of all habitat types and communities of species, have been little explored. Published versions of the index[34,35] have presented species data as potential indicators of ecosystem health in marine and inland water systems. The 1999 marine ecosystems index is based on population data for 102 fishes, reptiles, birds and mammals. The 1999 freshwater index is similarly based on 102 species and incorporates data on amphibians. These indices provide the first-ever quantitative guide to change in biodiversity over time, in this case using a global sample of species representing aquatic biodiversity. Although several populations and species have undergone dramatic increase in numbers as a result of intense conservation efforts, decline is the prevailing pattern. Population levels in the marine species sampled have declined by around 35% in the period 1970-1995, and by more than 40% in inland water species (see Chapters 6 and 8, and Figures 6.5 and 8.1 for further information).

The overall WWF Living Planet Index for 1999 is shown in Figure 5.2. This index combines the two aquatic species indices with data on change in area of forest ecosystems to represent the terrestrial biosphere. Work in progress is greatly expanding the set of species sampled, and will generate a separate terrestrial index in the year 2000.

Figure 5.2

The 1999 WWF Living Planet Index

Note: This figure indicates change in condition of world ecosystems, 1970-1995.

Source: index developed and produced by WCMC for WWF International; graphic reproduced with permission from WWF *Living Planet Report 1999* (Loh et al.[35]).

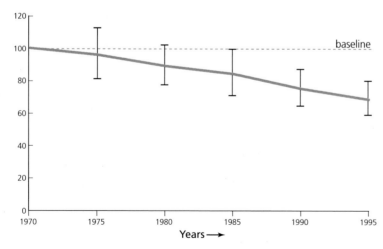

Table 5.4

Recent vertebrate extinctions

Notes: This table lists the vertebrate species that on good evidence are thought to be globally extinct. It is based in part on lists produced using the IUCN threat category system but mammals and fishes have been revised following criteria developed by the Committee on Recently Extinct Organisms (CREO[39]). The CREO system lists extinctions since AD 1500, the other groups start at AD 1600. Where information meets all CREO criteria, the extinction event is considered fully resolved; these species appear on a darker background in the table. Most other species listed are believed probably extinct, but key data (eg. on taxonomy or range) are missing. The many Lake Victoria cichlid fishes in the genus *Haplochromis* (*sensu lato*) at the end of the list are believed by many to be extinct, but because available data do not meet rigorous criteria, these fishes are listed as questionably extinct and are distinguished in the table by a pale background. The Date column indicates, in most cases very approximately, when the species was last recorded; for some species an entry such as "post-1500" indicates that the species persisted at least until the arrival of European colonists but no record of subsequent observation is available.

Sources: mammals[37], birds[1,27] based largely on assessments made by BirdLife International, reptiles[1,27], amphibians[1,27], fishes[39].

Species	Common Name	Place	Date
Class MAMMALIA			
Order DASYUROMORPHIA			
Family Thylacinidae			
Thylacinus cynocephalus	Thylacine	Tasmania	1936
Order PERAMELEMORPHIA			
Family Peramelidae			
Chaeropus ecaudatus	Pig-footed Bandicoot	Australia	1960s
Macrotis leucura	Lesser Bilby	Australia	1960s
Perameles eremiana	Desert Bandicoot	Australia	1960s
Order DIPROTODONTIA			
Family Macropodidae			
Caloprymnus campestris	Desert Rat-kangaroo	Australia	1932
Lagorchestes asomatus	Central Hare-wallaby	Australia	1960s
Lagorchestes leporides	Eastern Hare-wallaby	Australia	1890
Macropus greyi	Toolache Wallaby	Australia	1972
Onychogalea lunata	Crescent Nailtail Wallaby	Australia	1956
Potorous platyops	Broad-faced Potoroo	Australia	1875
Order INSECTIVORA			
Family Solenodontidae			
Solenodon marcanoi	Marcano's Solenodon	Hispaniola	post-1500
Family Nesophontidae			
Nesophontes hypomicrus	Atalaye Island-shrew	Hispaniola	post-1500
Nesophontes longirostris	Long-nosed Island-shrew	Cuba	post-1500
Nesophontes major		Cuba	post-1500
Nesophontes micrus	Western Cuban Island-shrew	Cuba	post-1500
Nesophontes paramicrus	St Michel Island-shrew	Hispaniola	post-1500
Nesophontes submicrus		Cuba	post-1500
Nesophontes superstes		Cuba	post-1500
Nesophontes zamicrus	Haitian Island-shrew	Hispaniola	post-1500
Nesophontes sp. 1	Grand Cayman Island-shrew	Grand Cayman	post-1500
Nesophontes sp. 2	Cayman Brac Island-shrew	Cayman Brac	post-1500

Species	Common Name	Place	Date
Order CHIROPTERA			
Family Pteropodidae			
Dobsonia chapmani	Dobson's Fruit Bat	Negros I. (Philippines)	1970s
Nyctimene sanctacrucis	Nendo Tube-nosed Fruit Bat	Santa Cruz (Solomons)	pre-1892
Pteropus brunneus	Dusky Flying Fox	Percy I. (Australia)	1874
Pteropus pilosus	Large Palau Flying Fox	Palau	1874
Pteropus subniger	Réunion Flying Fox	Mauritius, Réunion	pre-1866
Pteropus tokudae	Guam Flying Fox	Guam	1968
Family Vespertilionidae			
Pharotis imogene	Large-eared Nyctophilus	Papua New Guinea	1890
Kerivoula africana	Dobson's Painted Bat	Tanzania	pre-1878
Family Molossidae			
Mystacina robusta	New Zealand Lesser Short-tailed Bat	New Zealand	1965
Order PRIMATES			
Family Paleopropithecidae			
Palaeopropithecus cf. *P. ingen*	Sloth Lemur	Madagascar	post-1500
Family Megaladapidae			
Megaladapis cf. *M. edwardsi*	Tretretretre	Madagascar	post-1500
Family Xenotrichidae			
Xenothrix mcgregori	Jamaican Monkey	Jamaica	post-1500
Order CARNIVORA			
Family Canidae			
Dusicyon australis	Falkland Island Dog	Falkland Is.	1876
Family Phocidae			
Monachus tropicalis	Caribbean Monk Seal	Caribbean Sea	1962
Order SIRENIA			
Family Dugongidae			
Hydrodamalis gigas	Steller's Sea Cow	Russia, USA	1768

Species	Common Name	Place	Date
Order ARTIODACTYLA			
Family Hippopotamidae			
Hippopotamus lemerlei	Lemerle's Malagasy Hippo	Madagascar	post-1500
Hippopotamus madagascariensis	Common Malagasy Hippo	Madagascar	post-1500
Family Bovidae			
Gazella rufina	Red Gazelle	Algeria	pre-1894
Hippotragus leucophaeus	Bluebuck	South Africa	1800
Order RODENTIA			
Family Muridae			
Conilurus albipes	White Footed Rabbit Rat	Australia	1875
Crateromys paulus	Ilin Bushy-tailed Cloud-rat	Philippines	1953
Leimacomys buettneri	Groove-toothed Forest Mouse	Ghana	1890
Leporillus apicalis	Lesser Stick-nest Rat	Australia	1970
Oligoryzomys victus	St. Vincent Pygmy Rice Rat	Saint Vincent	1892
Malpaisomys insularis	Volcano Mouse	Canary Is.	post-1500
Microtus bavaricus	Bavarian Pine Vole	Germany	1962
Megalomys audreyae	Barbuda Giant Rice Rat	Barbuda and Antigua	pre-1890
Megalomys desmarestii	Martinique Rice Rat	Martinique	1902
Megalomys luciae	St Lucia Giant Rice Rat	Saint Lucia	pre-1881
Megaoryzomys curioi	Curio's Giant Rice Rat	Santa Cruz, Galápagos	post-1500
Megaoryzomys sp. 1	Isabela Giant Rice Rat	Isabela, Galápagos	post-1500
Nesoryzomys darwini	Galapagos Rice Rat	Santa Cruz, Galápagos	post-1500
Nesoryzomys swarthi	San Salvador Rice Rat	San Salvador, Galápagos	post-1500
Nesoryzomys sp. 1	Isabela Island Rice Rat "A"	Isabela, Galápagos	post-1500
Nesoryzomys sp. 2	Isabela Island Rice Rat "B"	Isabela, Galápagos	post-1500
Notomys amplus	Short-tailed Hopping-mouse	Australia	pre-1896
Notomys longicaudatus	Long-tailed Hopping-mouse	Australia	1901
Notomys macrotis	Big-eared Hopping-mouse	Australia	1843
Notomys mordax	Darling Downs Hopping-mouse	Australia	1846
Notomys sp. 1	Great Hopping-mouse	Australia	pre-1900
Oryzomin sp. 1	Vespucci's Rice Rat	Fernando da Noronha I. (Brazil)	post-1500
Oryzomys antillarum	Jamaican Rice Rat	Jamaica	1877
Oryzomys hypenemus	Barbuda Rice Rat	Barbuda and Antigua	post-1500
Oryzomys nelsoni	Nelson's Rice Rat	Maria Madre I. (Mexico)	1897
Oryzomys sp. 1	Barbados Rice Rat	Barbados	pre-1890

Species	Common Name	Place	Date
Peromyscus pembertoni	Pemberton's Deer Mouse	San Pedro Nolasco I. (Mexico)	1931
Pseudomys gouldii	Gould's Mouse	Australia	1930
Rattus macleari	Maclear's Rat	Christmas I (Australia)	1908
Rattus nativitatis	Bulldog Rat	Christmas I (Australia)	1908
Uromys imperator	Giant Naked-tailed Rat	Guadalcanal (Solomons)	1960s
Uromys porculus	Little Pig Rat	Guadalcanal (Solomons)	1887
Family Echimyidae			
Boromys offella	Cuban Esculent Spiny Rat	Cuba	post-1500
Boromys torrei	De la Torre's Esculent Spiny Rat	Cuba	post-1500
Brotomys voratus	Muhoy	Hispaniola	post-1500
Family Capromyidae			
Capromys sp. 1	Cayman Hutia	Cayman Is.	post-1500
Geocapromys colombianus	Cuban Coney	Cuba Little	post-1500
Geocapromys thoractus		Swan I. (Honduras)	1950s
Geocapromys sp. 1	Great Cayman Coney	Grand Cayman	post-1500
Geocapromys sp. 2	Cayman Brac Coney	Cayman Brac	post-1500
Hexolobodon phenax		Hispaniola	post-1500
Isolobodon montanus	Montane Hutia	Hispaniola	post-1500
Isolobodon portoricensis	Allen's Hutia	Hispaniola	post-1500
Plagiodontia ipnaeum	Johnson's Hutia	Hispaniola	post-1500
Rhizoplagiodontia lemkei		Hispaniola	post-1500
Family Heptaxodontidae			
Quemisia gravis	Quemi	Hispaniola	post-1500
Order LAGOMORPHA			
Family Ochotonidae			
Prolagus sardus	Sardinian Pika	Corsica, Sardinia	1777
Family Leporidae			
Sylvilagus insonus	Omilteme Cottontail	Mexico	1991
Class AVES			
Order STRUTHIONIFORMES			
Family Aepyornithidae			
Aepyornis maximus	Great Elephantbird	Madagascar	1650

Species	Common Name	Place	Date
Family Anomalopterygidae			
Dinornis torosus	Brawny Great Moa	New Zealand	1670
Eurapteryx gravis	Burly Lesser Moa	New Zealand	1640
Megalaperyx didinus	South Island Tokoweka	New Zealand	1785
Order CASUARIIFORMES			
Family Dromaiidae			
Dromaius ater	King Island Emu	King I. (Australia)	post-1600
Dromaius baudinianus	Kangaroo Island Emu	Kangaroo I. (Australia)	post-1600
Order PODICIPEDIFORMES			
Family Podicipedidae			
Podiceps andinus	Colombian Grebe	Colombia	1977
Podilymbus gigas	Atitán Grebe	Guatemala	1980-1986/7
Order PROCELLARIIFORMES			
Family Procellariidae			
Pterodroma sp.		Rodrigues	1726
Order PELECANIFORMES			
Family Phalacrocoracidae			
Phalacrocorax perspicillatus	Pallas's Cormorant	Bering Straits (Russia)	1852
Family Ardeidae			
Ixobrychus novaezelandia	Black-backed Bittern	New Zealand	1900
Nycticorax mauritianus	Mauritius Night-heron	Mauritius	by 1700
Nycticorax megacephalus	Rodrigues Night-heron	Rodrigues	1761
Nycticorax sp.		Réunion	by 1700
Family Ciconiidae			
Ciconia sp.		Réunion	1674
Family Threskiornithidae			
Borbonibis latipes	Réunion Flightless Ibis	Réunion	1773
Order ANSERIFORMES			
Family Anatidae			
Alopochen mauritiania	Mauritian Shelduck	Mauritius	1698

Species	Common Name	Place	Date
Anas marecula	Amsterdam Island Duck	Amsterdam I. (France)	1793
Anas theodori	Mauritian Duck	Mauritius, ?Réunion	1696
Camptorhynchus labradorius	Labrador Duck	Canada, USA	1878
Cygnus sumnerensis	Chatham Islands Swan	Chatham Is. (New Zealand)	1590-1690
Mascarenachen kervazoi	Réunion Shelduck	Réunion	1674
Mergus australis	Auckland Island Merganser	New Zealand	1905
Pachyanas chathamica	Chatham Islands Shelduck	Chatham Is. (New Zealand)	post 1600

Order FALCONIFORMES

Species	Common Name	Place	Date
Family Falconidae			
Falco sp.		Réunion	1674
Polyborus lutosus	Guadalupe Caracara	Guadalupe (Mexico)	1900

Order GALLIFORMES

Species	Common Name	Place	Date
Family Phasianidae			
Coturnix novaezelandiae	New Zealand Quail	New Zealand	1875

Order GRUIFORMES

Species	Common Name	Place	Date
Family Rallidae			
Aphanapteryx bonasia	Red Rail	Mauritius	1700
Aphanapteryx leguati	Rodrigues Rail	Rodrigues	1761
Atlantisia elpenor	Ascension Flightless Crake	Ascension I.	1650
Fulica newtoni	Mascarene Coot	Mauritius, Réunion	1693
Gallirallus dieffenbachii	Dieffenbach's Rail	Chatham Is. (New Zealand)	1840
Gallirallus modestus	Chatham Islands Rail	Chatham Is. (New Zealand)	1900
Gallirallus pacificus	Tahiti Rail	French Polynesia	1773-4
Gallirallus wakensis	Wake Island Rail	Wake I. (USA)	1945
Nesoclopeus poecilopterus	Bar-winged Rail	Fiji	c. 1973?
Porphyrio albus	Lord Howe Swamphen	Lord Howe I. (Australia)	1834
Porzana monasa	Kosrae Crake	Fed. States Micronesia	1827
Porzana palmeri	Laysan Rail	Hawaii	1944
Porzana sandwichensis	Hawaiian Crake	Hawaii	1898

Order CHARADRIIFORMES

Species	Common Name	Place	Date
Family Charadriidae			
Vanellus macropterus	Javanese Lapwing	Java	1940
Haematopus meadewaldoi	Canary Islands Oystercatcher	Canary Is.	1913

Species	Common Name	Place	Date
Family Scolopacidae			
Prosobonia leucoptera	Tahitian Sandpiper	Tahiti, Mooréa (F. Polynesia)	1773
Family Laridae			
Alca impennis	Great Auk	North Atlantic coasts	1844

Order COLUMBIFORMES

Species	Common Name	Place	Date
Family Raphidae			
Raphus cucullatus	Dodo	Mauritius	1665
Raphus solitarius	Réunion Solitaire	Réunion	1710-1715
Pezophaps solitaria	Rodrigues Solitaire	Rodrigues	1765
Family Columbidae			
Alectroenas nitidissima	Mauritius Blue Pigeon	Mauritius	1835
'Alectroenas' rodericana	Rodrigues Pigeon	Rodrigues	1726
Columba jouyi	Ryukyu Pigeon	Nansei-shoto (Japan)	1936
Columba versicolor	Bonin Wood Pigeon	Ogasawara-shoto (Japan)	1889
Ectopistes migratorius	Passenger Pigeon	USA	1914
Microgoura meeki	Choiseul Pigeon	Choiseul (Solomon Is.)	1904
Ptilinopus mercierii	Red-moustached Fruit-dove	Marquesas Is. (F. Polynesia)	1922

Order PSITTACIFORMES

Species	Common Name	Place	Date
Family Psittacidae			
Anodorhynchus glaucus	Glaucous Macaw	Brazil, Uruguay	1955
Ara tricolor	Cuban Red Macaw	Cuba	1885
Conuropsis carolinensis	Carolina Parakeet	USA	1914
Cyanoramphus ulietanus	Raiatea Parakeet	Raiatea (F. Polynesia)	1773
Cyanoramphus zealandicus	Black-fronted Parakeet	Tahiti (F. Polynesia)	1844
'Lophopsittacus' bensoni	Mauritius Grey Parrot	Mauritius	1765
Lophopsittacus mauritianus	Mauritius Parrot	Mauritius	1675
Mascarinus mascarinus	Mascarene Parrot	Réunion	1834
'Necropsittacus' rodericanus	Rodrigues Parrot	Rodrigues	1761
Nestor productus	Norfolk Island Kaka	Phillip I. (Australia)	1851
Psephotus pulcherrimus	Paradise Parrot	Australia	1927
Psittacula exsul	Newton's Parakeet	Rodrigues	1876
Psittacula wardi	Seychelles Parrot	Seychelles	1870

Species	Common Name	Place	Date
Order CUCULIFORMES			
Family Cuculidae			
Coua delalandei	Snail-eating Coua	Madagascar	1930
Order STRIGIFORMES			
Family Strigidae			
'Athene' murivora	Rodrigues Little Owl	Rodrigues	1726
Sauzieri sp.	Mauritian Owl	Mauritius	post-1600
Sceloglaux albifacies	Laughing Owl	New Zealand	1914
'Scops' commersoni	Mauritian Owl	Mauritius	1836
Order APODIFORMES			
Family Trochilidae			
Chlorostilbon bracei	Grace's Emerald	Bahamas	1877
Order PICIFORMES			
Family Picidae			
Campephilus principalis	Ivory-billed Wookpecker	Cuba, USA	1991
Order PASSERIFORMES			
Family Acanthisittidae			
Xenicus longipes	Bush Wren	New Zealand	1972
Xenicus lyalli	Stephens Island Wren	Stephens I. (New Zealand)	1874
Family Pycnonotidae			
Hypsipetes sp.		Rodrigues	1600s?
Family Muscicapidae			
Gerygone insularis	Lord Howe Gerygone	Australia	1938
Myiagra freycineti	Guam Flycatcher	Guam	1983
Nesillas aladabrana	Aldabra Bush-warbler	Aldabra (Seychelles)	1986
Turnagra capensis	Piopio	New Zealand	1955
Turdus ravidus	Grand Cayman Thrush	Cayman Is.	1938
Zoothera terrestris	Bonin Thrush	Ogasawara-shoto (Japan)	1928
Family Zosteropidae			
Zosterops strenuus	Robust White-eye	Lord Howe I. (Australia)	1920
Family Sylviidae			
Megalurus rufescens	Chatham Island Fernbird	Chatham Is. (New Zealand)	1900
Family Meliphagidae			
Chaetoptila angustipluma	Kioea	Hawaii	1860
Moho apicalis	Oahu Oo	Hawaii	1837
Moho nobilis	Hawaii Oo	Hawaii	1934
Family Drepanididae			
Ciridops anna	Ula-ai-hawane	Hawaii	1892
Drepanis funerea	Black Mamo	Hawaii	1907
Drepanis pacifica	Hawaii Mamo	Hawaii	1899
Hemignathus obscurus	Akialoa	Hawaii	1969
Paroreomyza flammea	Kakawihie	Hawaii	1963
Chloridops kona	Kona Grosbeak	Hawaii	1894
Rhodacanthis flaviceps	Lesser Koa-finch	Hawaii	1891
Rhodacanthis palmeri	Greater Koa-finch	Hawaii	1896
Viridonia sagittirostris	Greater Amakihi	Hawaii	1900
Family Icteridae			
Quiscalus palustris	Slender-billed Grackle	Mexico	1910
Family Ploceidae			
Foudia sp.	Réunion Fody	Réunion	1671
Family Fringillidae			
Chaunoproctus ferreorostris	Bonin Grosbeak	Ogasawara-shoto (Japan)	1890
Spiza townsendi	Townsend's Finch	USA	1833
Family Sturnidae			
Aplonis corvina	Kosrae Mountain Starling	Kosrae (Fed. States Micronesia)	1828
Aplonis fusca	Norfolk Island Starling	Norfolk I. (Australia)	1923
Aplonis mavornata	Mysterious Starling	Cook Is.	1825
Fregilupus varius	Réunion Starling	Réunion	1854
Necrospar rodericanus	Rodrigues Starling	Rodrigues	1726
Family Callaeidae			
Heteralocha acutirostris	Huia	New Zealand	1907

Species	Common Name	Place	Date
Class **REPTILIA**			
Order SAURIA			
Family Anguidae			
Celestus occiduus	Jamaican Giant Galliwasp	Jamaica	1840
Family Gekkonidae			
Hoplodactylus delcourti		New Zealand (?)	mid 19th C?
Phelsuma gigas	Giant Day Gecko	Rodrigues	end 19th C
Family Iguanidae			
Leiocephalus eremitus		Navassa I. (USA)	1900
Leiocephalus herminieri		Martinique	1830s
Family Scincidae			
Leiolopisma mauritiana		Mauritius	1600
Macroscincus coctei	Cape Verde Giant Skink	Cape Verde	early 20th C
Tachygia microlepis	Tongan Ground Skink	Tonga	post 1600
Tetradactylus eastwoodae	Eastwood's Longtailed Seps	South Africa	post 1900
Family Teiidae			
Ameiva cineracea		Guadeloupe	early 20th C
Ameiva major	Martinique Giant Ameiva	Martinique	post-1600
Order SERPENTES			
Family Boidae			
Bolyeria multocarinata		Round I. (Mauritius)	1975
Family Colubridae			
Alsophis sancticrucis	St Croix Racer	Virgin Is. (US)	20th C
Family Typhlopidae			
Typhlops cariei		Mauritius	17th C
Order TESTUDINES			
Family Testudinidae			
Cylindraspis borbonica		Réunion	1800
Cylindraspis indica		Réunion	1800
Cylindraspis inepta		Mauritius	early 18th C
Cylindraspis peltastes		Rodrigues	1800
Cylindraspis triserrata		Mauritius	early 18th C
Cylindraspis vosmaeri		Rodrigues	1800
Class **AMPHIBIA**			
Order ANURA			
Family Discoglossidae			
Discoglossus nigriventer	Israel Painted Frog	Israel	1940
Family Myobatrachidae			
Uperoleia marmorata	Marbled Toadlet	Australia	post-1840
Family Ranidae			
Arthroleptides dutoiti		Kenya	post-1900
Rana fisheri	Relict Leopard Frog	USA	1960
Rana tlaloci		Mexico	1990s
Class **CEPHALASPIDOMORPHI**			
Order PETROMYZONTIFORMES			
Family Petromyzontidae			
Lampetra minima	Miller Lake Lamprey	USA	1952-1959
Class **ACTINOPTERYGII**			
Order CYPRINIFORMES			
Family Cyprinidae			
Acanthobrama hulensis		Lake Huleh (Israel)	1957-1996
Anabarilius alburnops		Lake Dianchi (China)	1904-1993
Anabarilius polylepis		Lake Dianchi (China)	1904-1993
Barbus microbarbis		Lake Luhondo (Rwanda)	1934-1952
Cephalakompsus pachycheilus		Lake Lanao (Philippines)	1921-1982
Chondrostoma scodrensis		Lake Skadar (Albania, Yugoslavia)	1881-1990
Cyprinus yilongensis		Lake Yilong (China)	1981-1993

Species	Common Name	Place	Date
Evarra bustamantei	Mexican Chub	Mexico	1957-1983
Evarra eigenmanni	Plateau Chub	Mexico	1954-1983
Evarra tlahuacensis	Endorheic Chub	Mexico	1902-1954
Gila crassicauda	Thicktail Chub	USA	1957-1989
Hybopsis amecae	Ameca Shiner	Mexico	1969
Hybopsis aulidion	Durango Shiner	Mexico	1961-1985
Lepidomeda altivelis	Pahranagat Spinedace	USA	1938-1959
Mandibularca resinus	Bagangan	Lake Lanao (Philippines)	1922-1982
Notropis orca	Phantom Shiner	Mexico, USA	1975-1981
Ospatulus palaemorphagus		Lake Lanao (Philippines)	1924-1982
Ospatulus trunculatus	Bitungu	Lake Lanao (Philippines)	1921-1982
Phoxinellus egridiri	Yag Baligi	Lake Egridir (Turkey)	1955-1995
Phoxinellus handlirschi	Ciçek	Lake Egridir (Turkey)	1955-1995
Pogonichthys ciscoides	Clear Lake Splittail	USA	1970-1976
Puntius amarus	Pait	Lake Lanao (Philippines)	1910-1982
Puntius baoulan	Baolan	Lake Lanao (Philippines)	1926-1982
Puntius clemensi	Bagangan	Lake Lanao (Philippines)	1921-1982
Puntius disa	Disa	Lake Lanao (Philippines)	1932-1982
Puntius flavifuscus	Katapa-tapa	Lake Lanao (Philippines)	1921-1982
Puntius herrei		Lake Lanao (Philippines)	1908-1982
Puntius lanaoensis	Kandar	Lake Lanao (Philippines)	1922-1982
Puntius manalak	Manalak	Lake Lanao (Philippines)	1924-1982
Puntius sirang	Sirang	Lake Lanao (Philippines)	1932-1982
Puntius tras	Tras	Lake Lanao (Philippines)	1925-1982
Rhinichthys deaconi	Las Vegas Dace	USA	1940-1965
Sprattlicypris palata	Palata	Lake Lanao (Philippines)	1921-1982
Stypodon signifer	Stumptooth Minnow	Mexico	1903-1953

Family Catostomidae

Species	Common Name	Place	Date
Chasmistes muriei	Snake River Sucker	USA	1927-1970
Moxostoma lacerum	Harelip Sucker	USA	1893-1940

Order CHARACIFORMES

Family Characidae

Species	Common Name	Place	Date
Brycyon acuminatus		Brazil	1900-1982
Henochilus wheatlandi		Brazil	1865-1990

Order SILURIFORMES

Family Schilbeidae

Species	Common Name	Place	Date
Platytropius siamensis		Thailand	1966-1997

Family Trichomycteridae

Species	Common Name	Place	Date
Rhizosomichthys totae		Lake Tota (Colombia)	1957-1990

Order SALMONIFORMES

Family Retropinnidae

Species	Common Name	Place	Date
Prototroctes oxyrhynchus	New Zealand Grayling	New Zealand	1920s

Family Salmonidae

Species	Common Name	Place	Date
Coregonus alpenae	Longjaw Cisco	Great Lakes (Canada, USA)	1975-1989
Coregonus confusus	Pärrig	Lake Morat (Switzerland)	1885-1950
Coregonus fera	Féra	Lake Geneva (Switzerland)	1950-1958
Coregonus gutturosus	Kilch	Lake Constance (Switzerland)	1972-1992
Coregonus hiemalis	Gravenche	Lake Geneva (Switzerland)	1900-1950
Coregonus johannae	Deepwater Cisco	Great Lakes (Canada, USA)	1952-1980
Coregonus restrictus	Férit	Lake Morat (Switzerland)	1885-1950
Salvelinus agassizi	Silver Trout	Dublin Pond (USA)	1930-1939
Salvelinus inframundis	Orkney Char	Hoy I. (UK)	1862-1909
Salvelinus scharffi	Scharff's Char	Lough Owel (Ireland)	1908

Order ATHERINIFORMES

Family Atherinidae

Species	Common Name	Place	Date
Chirostoma compressum	Cuitzeo Silverside	Lake Cuitzeo (Mexico)	1940-1974
Rheocles sikorae	Zona	Madagascar	1966-1990

Order CYPRINODONTIFORMES

Family Fundulidae

Species	Common Name	Place	Date
Fundulus albolineatus	Whiteline Topminnow	USA	1899-1966

Family Goodeidae

Species	Common Name	Place	Date
Characodon garmani	Parras Characodon	Mexico	1903-1953
Empetrichthys merriami	Ash Meadows Killifish	USA	1948-1953

Family Poeciliidae

Species	Common Name	Place	Date
Gambusia amistadensis	Amistad Gambusia	USA	1974-1987

Species	Common Name	Place	Date
Family Cyprinodontidae			
Cyprinodon ceciliae	Cachorrito de la Presa	Mexico	1988-1990
Cyprinodon inmemoriam	Cachorrito de la Trinidad	Mexico	1985
Cyprinodon latifasciatus	Perrito de Parras	Mexico	1903-1953
Cyprinodon longidorsalis	Cachorrito de Charco Palma	Mexico	1985-1994
Cyprinodon sp. ?	Monkey Spring Pupfish	USA	1971
Leptolebias marmoratus	Ginger Pearlfish	Brazil	1944
Orestias cuvieri	Lake Titicaca Orestias	Lake Titicaca (Bolivia, Peru)	1937-1962
Order BELONIFORMES			
Family Adrianichthyidae			
Adrianichthys kruyti	Duck-billed Buntingi	Lake Poso (Indonesia)	1983-1987
Order GASTEROSTEIFORMES			
Family Gasterosteidae			
Pungitius kaibarae	Kyoto Ninespine Stickleback	Japan	1959-1993
Order SCORPAENIFORMES			
Family Cottidae			
Cottus echinatus	Utah Lake Sculpin	Utah Lake (USA)	1928-1963
Order PERCIFORMES			
Family Gobiidae			
Weberogobius amadi	Poso Bungu	Lake Poso, Sulawesi (Indonesia)	1985-1987
Family Cichlidae			
Tristramella magdalenae		Israel	1997
Haplochromis altigenis		Lake Victoria	1989-1996
Haplochromis apogonoides		Lake Victoria	1982-1987
Haplochromis arcanus		Lake Victoria	1969-1996
Haplochromis argenteus		Lake Victoria	1982-1987
Haplochromis artaxerxes		Lake Victoria	1962-1996
Haplochromis barbarae		Lake Victoria	1967-1989
Haplochromis bareli		Lake Victoria	1982-1987
Haplochromis bartoni		Lake Victoria	1962-1996
Haplochromis bayoni		Lake Victoria	1962-1996

Species	Common Name	Place	Date
Haplochromis boops		Lake Victoria	1962-1996
Haplochromis cassius		Lake Victoria	1982-1987
Haplochromis cinctus		Lake Victoria	1982-1987
Haplochromis cnester		Lake Victoria	1982-1987
Haplochromis decticostoma		Lake Victoria	1969-1996
Haplochromis dentex group		Lake Victoria	1982-1987
Haplochromis diplotaenia		Lake Victoria	1928-1996
Haplochromis estor		Lake Victoria	1962-1996
Haplochromis flavipinnis		Lake Victoria	1962-1989
Haplochromis gilberti		Lake Victoria	1978-1996
Haplochromis gowersi		Lake Victoria	1962-1996
Haplochromis guiarti		Lake Victoria	1982-1987
Haplochromis heusinkveldi		Lake Victoria	1990-1991
Haplochromis hiatus		Lake Victoria	1982-1987
Haplochromis iris		Lake Victoria	1982-1987
Haplochromis longirostris		Lake Victoria	1982-1987
Haplochromis macrognathus		Lake Victoria	1982-1987
Haplochromis maculipinna		Lake Victoria	1967-1996
Haplochromis mandibularis		Lake Victoria	1962-1996
Haplochromis martini		Lake Victoria	1960-1989
Haplochromis megalops		Lake Victoria	1982-1987
Haplochromis michaeli		Lake Victoria	1982-1987
Haplochromis microdon		Lake Victoria	1982-1987
Haplochromis mylergates		Lake Victoria	1978-1995
Haplochromis nanoserranus		Lake Victoria	1978-1996
Haplochromis nigrescens		Lake Victoria	1912?-1996
Haplochromis nyanzae		Lake Victoria	1962-1996
Haplochromis obtusidens		Lake Victoria	1960-1996
Haplochromis pachycephalus		Lake Victoria	1967-1996
Haplochromis paraguiarti		Lake Victoria	1967-1996
Haplochromis paraplagiostoma		Lake Victoria	1969-1996
Haplochromis parorthostoma		Lake Victoria	1967-1996
Haplochromis percoides		Lake Victoria	1962-1996
Haplochromis pharyngomylus		Lake Victoria	1960-1989
Haplochromis prognathus		Lake Victoria	1967-1996
Haplochromis pseudopellegrini		Lake Victoria	1967-1996
Haplochromis pyrrhopteryx		Lake Victoria	1982-1987

Species	Common Name	Place	Date
Haplochromis spekii		Lake Victoria	1967-1996
Haplochromis teegelaari		Lake Victoria	1982-1987
Haplochromis thuragnathus		Lake Victoria	1967-1996
Haplochromis tridens		Lake Victoria	1967-1989
Haplochromis victorianus		Lake Victoria	1962-1996
Haplochromis xenostoma		Lake Victoria	1982-1987
Haplochromis "bartoni-like"		Lake Victoria	1982-1987
Haplochromis "bicolor"		Lake Victoria	1982-1987
Haplochromis "big teeth"		Lake Victoria	1982-1987
Haplochromis "black cryptodon"		Lake Victoria	1982-1987
Haplochromis "black pectoral"		Lake Victoria	1982-1987
Haplochromis "chlorocephalus"		Lake Victoria	1982-1987
Haplochromis "citrus"		Lake Victoria	1982-1987
Haplochromis "coop"		Lake Victoria	1982-1987
Haplochromis "elongate rockpicker"		Lake Victoria	1993-1995
Haplochromis "filamentus"		Lake Victoria	1982-1987
Haplochromis "fleshy lips"		Lake Victoria	1993-1996
Haplochromis "grey pseudo-nigricans"		Lake Victoria	1979-1990
Haplochromis "large eye guiarti"		Lake Victoria	1982-1987
Haplochromis "lividus-frels"		Lake Victoria	1982-1987
Haplochromis "longurius"		Lake Victoria	1979-1990
Haplochromis "macrops like"		Lake Victoria	1982-1987
Haplochromis "micro-obesus"		Lake Victoria	1982-1987
Haplochromis "morsei"		Lake Victoria	1990-1991
Haplochromis "orange cinereus"		Lake Victoria	1990-1991
Haplochromis "orange macula"		Lake Victoria	1982-1987
Haplochromis "orange yellow big teeth"		Lake Victoria	1982-1987
Haplochromis "orange yellow small teeth"		Lake Victoria	1982-1987
Haplochromis "paropius-like"		Lake Victoria	1982-1987
Haplochromis "pink paedophage"		Lake Victoria	1982-1987
Haplochromis "pseudo-morsei"		Lake Victoria	1982-1987
Haplochromis "purple head"		Lake Victoria	1990-1991
Haplochromis "purple miller"		Lake Victoria	1982-1987
Haplochromis (?) "purple rocker"		Lake Victoria	1990-1991
Haplochromis "red empodisma"		Lake Victoria	1982-1987
Haplochromis "red eye scraper"		Lake Victoria	1982-1987
Haplochromis "reginus"		Lake Victoria	1982-1987
Haplochromis "regius"		Lake Victoria	1989
Haplochromis "short supramacrops"		Lake Victoria	1982-1987
Haplochromis "small blue zebra"		Lake Victoria	1991-1993
Haplochromis "small empodisma"		Lake Victoria	1982-1987
Haplochromis "smoke"		Lake Victoria	1979-1990
Haplochromis "soft grey"		Lake Victoria	1982-1987
Haplochromis "stripmac"		Lake Victoria	1990-1991
Haplochromis "supramacrops"		Lake Victoria	1982-1987
Haplochromis "theliodon-like"		Lake Victoria	1982-1987
Haplochromis "tigrus"		Lake Victoria	1982-1987
Haplochromis "too small"		Lake Victoria	1982-1987
Haplochromis "twenty"		Lake Victoria	1982-1987
Haplochromis "two stripe white lip"		Lake Victoria	1982-1987
Haplochromis "wyber"		Lake Victoria	1982-1987
Haplochromis "xenognathus-like"		Lake Victoria	1982-1987
Haplochromis "yellow"		Lake Victoria	1982-1987
Haplochromis "yellow-blue"		Lake Victoria	1991-1996
Hoplotilapia retrodens		Lake Victoria	1991-1996
Psammochromis cryptogramma group		Lake Victoria	1982-1987

Table 5.5

Biodiversity at country level

Note: This table includes estimates of the number of mammals, breeding birds and plants in each country of the world, together with estimates of the number of these endemic to each country, and the number and percentage of species globally threatened in each country. The columns headed DI and AI, respectively contain the unweighted national diversity indices, and the diversity indices adjusted for area, against the value expected by regressing diversity against country area. Missing values refer to countries with land area below 5000 km², for which the methodology[19] is invalid. The index is based on data for the groups shown here (0 = zero, – = no data) plus reptiles and amphibians, not included here.

Source: WCMC database; data derived from a very large number of published and unpublished sources, including country reports and regional checklists. Numbers of threatened species after animal and plant Red Lists[27,36].

Country	Area	DI	AI	Mammals total	Mammals endemic	Mammals no. thr.	Mammals % total	Birds total br	Birds endemic	Birds no. thr.	Birds % total	Plants total	Plants endemic	Plants no. thr.	Plants % total
Afghanistan	652225	0.063	–0.296	123	2	11	9	235	0	13	6	4000	800	4	0
Albania	28750	0.035	–0.019	68	0	2	3	230	0	7	3	3031	24	79	3
Algeria	2381745	0.045	–1.003	92	2	15	16	192	1	8	4	3164	250	141	4
American Samoa	197			3	0	2	67	34	0	2	6	471	15	9	2
Andorra	465			44	0	0	0	113	0	0	0	1350	–	–	–
Angola	1246700	0.176	0.544	276	7	17	6	765	12	13	2	5185	1260	30	1
Anguilla	91			3	0	0	0	–	0	0	–	321	1	–	–
Antigua & Barbuda	442			7	0	0	0	49	0	1	2	845	–	3	0
Argentina	2777815	0.196	0.423	320	49	27	8	897	19	41	5	9372	1100	247	3
Armenia	29800	0.042	0.153	84	3	4	5	242	0	5	2	–	–	31	–
Aruba	193			–	0	1	–	48	0	1	2	460	25	–	–
Australia	7682300	0.608	1.268	252	206	58	23	649	350	45	7	15638	14074	2245	14
Austria	83855	0.036	–0.293	83	0	7	8	213	0	5	2	3100	35	23	1
Azerbaijan	86600	0.050	0.027	99	0	11	11	248	0	8	3	4300	240	28	1
Bahamas	13865	0.017	–0.503	12	3	4	33	88	3	4	5	1111	118	31	3
Bahrain	661			17	0	1	6	28	0	1	4	195	–	–	–
Bangladesh	144000	0.059	0.058	109	0	18	17	295	0	30	10	5000	–	24	0
Barbados	430			6	0	0	0	24	0	1	4	572	3	2	0
Belarus	207600	0.029	–0.771	74	0	4	5	221	0	4	2	2100	–	1	0
Belgium	30520	0.023	–0.441	58	0	6	10	180	0	3	2	1550	1	2	0
Belize	22965	0.056	0.526	125	0	5	4	356	0	1	0	2894	150	57	2
Benin	112620	0.080	0.437	188	0	9	5	307	0	1	0	2201	–	4	0
Bermuda	54			3	0	0	0	8	1	2	25	167	15	10	6
Bhutan	46620	0.058	0.366	99	0	20	20	448	0	14	3	5468	75	23	0
Bolivia	1098575	0.239	0.882	316	16	24	8	–	18	27	–	17367	4000	227	1
Bosnia & Herzegovina	51129	0.034	–0.200	72	0	10	14	218	0	2	1	–	–	64	–
Botswana	575000	0.062	–0.287	164	0	5	3	386	1	7	2	2151	17	7	0
Brazil	8511965	0.740	1.436	394	119	71	18	1492	185	103	7	56215	–	1358	2
British Ind. Oc. Terr.				–	0	0	–	14	0	0	0	101	–	1	1
Brunei	5765	0.071	1.145	157	0	9	6	359	0	14	4	6000	7	25	0
Bulgaria	110910	0.044	–0.167	81	0	13	16	240	0	12	5	3572	320	106	3
Burkina Faso	274122	0.068	0.011	147	0	6	4	335	0	1	0	1100	–	–	–
Burundi	27835	0.072	0.723	107	0	5	5	451	0	6	1	2500	–	–	–
Cambodia	181000	0.059	0.001	123	0	23	19	307	0	18	6	–	–	5	–
Cameroon	475500	0.167	0.762	409	14	32	8	690	8	14	2	8260	156	89	1

Country	Area	DI	AI	Mammals total	Mammals endemic	Mammals no. thr.	Mammals % total	Birds total br	Birds endemic	Birds no. thr.	Birds % total	Plants total	Plants endemic	Plants no. thr.	Plants % total
Canada	9922385	0.067	−1.014	193	7	7	4	426	5	5	1	3270	147	278	9
Cape Verde	4035			5	0	1	20	38	4	3	8	774	86	1	0
Cayman Islands	259			8	0	0	0	45	0	1	2	539	19	13	2
Central African Republic	624975	0.080	−0.058	209	2	11	5	537	1	2	0	3602	100	1	0
Chad	1284000	0.049	−0.739	134	1	14	10	370	0	3	1	1600	–	12	1
Chile	751625	0.112	0.229	91	16	16	18	296	16	18	6	5284	2698	329	6
China	9597000	0.392	0.767	394	83	75	19	1100	70	90	8	32200	18000	312	1
Colombia	1138915	0.538	1.685	359	34	35	10	1695	67	64	4	51220	1500	712	1
Comoros	1860			12	2	3	25	50	14	6	12	721	136	4	1
Congo, D.R.	2345410	0.218	0.579	450	28	38	8	929	24	26	3	11007	1100	78	1
Congo, Republic	342000	0.128	0.589	200	2	10	5	449	0	3	1	6000	1200	30	
Cook Islands	233			1	0	0	0	27	6	6	22	284	3	12	4
Costa Rica	50900	0.162	1.358	205	7	14	7	600	6	13	2	12119	950	527	4
Côte d'Ivoire	322465	0.116	0.507	230	0	16	7	535	2	12	2	3660	62	94	3
Croatia	56538	0.036	−0.169	76	0	10	13	224	0	4	2	–	–	6	–
Cuba	114525	0.120	0.829	31	12	9	29	137	21	13	9	6522	3229	888	14
Cyprus	9250	0.017	−0.429	21	1	3	14	79	2	4	5	1682	–	51	3
Czech Republic	78864	0.033	−0.356	81	0	7	9	199	0	6	3	–	–	–	
Denmark	43075	0.021	−0.643	43	0	3	7	196	0	2	1	1450	1	2	0
Djibouti	23000	0.020	−0.528	61	0	3	5	126	1	3	2	826	6	2	0
Dominica	751			12	0	1	8	52	2	2	4	1228	11	57	5
Dominican Republic	48440	0.076	0.625	20	0	4	20	136	0	11	8	5657	1800	1362	
Ecuador	461475	0.353	1.519	302	25	28	9	1388	37	53	4	19362	4000	824	4
Egypt	1000250	0.038	−0.936	98	7	15	15	153	0	11	7	2076	70	82	4
El Salvador	21395	0.048	0.393	135	0	2	1	251	0	0	0	2911	17	42	1
Equatorial Guinea	28050	0.084	0.869	184	1	12	7	273	3	4	1	3250	66	110	
Eritrea	117600	0.057	0.088	112	0	6	5	319	0	3	1	–	–	–	–
Estonia	45100	0.025	−0.483	65	0	4	6	213	0	2	1	1630	–	2	0
Ethiopia	1104300	0.145	0.383	277	31	35	13	626	28	20	3	6603	1000	163	2
F. S. Micronesia	702			6	3	6	100	40	18	6	15	1194	293	40	
Faeroe Islands				–	0	1	–	71	0	0	0	236	1	–	–
Falkland Islands	15931	0.004	−2.040	0	0	0	–	64	4	1	2	165	14	64	
Fiji	18330	0.028	−0.100	4	1	4	100	74	24	9	12	1518	760	74	5
Finland	337030	0.023	−1.145	60	0	4	7	248	0	4	2	1102	–	6	1
France	543965	0.051	−0.473	93	0	13	14	269	1	7	3	4630	133	195	4
French Guiana	91000	0.079	0.483	150	3	9	6	–	1	1	–	5625	144	98	2
French Polynesia	3940			0	0	0	–	60	25	22	37	959	560	187	19
French S. & Antarctic	7241	0.001	−3.261	–	0	0	–	48	3	3	6	–	–	–	–

Country	Area	DI	AI	Mammals total	Mammals endemic	Mammals no. thr.	Mammals % total	Birds total br	Birds endemic	Birds no. thr.	Birds % total	Plants total	Plants endemic	Plants no. thr.	Plants % total
Gabon	267665	0.116	0.560	190	3	12	6	466	1	4	1	6651	–	91	1
Gambia	10690	0.036	0.308	117	0	4	3	280	0	1	0	974	–	1	0
Georgia	69700	0.051	0.111	107	2	10	9	–	0	5	–	4350	380	29	1
Germany	356840	0.033	-0.770	76	0	8	11	239	0	5	2	2682	6	14	1
Ghana	238305	0.114	0.571	222	1	13	6	529	0	10	2	3725	43	103	3
Gibraltar	7			7	0	1	14	34	0	1	3	600	–	4	1
Greece	131985	0.062	0.129	95	3	13	14	251	0	10	4	4992	742	571	11
Greenland	2175600	0.007	-2.821	9	0	2	22	62	0	0	0	529	15	5	1
Grenada	345			15	0	0	0	50	1	1	2	1068	4	8	1
Guadeloupe	1780			11	4	5	45	52	2	0	0	1400	26	26	2
Guam	450			2	0	2	100	18	2	3	17	330	69	18	5
Guatemala	108890	0.142	1.014	250	3	8	3	458	1	4	1	8681	1171	355	4
Guinea	245855	0.094	0.373	190	1	11	6	409	0	12	3	3000	88	39	1
Guinea–Bissau	36125	0.050	0.289	108	0	4	4	243	0	1	0	1000	12	–	–
Guyana	214970	0.133	0.758	193	1	10	5	678	0	3	0	6409	–	152	2
Haiti	27750	0.071	0.710	3	0	4	100	75	1	11	15	5242	1623	100	2
Honduras	112085	0.094	0.597	173	2	7	4	422	1	4	1	5680	148	96	2
Hungary	93030	0.031	-0.457	83	0	8	10	205	0	10	5	2214	38	30	1
Iceland	102820	0.006	-2.080	11	0	1	9	88	0	0	0	377	1	1	0
India	3166830	0.326	0.896	316	44	75	24	923	58	73	8	16000	5000	1236	8
Indonesia	1919445	0.731	1.844	436	222	128	29	1519	408	104	7	29375	17500	264	1
Iran	1648000	0.091	-0.194	140	6	20	14	323	1	14	4	8000	–	2	0
Iraq	438445	0.041	-0.629	81	2	7	9	172	1	12	7	–	–	–	–
Ireland	68895	0.013	-1.248	25	0	2	8	142	0	1	1	950	–	1	0
Israel	20770	0.043	0.285	116	4	13	11	180	0	8	4	2317	–	32	1
Italy	301245	0.065	-0.056	90	3	10	11	234	0	7	3	5599	712	311	6
Jamaica	11425	0.051	0.619	24	2	4	17	113	26	7	6	3308	923	744	22
Japan	369700	0.124	0.536	188	42	29	15	250	21	33	13	5565	2000	707	13
Jordan	96000	0.036	-0.310	71	0	7	10	141	0	4	3	2100	–	9	0
Kazakhstan	2717300	0.071	-0.581	178	4	15	8	396	0	15	4	–	–	71	–
Kenya	582645	0.145	0.560	359	23	43	12	844	9	24	3	6506	265	240	4
Kiribati	684			–	0	0	–	26	1	4	15	60	2	–	–
Korea, D.P.R.	122310	0.025	-0.775	–	0	7	–	115	1	19	17	2898	107	4	0
Korea, Republic	98445	0.030	-0.518	49	0	6	12	112	0	19	17	2898	224	662	
Kuwait	24280	0.007	-1.564	21	0	1	5	20	0	3	15	234	–	–	–
Kyrgyzstan	198500	0.036	-0.537	83	1	6	7	–	0	5	–	3786	–	34	1
Lao P.D.R.	236725	0.081	0.229	172	0	30	17	487	1	27	6	–	–	2	–
Latvia	63700	0.025	-0.553	83	0	4	5	217	0	6	3	1153	–	–	–

Country	Area	DI	AI	Mammals total	Mammals endemic	Mammals no. thr.	Mammals % total	Birds total br	Birds endemic	Birds no. thr.	Birds % total	Plants total	Plants endemic	Plants no. thr.	Plants % total
Lebanon	10400	0.031	0.145	57	0	5	9	154	0	5	3	3000	–	5	0
Lesotho	30345	0.025	–0.354	33	0	2	6	58	0	5	9	1591	2	21	1
Liberia	111370	0.059	0.132	193	0	11	6	372	1	13	3	2200	103	25	1
Libya	1759540	0.029	–1.343	76	5	11	14	91	0	2	2	1825	134	57	3
Liechtenstein	160			64	0	0	0	124	0	1	1	1410	–	3	0
Lithuania	65200	0.026	–0.544	68	0	5	7	202	0	4	2	1796	–	1	0
Luxembourg	2585			55	0	3	5	126	0	1	1	1246	–	1	0
Macedonia Y.R.	25713	0.037	0.077	78	0	10	13	210	0	3	1	3500	–	–	
Madagascar	594180	0.298	1.277	141	93	46	33	202	105	28	14	9505	6500	306	3
Malawi	94080	0.079	0.473	195	0	7	4	521	0	9	2	3765	49	61	2
Malaysia	332965	0.254	1.280	300	36	42	14	501	18	34	7	15500	3600	490	3
Maldives	298			3	0	0	0	23	0	1	4	–	–	–	–
Mali	1240140	0.053	–0.658	137	0	13	9	397	0	6	2	1741	11	15	1
Malta	316			22	0	0	0	26	0	2	8	914	5	15	2
Marshall Islands	181			0	0	0	–	17	0	1	6	100	5	–	–
Martinique	1079			9	0	0	0	52	1	2	4	1287	30	44	3
Mauritania	1030700	0.041	–0.856	61	1	14	23	273	0	3	1	1100	–	3	0
Mauritius	1865			4	1	4	100	27	8	10	37	750	325	294	39
Mayotte	376			–	0	0	–	27	2	2	7	–	–	–	–
Mexico	1972545	0.589	1.621	491	140	64	13	769	92	36	5	26071	12500	1593	6
Moldova	33700	0.025	–0.396	68	0	2	3	177	0	7	4	1752	–	5	0
Monaco	2			–	0	0	–	–	0	0	–	–	–	–	–
Mongolia	1565000	0.051	–0.767	133	0	12	9	426	0	14	3	2823	229	0	0
Montserrat	104			7	0	1	14	37	1	0	0	671	2	2	0
Morocco	458730	0.057	–0.304	105	4	18	17	210	0	11	5	3675	625	186	5
Mozambique	784755	0.090	0.005	179	2	13	7	498	0	14	3	5692	219	89	2
Myanmar	678030	0.141	0.493	251	6	31	12	867	4	44	5	7000	1071	32	0
Namibia	824295	0.102	0.116	250	3	11	4	469	3	8	2	3174	687	75	2
Nauru				–	0	0	–	9	1	2	22	50	1	–	–
Nepal	141415	0.096	0.549	181	2	28	15	611	2	27	4	6973	315	20	0
Netherlands	41160	0.022	–0.599	55	0	6	11	191	0	3	2	1221	–	1	0
Netherlands Antilles	800			–	0	1	–	77	0	1	1	–	–	2	–
New Caledonia	19105	0.078	0.904	11	3	5	45	107	22	10	9	3250	3200	480	15
New Zealand	265150	0.065	–0.017	2	2	3	100	150	74	44	29	2382	1942	211	9
Nicaragua	148000	0.098	0.555	200	2	4	2	482	0	3	1	7590	40	98	1
Niger	1186410	0.061	–0.512	131	0	11	8	299	0	2	1	1170	–	–	–
Nigeria	923850	0.107	0.131	274	4	26	9	681	2	9	1	4715	205	37	1
Niue	259			1	0	0	0	15	0	1	7	178	1	1	1

Country	Area	DI	AI	Mammals total	Mammals endemic	Mammals no. thr.	Mammals % total	Birds total br	Birds endemic	Birds no. thr.	Birds % total	Plants total	Plants endemic	Plants no. thr.	Plants % total
Northern Marianas	477			–	0	1	–	28	2	7	25	315	81	11	3
Norway	386325	0.024	−1.107	54	0	4	7	243	0	3	1	1715	1	12	1
Oman	271950	0.030	−0.812	56	2	9	16	107	0	5	5	1204	73	30	2
Pakistan	803940	0.080	−0.121	151	4	13	9	375	0	25	7	4950	372	14	0
Palau	492			2	0	3	100	45	10	2	4	–	–	–	–
Panama	78515	0.162	1.236	218	16	17	8	732	9	10	1	9915	1222	1302	13
Papua New Guinea	462840	0.271	1.254	214	65	57	27	644	94	31	5	11544	–	92	1
Paraguay	406750	0.115	0.429	305	2	10	3	556	0	26	5	7851	–	129	2
Peru	1285215	0.396	1.344	460	49	46	10	1538	112	64	4	17144	5356	906	5
Philippines	300000	0.225	1.188	153	102	49	32	196	186	86	44	8931	3500	360	4
Pitcairn Islands				0	0	0	–	19	5	5	26	76	14	14	18
Poland	312685	0.032	−0.761	84	0	10	12	227	0	6	3	2450	3	27	1
Portugal	92390	0.045	−0.088	63	1	13	21	207	2	7	3	5050	150	269	5
Puerto Rico	8960	0.033	0.259	16	0	3	19	105	12	11	10	2493	235	223	9
Qatar	11435	0.005	−1.770	11	0	0	0	23	0	1	4	220	–	–	–
Réunion	2510			2	0	2	100	18	4	3	17	546	165	99	18
Romania	237500	0.039	−0.490	84	0	16	19	247	0	11	4	3400	41	99	3
Russia	17075400	0.179	−0.179	269	22	31	12	628	13	38	6	–	–	214	–
Rwanda	26328	0.087	0.925	151	0	9	6	513	0	6	1	2288	26	–	–
Saint Helena & dep.	411			2	0	0	0	53	9	9	17	165	50	68	41
San Marino				13	0	–	–	–	0	–	–	–	–	–	–
São Tomé & Príncipe	964			8	4	3	38	63	25	9	14	895	134	3	0
Saudi Arabia	2400900	0.040	−1.129	77	0	9	12	155	0	11	7	2028	–	7	0
Senegal	196720	0.057	−0.065	192	0	13	7	384	0	6	2	2086	26	31	1
Seychelles	404			6	2	2	33	38	11	9	24	250	182	78	31
Sierra Leone	72325	0.083	0.588	147	0	9	6	466	1	12	3	2090	74	29	1
Singapore	616			85	1	6	7	118	0	9	8	2282	2	29	1
Slovakia	14035	0.037	0.252	85	0	8	9	209	0	4	2	3124	92	–	–
Slovenia	20251	0.036	0.106	75	0	10	13	207	0	3	1	3200	22	13	0
Solomon Islands	29790	0.049	0.316	53	21	20	38	163	43	18	11	3172	30	42	1
Somalia	630000	0.087	0.025	171	12	18	11	422	11	8	2	3028	500	103	3
South Africa	1184825	0.252	0.915	247	35	33	13	596	8	16	3	23420	–	2215	9
Spain	504880	0.067	−0.172	82	4	19	23	278	5	10	4	5050	941	985	20
Sri Lanka	65610	0.082	0.606	88	15	14	16	250	24	11	4	3314	890	455	14
St Kitts–Nevis	261			7	0	0	0	32	0	1	3	659	1	4	1
St Lucia	619			9	0	0	0	50	4	3	6	1028	11	6	1
St Vincent	389			8	1	1	13	108	2	2	2	1166	–	9	1
Sudan	2505815	0.137	0.093	267	11	21	8	680	1	9	1	3137	50	10	0

Country	Area	DI	AI	Mammals total	Mammals endemic	Mammals no. thr.	Mammals % total	Birds total br	Birds endemic	Birds no. thr.	Birds % total	Plants total	Plants endemic	Plants no. thr.	Plants % total
Suriname	163820	0.092	0.471	180	2	10	6	603	0	2	0	5018	–	103	2
Swaziland	17365	0.044	0.353	47	0	5	11	364	0	6	2	2715	4	42	2
Sweden	440940	0.026	−1.067	60	0	5	8	249	0	4	2	1750	1	13	1
Switzerland	41285	0.033	−0.173	75	0	6	8	193	0	4	2	3030	1	30	1
Syria	185680	0.046	−0.265	63	2	4	6	204	0	7	3	3000	–	8	0
Taiwan	36960	0.058	0.418	63	11	10	16	160	14	13	8	3568	–	325	9
Tajikistan	143100	0.033	−0.536	84	1	5	6	–	0	9	–	–	–	50	–
Tanzania	939760	0.189	0.693	316	15	33	10	822	24	30	4	10008	1122	436	4
Thailand	514000	0.162	0.709	265	7	34	13	616	2	45	7	11625	–	385	3
Togo	56785	0.094	0.781	196	0	8	4	391	0	1	0	3085	–	4	0
Tokelau				0	0	0	–	5	0	1	20	26	–	–	–
Tonga	699			2	0	0	0	37	2	2	5	463	25	2	0
Trinidad & Tobago	5130	0.045	0.729	100	1	1	1	260	1	3	1	2259	236	21	1
Tunisia	164150	0.033	−0.572	78	1	11	14	173	0	6	3	2196	–	24	1
Turkey	779450	0.114	0.237	116	2	15	13	302	0	14	5	8650	2675	1876	22
Turkmenistan	488100	0.044	−0.572	103	0	11	11	–	0	12	–	–	–	17	–
Turks & Caicos Islands	430			–	0	0	–	42	0	3	7	448	9	2	0
Tuvalu	25			–	0	0	–	9	0	1	11	–	–	–	–
Uganda	236580	0.120	0.624	338	6	18	5	830	3	10	1	5406	–	15	0
Ukraine	603700	0.050	−0.509	108	1	15	14	263	0	10	4	5100	–	52	1
United Arab Emirates	75150	0.019	−0.883	25	0	3	12	67	0	4	6	–	–	–	–
United Kingdom	244880	0.024	−1.003	50	0	4	8	230	1	2	1	1623	16	18	1
Uruguay	186925	0.050	−0.186	81	1	5	6	237	0	11	5	2278	40	15	1
USA	9372614	0.342	0.638	428	105	35	8	650	67	50	8	19473	4036	4669	24
USA Pacific Islands	658			–	0	0	–	–	0	1	–	–	–	–	–
Uzbekistan	447400	0.051	−0.413	97	0	7	7	–	0	11	–	4800	400	41	1
Vanuatu	14765	0.014	−0.728	11	2	3	27	76	9	6	8	870	150	26	3
Venezuela	912045	0.379	1.398	323	19	24	7	1340	40	22	2	21073	8000	426	2
Viet Nam	329565	0.147	0.737	213	9	38	18	535	10	47	9	10500	1260	341	3
Virgin Islands (British)	153			3	0	1	33	70	0	2	3	–	–	14	–
Virgin Islands (US)	352			–	0	0	–	70	0	2	3	–	–	40	–
Wallis & Futuna	255			1	0	0	0	25	0	0	0	475	7	–	–
Western Sahara				32	1	7	22	60	0	1	2	330	–	–	–
Western Samoa	2840			3	0	2	67	40	8	6	15	737	–	18	2
Yemen	477530	0.041	−0.654	66	1	5	8	143	8	13	9	1650	135	149	9
Yugoslavia	102173	0.046	−0.086	96	0	12	13	224	0	8	4	4082	–	–	–
Zambia	752615	0.096	0.074	233	3	11	5	605	2	10	2	4747	211	12	0
Zimbabwe	390310	0.099	0.298	270	0	9	3	532	0	9	2	4440	95	100	2

References

1 World Conservation Monitoring Centre. 1992. *Global biodiversity: status of the Earth's living resources*. Chapman & Hall, London. xx +594pp.

2 United Nations Environment Programme. 1995. Heywood, V (ed). *Global biodiversity assessment*. Cambridge University Press, Cambridge. x + 1140pp.

3 Gaston, K.J. and Williams, P.H. 1996. Spatial patterns in taxonomic diversity. In, Gaston, K.J. (ed) *Biodiversity*. Pp. 202-229. Blackwell Science, Oxford.

4 http://www.nhm.ac.uk/science/projects/worldmap

5 Williams, P.H., Gaston, K.J. and Humphries, C.J. 1997. Mapping biodiversity value worldwide: combining higher-taxon richness from different groups. *Proc. R. Soc. Lond.* B 264:141-148.

6 Wright, D.H., Currie, D.J. and Maurer, B.A. 1993. Energy supply and patterns of species richness on local and regional scales. In, Ricklefs, R.E. and Schluter, D. (eds). *Species diversity in ecological communities*. Pp. 66-74. University of Chicago Press, Chicago.

7 Wright, D.H. 1983. Species-energy theory: an extension of species-area theory. *Oikos* 41:496-506.

8 Wright, D.H. 1987. Estimating human impacts on global extinction. *International Journal of Biometeorology* 31:293-299.

9 Adams, J.M. 1989. Species diversity and productivity of trees. *Plants today* Nov.-Dec. 183-187.

10 Currie, D.J. and Paquin, V. 1987. Large-scale biogeographical patterns of species richness of trees. *Nature* 329:326-327.

11 Kerr, J.T. and Packer, L. 1999. The environmental basis of North American species richness patterns among Epicaudata (Coleoptera, Meloidae). *Biodiversity and Conservation* 8:617-628.

12 Kerr, J.T. and Packer, L. 1997. Habitat heterogeneity as a determinant of mammal species richness in high-energy regions. *Nature* 385:252-254.

13 Burnett, M.R., August, P.V., Brown, J.H. and Killingbeck, K.T. 1998. The influence of geomorphological heterogeneity on biodiversity. I. A patch-scale perspective. *Conservation Biology* 12(2):363-370.

14 Nichols, W.F., Killingbeck, K.T. and August, P.V. 1998. The influence of geomorphological heterogeneity on biodiversity. II. A landscape perspective. *Conservation Biology* 12(2):371-379.

15 Mittermeier, R. 1988. Primate diversity and the tropical forest. Case studies from Brazil and Madagascar and the importance of megadiversity countries. In, Wilson, E.O. (ed) *Biodiversity*. Pp. 145-154. National Academic Press, Washington, DC.

16 Myers, N. 1988. Threatened biotas: 'hotspots' in tropical forests. *Environmentalist* 8:1-20.

17 Myers, N. 1990. The biodiversity challenge: expanded hot-spots analysis. *Environmentalist* 10:243-256.

18 http://www.conservation.org/web/fieldact/hotspots/hotspots.htm

19 WCMC. 1998. Development of a national biodiversity index. Version 2. Unpublished report for The World Bank.

20 WWF and IUCN 1994. *Centres of plant diversity. A guide and strategy for their conservation*. 3 vols. IUCN Publications Unit, Cambridge.

21 Stattersfield, A.J., Crosby, M.J., Long, A.J. and Wege, D.C. 1998. *Endemic bird areas of the world*. BirdLife International, Cambridge, UK.

22 Reid, W.V. 1992. How many species will there be? In, Whitmore, T.C. and Sayer, J.A. (eds) *Tropical deforestation and species extinction*. Pp. 55-73. Chapman & Hall, London.

23 IUCN. 1994. IUCN *Red List categories*. IUCN, Gland, Switzerland.

24 Collar, N.J., Crosby, M.J. and Stattersfield, A.J. 1994. *Birds to watch 2. The world list of threatened birds*. BirdLife Conservation Series No.4. BirdLife International, Cambridge.

25 Master, L.L., Flack, S.R. and Stein, B.A. (eds). 1998. *Rivers of life: critical watersheds for protecting freshwater biodiversity*. The Nature Conservancy, Arlington, Virginia.

26 Richards, J.F. 1990. Chapter 10, Land Transformation, pp. 163-178. In, B.L. Turner II et al. (eds) *The Earth as transformed by human action*. Cambridge University Press, with Clark University.

27 IUCN. 1996. 1996 IUCN *Red List of threatened animals*. IUCN, Gland, Switzerland.

28 UNEP. 1997. Recommendations for a core set of indicators of biological diversity. Note by the Executive Secretary. UNEP/CBD/SBSTTA/3/9. (available from the CBD Secretariat website, http://www.biodiv.org/).

29 UNEP. 1997. Recommendations for a core set of indicators of biological diversity. Background paper prepared by the liaison group on indicators of biological diversity. UNEP/CBD/SBSTTA/3/Inf.13. (available from the CBD Secretariat website, http://www.biodiv.org/).

30 Brink, B.J.E. ten, Hosper, S.H. and Colijn, F. 1991. A quantitative method for description and assessment of ecosystems: the AMOEBA-approach. *Marine Pollution Bulletin* 23:265-270.

31 OECD. 1993. OECD core set of indicators for environmental performance reviews. OCDE/GD(93)179. *Environment Monographs No.83*. Organisation for Economic Co-operation and Development, Paris. Available from http://www.oecd.org/env/soe/.

32 OECD. 1998. Sustainable development indicators. Proceedings of an OECD workshop. Organisation for Economic Co-operation and Development, Paris. Available from http://www.oecd.org/env/soe/.

33 Collar, N.J. and Andrew, P. 1988. *Birds to watch*. ICBP Technical Publication No. 8. ICBP, Cambridge.

34 Loh, J., Randers, J., MacGillivray, A., Kapos, V., Jenkins, M., Groombridge, B. and Cox, N. 1998. *Living planet report 1998*. WWF International, Gland, Switzerland.

35 Loh, J., Randers, J., MacGillivray, A., Kapos, V., Jenkins, M., Groombridge, B., Cox, N. and Warren, B. 1999. *Living planet report 1999*. WWF International, Gland, Switzerland.

36 Walter, K.S. and Gillet, H.J. 1998. 1997 IUCN *Red List of threatened plants*. Compiled by WCMC. IUCN: Gland, Switzerland and Cambridge, UK.

37 MacPhee R.D.E. and Flemming, C. 1999. Requiem Æternam: the last five hundred years of mammalian species extinctions. In, R.D.E. MacPhee (ed.), *Extinctions in near time: causes, contexts, and consequences.* pp. 333-372. Kluwer Academic/Plenum Publishers, New York.

38 Harrison I.J. and Stiassny, M.L.J. 1999. The quiet crisis: a preliminary listing of the freshwater fishes of the World that are extinct or "missing in action". In, R.D.E. MacPhee (ed.), *Extinctions in near time: causes, contexts, and consequences,* pp.271-332. Kluwer Academic/Plenum Publishers, New York.

39 Website of Committee on Recently Extinct Organisms (CREO). http://www.creo.org/

40 Berra, T.M. 1981. *An atlas of distribution of the freshwater fish families of the world.* University of Nebraska Press, Lincoln and London.

41 http://www.botanik.uni-bonn.de/biodiv/biomaps.htm#introduction

42 O'Brien, E.M. 1998. Water-energy dynamics, climate, and prediction of woody plant species richness: an interim general model. *Journal of Biogeography* 25:379-398.

43 Prendergast, J.R. *et al.* 1993. Rare species, the coincidence of biodiversity hotspots and conservation strategies. *Nature* 365:335-337.

44 Prendergast, J.R. and Eversham, B. 1997. Species richness covariance in higher taxa: empirical tests of the biodiversity indicator concept. *Ecography* 20:210-216.

45 Lawton, J.H. *et al.* 1998. Biodiversity inventories, indicator taxa and effects of habitat modification in tropical forest. *Nature* 391:72-76.

46 Van Jaarsveld, A.S. *et al.* 1997. Biodiversity assessment and conservation strategies. *Science.* 279(5329):2106-2108.

47 Guégan, J.-F, Lek, S. and Oberdorff, T. 1998. Energy availability and habitat heterogeneity predict global riverine fish diversity. *Nature* 391, 22 January, 382-384.

48 MacGillivray, A. *et al.* 1994. *Environmental measures. Indicators for the UK environment.* Environment Challenge Group (New Economics Foundation, FoE, IIED, RSNC, RSPB, Wildlife and Countryside Link, WWF-UK).

49 Hammond, A., Adriaanse, A., Rodenburg, E., Bryant, D. and Woodward, R. 1995. *Environmental Indicators.* World Resources Institute. Washington, DC.

50 For example, for Canada: http:www1.ncr.ec.gc.ca/~ind/

51 Barthlott, W., Lauer, W. and Placke, A. 1996: Global distribution of species diversity in vascular plants: towards a world map of phytodiversity. In, *Erdkunde* 50(4):317-327.

52 Barthlott, W., Biedinger, N., Braún, G., Feig, F., Kier, G. and J. Mutke. 1999: Terminological and methodological aspects of the mapping and analysis of global biodiversity. In *Acta Botanica Fennica* 162:103-110.

Suggested introductory sources

Gaston, K.J. and Spicer, J.I. 1998. *Biodiversity: an introduction.* Blackwell Science Ltd., Oxford.

6

MARINE BIODIVERSITY

Most of the planet is covered by ocean waters whose average depth is four times the average elevation of the land, making the open sea by far the largest eco-system on Earth. Despite this volume, marine net primary production remains similar to or less than that on land because photosynthesis in the sea is carried out by microscopic bacteria and algae restricted to the sunlit surface layers (plants are virtually absent).

The diversity of major lineages (phyla and classes) is much greater in the sea than on land or in freshwaters, and many phyla of invertebrate animals occur only in marine waters. Species diversity appears to be far lower, perhaps because marine waters are physically much less variable in space and time than the terrestrial environment. Planktonic organisms and others with planktonic larvae tend to be widely distributed and relatively resistant to extinction.

Marine fisheries are the largest source of wild protein, derived from fishes, molluscs and crustaceans. The world catch from capture fisheries has grown five-fold over the past five decades, but despite increasing effort has levelled off dur-ing the 1990s. Half of the world's major fishery resources are now in urgent need of remedial management, mainly because of excess exploitation, and pollution in semi-enclosed seas.

THE SEAS

Oceans cover 71% of the world's surface. They are on average around 3.8 km deep and have an overall volume of some 1370 million cubic kilometres. The whole of the world ocean* is theoretically capable of supporting life, so that the marine part of the biosphere is far larger than the terrestrial part. However, as on land, life in the oceans is very unevenly distributed – some parts are astonishingly productive and diverse, while others are virtually barren (Table 6.1).

Although knowledge of the functioning of the marine biosphere has increased enormously in the past few decades, overall it remains far less well known and understood than the terrestrial part of the globe. The main reason for this is, quite simply, that much of it is inaccessible to humans. Study of any part below the top few metres requires specialised equipment, is expensive and time-consuming. Knowledge of most of the sea is thus based largely on a range of remote sensing and sampling techniques and remains often sketchy. As these techniques beome more sophisticated, so our understanding of marine ecosystems, particularly those away from the coastal zone, is undergoing constant revision.

THE COMPOSITION AND MOVEMENT OF SEAWATER

Seawater is a complex but relatively uniform mixture of chemicals. Most of the 92 naturally occurring elements can be detected dissolved in it, but most only in trace concentrations; the most abundant are sodium (as Na^+) and chlorine (as Cl^-), which occur at a concentration some ten times higher than the next most abundant element, magnesium. The term used to quantify the total amount of dissolved salts in seawater is salinity, a dimensionless ratio, which generally ranges between 33 and 37 and averages 35. Most of the substances dissolved in seawater are unreactive and remain at relatively stable concentrations; however those that play a part in biological systems can be highly variable in time and space.

The world's seawaters are constantly in motion, at all scales from the molecular to the oceanic. Large-scale ocean circulation plays a vital rôle in mediating global climate as well as influencing the functioning of marine ecosystems. It is driven by complex interactions between a number of physical variables, notably latitudinal variations in solar radiation (and consequent heating and cooling), precipitation and evaporation, transfer of frictional energy across the ocean surface from winds, and forces resulting from the rotation of the planet.

Table 6.1

Area and maximum depth of the world's oceans and seas

Source: Times Atlas of the Oceans

Ocean/Sea	Area (km²)	Depth (metres)
Pacific Ocean	165 384 000	11 524
Atlantic Ocean	82 217 000	9560
Indian Ocean	73 481 000	9000
Arctic Ocean	14 056 000	5450
Mediterranean Sea	2 505 000	4846
South China Sea	2 318 000	5514
Bering Sea	2 269 000	5121
Caribbean Sea	1 943 000	7100
Gulf of Mexico	1 544 000	4377
Sea of Okhotsk	1 528 000	3475
East China Sea	1 248 000	2999
Yellow Sea	1 243 000	91
Hudson Bay	1 233 000	259
Sea of Japan	1 008 000	3743
North Sea	575 000	661
Black Sea	461 000	2245
Red Sea	438 000	2246
Baltic Sea	422 000	460

* **Note**: The world ocean refers to the contiguous seas and oceans of the world; because all are linked to each other, eventually the same body of seawater and its contents could at least theoretically flow through each in turn. The world ocean includes semi-enclosed areas such as the Mediterranean, Black and Arctic Seas but excludes the Caspian and now largely destroyed Aral Sea; the former has some of the characteristics of the sea and some of an inland water.

◄— Surface currents   ◄— Deep currents   Source areas of cold bottom water   Zones of upwelling

Map 6.1

Ocean circulation

This map shows in highly schematic form some major features of oceanic circulation, including source regions of dense cold bottom-flowing water and regions of upwelling nutrient-rich waters important for fishery maintenance.

Source: compiled from multiple sources.

Surface currents are largely driven by the prevailing winds. The most important features are vast, anticyclonic gyres in the subtropical regions of the world's oceans. These are primarily driven by the westerly trade winds in the 'roaring forties' and circulate clockwise in the northern hemisphere and anti-clockwise in the southern hemisphere. Also important are the eastward flowing Antarctic circumpolar current and equatorial current. At smaller scales, eddies and rings are ubiquitous and are analogous to weather systems in the atmosphere. Typical oceanic eddies may be 100 km across and may persist for one or two years (Map 6.1).

Vertical gradients and vertical mixing have a profound effect on the marine biosphere and are mostly driven by differences in water density. Density of seawater increases with depth (slightly), with increasing salinity and with decreasing temperature until it freezes at around −1.9°C, unlike freshwater, which shows maximum density at 4°C and freezes at 0°C.

Because seawater of different density does not mix readily, the oceans tend to be well stratified vertically, with bodies of less dense seawater (warmer and less saline) sitting on top of cooler, more saline and denser bodies. The stratification is rarely stable over the long-term, however, as the influence of climate changes the properties of surface waters and causes various forms of vertical mixing. The single most important factor controlling large-scale deep water circulation appears to be the generation of cold, high salinity water near the surface in the Weddell Sea and off Greenland in the north Atlantic. Here, during winter, the sea freezes. The ice, which forms a floating ice-sheet, is virtually free of salt, so that the underlying water is more saline, cold and dense. This sinks to the bottom and moves south along the Atlantic floor, some passing south of the equator and circulating throughout the world ocean, producing what is known as the Great Conveyor – the interlocking system of major circulation currents in the deep sea. Because it originates near the surface it is well oxygenated as well as cold, and is the major reason that aerobic organisms can thrive in the deep sea.

Vertical movement of seawater elsewhere is largely determined by two factors: surface mixing caused by winds and the relative influence of precipitation and evaporation. When winds are strong (eg. during winter at high latitudes) surface mixing may extend several hundred metres into the water column; when winds are weak, surface mixing may have negligible impact beyond the top few tens of metres. Precipitation and evaporation affect surface water salinity and density. Where precipitation exceeds evaporation, surface waters decrease in salinity and density and therefore remain at the surface, tending to stabilize the water column and reduce vertical mixing. Where the reverse is the case, surface waters increase in salinity and density and therefore sink through the water column, increasing mixing. In some areas (eg. the Arctic Ocean), major inflow of river water may reduce the salinity and density of surface waters and stabilize the upper water column.

Because many of the factors affecting mixing are likely to change seasonally, there may be marked differences in the behaviour of seawater throughout the year. In particular, at latitudes above 40° winter cooling and winter storms break down vertical stratification leading to considerable vertical mixing. In the tropics and subtropics, stratification tends to persist throughout the year, other than in zones of upwelling. These occur along the western boundaries of continents where trade winds blow towards the equator, causing surface waters to be pushed offshore, to be replaced by cooler deep waters. There are five major upwelling areas: the Humboldt Current region off the Chilean and southern Peruvian coast of South America; the California Current region off western North America; the Canary Current off the coast of Mauritania in north-west Africa; the Benguela Current region off southern Africa; and the northwest Arabian Sea.

THE MAJOR ZONES OF THE OCEANS

The continental shelf

Marine waters around major landmasses are typically shallow, lying over a continental shelf which may be anything from a few kilometres to several hundred kilometres wide. The most landward part is the *littoral* or intertidal zone where the bottom is subject to periodic exposure to the air. Water depth here varies from zero to several metres. Seaward of this the shelf slopes gently from shore to depths of one to several hundred metres, forming the sublittoral or shelf zone. Waters below low tide mark in this region are referred to as *neritic*.

The extent, gradient and superficial geology of continental shelf areas are determined by many factors, including levels of tectonic activity in the Earth's crust. A small number of mainly tropical rivers dominate transport of sediment from land to sea. More than 80% of the global volume of river-borne sediment is deposited in the tropics (and an estimated 40% of it by just two river systems: the Huang He or Yellow River and the Ganges-Brahmaputra)[1], and this is reflected in the extent of shelf areas in parts of the tropics (especially the east Indian Ocean and west Pacific), and in the high turbidity of coastal waters in monsoon regions. Most shelf areas in the tropics are overlain by sands or muds composed of sediment of terrestrial origin (terrigenous deposits).

With the present configuration of landmasses, a major part (37%) of the world ocean is within the tropics, and about 75% lies between the 45° latitudes. The largest continental shelf areas are in high northern latitudes (Table 6.2), but about 30% of the total shelf area is in the tropics. Within the tropics, the shelf is most extensive in the western Pacific (China Seas south to north Australia).

	Open ocean	Continental shelf
Total area (million km²)	360.3	26.7
Latitude bands (% of total)		
Polar and boreal (45-90°)	26.6	40.9
Temperate (20-45°)	36.8	28.8
Tropical (0-20°)	36.6	30.3

The deep-sea bottom

At the outer edge of the shelf there is an abrupt steepening of the sea bottom, forming the continental slope which descends to depths of 3-5 km. The sea-bottom along the slope is referred to as the bathyal zone. At the base of the continental slope are huge *abyssal* plains which form the floor of much of the world's oceans. The plains are punctuated by numerous submarine ridges and sea-mounts which may break the surface to form islands. There are also a number of narrow trenches which have depths of from 7000 to 11 000 m. These constitute the *hadal* zone. All marine areas beyond the continental shelf are referred to as *oceanic*.

Ocean trenches

Ocean trenches are formed as a consequence of plate tectonic processes where sectors of expanding ocean floor are compressed against an unyielding continental mass or island arc, resulting in the crust buckling downwards (subducting) and being destroyed within the hot interior of the Earth. As oceanic crust ages and cools, it becomes denser and stiffer, resulting in a steeper angle of subduction and a deepening trench. Trenches along the western edge of the Pacific are deepest and oldest. Seismically, ocean trenches are highly active, as subduction is an episodic rather than continuous process.

The oceanic pelagic zone

The term pelagic refers to open water, away from the sea bed; the pelagic zone includes continental shelf waters (neritic; see above) and the remainder, the oceanic pelagic zone. Given that the oceans cover some 71% of the globe, and that the shelf area is relatively narrow, the oceanic pelagic zone is by far the most extensive ecosystem on Earth.

THE BASIS OF LIFE IN THE SEAS

In the sea, as on the land, photosynthesis is the driving force behind maintenance of life. Because photosynthesis in nature depends on sunlight, with few exceptions primary productivity is confined to those parts of the ocean which are sunlit. Water absorbs sunlight strongly so that light intensity decreases rapidly with increasing depth. Red wavelengths are most rapidly absorbed, except where turbitidy is high, while blue-green wavelengths penetrate the deepest. Even in the clearest waters the latter are completely absorbed by around 1 km depth, this marking the extreme limit of the so-called *photic* zone. Photosynthesis is thus limited to the continental shelf area and the first few hundred metres of surface waters (and often very much less) of the open ocean which together make up a very small proportion of the total volume of the oceans. The portion of the photic zone where sunlight is strong enough to support appreciable amounts of photosynthesis is the *euphotic* zone.

Virtually all other marine organisms, including those of the unlit middle depths and the deep sea are dependent ultimately on growth of primary producers in areas that may be widely distant from them in time and space. The most important exceptions to this are the bacteria living around hydrothermal vents associated with rift zones in the ocean floor. The water here can be 10°C warmer than adjacent areas and the bacteria are able to grow using as an energy source hydrogen sulphide gas emitted at the vents, and they in turn are used by other organisms.

With some exceptions primary production per unit area tends to be lower in marine environments than terrestrial ones, especially if highly-managed terrestrial agricultural systems are considered. This is because over the vast majority of the ocean the euphotic zone is far distant from the lithosphere. The latter provides essential nutrients for life, and these therefore have to be transported to the euphotic zone to allow life processes to continue. There is also usually a steady loss of nutrients and organic compounds from the euphotic zone, owing to the sinking of particles and bodies into the dark regions of the sea where no photosynthesis can take place. Continued productivity in the open sea is contingent on the replacement of the lost nutrients. The latter may originate either on land, in the form of river outflow, or from marine sediments. Their replacement in the pelagic euphotic zone is dependent on the mixing or vertical movement of the water column. As noted above, at latitudes higher than 40°, winter mixing particularly in continental shelf areas, allow replenishment of the euphotic zone. However, in permanently stratified subtropical and tropical oceanic waters there is little vertical mixing and therefore little influx of new nutrients. Productivity in these areas is correspondingly low. However, in zones of upwelling, surface waters are regularly replaced by nutrient rich bottom waters and very high levels of productivity can be achieved, at least seasonally. Unsurprisingly, several major upwelling zones play a very important rôle in world fisheries production, discussed below.

Until the 1980s it had been believed that photosynthesis in the pelagic ocean was carried out only by single-celled phytoplankton, between 1 and 100 microns in diameter (1 micron = .001 mm), and also that vast expanses of open ocean where phytoplankton could not be detected were, in terms of productivity, the marine equivalent of deserts. New observational techniques have since revealed the presence in great abundance of exceptionally small and previously unknown organisms, collectively termed *picoplankton*. These appear to be predominantly photosynthesising unicellular cyanobacteria 0.6-1 micron in size, such as *Prochlorococcus*. Because of their extra-ordinary abundance (some 100 million cells may be present in one litre), and despite their minute size, these organisms play a crucial role in productivity of open ocean waters[2] and have led to marked upward revisions in estimates of overall marine productivity.

The major rôle played by microscopic organisms in marine productivity has a number of major implications for marine ecology. Although much remains unknown about the structure and dynamics of pelagic food-webs its seems that a high proportion of marine primary production is used directly by microscopic organisms (both autotrophic and heterotrophic) and cycled back into non-living forms (dissolved CO_2 and organic carbon) rather than supporting populations of larger organisms.

Oceanic primary producers divert a high proportion of their energy into reproduction rather than accumulating biomass, in contrast to terrestrial primary producers (plants). Because of this, average standing biomass per unit area in the oceans has been estimated at around one-thousandth that on land. Population turnover of oceanic primary producers is also several orders of magnitude higher than turnover of major terrestrial primary producers (plants).

The small size and rapid turnover of oceanic primary producers means that there are no organisms directly analogous to the terrestrial woody plants that so enrich terrestrial environments by providing structurally complex habitats for other organisms. The nearest equivalents are the large brown algae known as kelp (Phylum Phaeophyta), whose structure is less complex and which are much more narrowly distributed. Structurally complex habitats may in contrast be created by animals, particularly corals (Phylum Cnidaria) and, to a lesser extent, sponges (Phylum Porifera) molluscs and serpulid worms (Phylum Annelida).

BIOLOGICAL DIVERSITY IN THE SEAS

It is well known that diversity at higher taxonomic levels (Phyla and Classes) is much greater in the sea than on land or in freshwater. Of the 82 or so eucaryote phyla currently recognised (see Chapter 2), around 60 have marine representatives compared with around 40 found in freshwater and 40 on land. Amongst animals the preponderance is even higher, with 36 out of 38 phyla having marine representatives (Table 6.3).

Some 23 eucaryote phyla, of which 18 are animal phyla, are confined to marine environments. Most of these are relatively obscure and comprise few species. The major exceptions are the Echinodermata, of which some 6000 species are known, and the Foraminifera, with around 4000 known, extant species. A number of other important phyla including the coelenterates (Cnidaria), sponges (Porifera) and brown and red algae (Phaeophyta and Rhodophyta respectively) are very largely marine, each with only a small number of non-marine (usually freshwater) species.

The reason for this predominance of marine higher taxa (particularly amongst animals) is believed to be because most of the fundamental patterns of organisation and body plan, ie. the different basic kinds of organism that are distinguished as phyla, originated in the sea and remain there, but only a subset of them has spread to the land and into freshwaters. It is noteworthy that only a third or so of marine phyla are found in the pelagic realm, the remainder being confined to sea bottom (benthic) areas – the habitat where eucaryotic organisms are believed to have evolved.

In contrast, known species diversity in the sea is much lower than on land – some 250 000 species of marine organisms are currently known, compared with more than 1.5 million terrestrial ones. Much of this difference is because of the very large number of described terrestrial arthropods, for which there is no marine equivalent. Amongst fishes, almost as many freshwater species as marine are known, despite the fact that freshwater habitats account for only around one ten thousandth of the volume of marine ones. Similarly, the most diverse known marine habitats (coral reefs) are far less

Table 6.3

Marine diversity by phylum

Note: strictly marine groups shown in bold

Kingdom	Phyla		Estimated no. of described marine species
Bacteria	17 phyla		4800
Protoctista	27 phyla		
	Chlorophyta	Green algae	7000
	Phaeophyta	Brown algae	1500
	Rhodophyta	Red algae	4000
	others incl.		23 000
	Foraminifera		
Animalia			
	Placozoa		**1**
	Porifera	Sponges	10 000
	Cnidaria	Coelenterates	10 000
	Ctenophora	**Comb jellies**	**90**
	Platyhelminthes	Flatworms	15 000
	Nemertina	Nemertines	750
	Gnathostomulida	**Gnathostomulids**	**80**
	Rhombozoa	**Rhombozoans**	**65**
	Orthonectida	**Orthonectids**	**20**
	Gastrotricha	Gastrotrichs	400
	Rotifera	Rotifers	50
	Kinorhyncha	**Kinorhynchs**	**100**
	Loricifera	**Loriciferans**	**10**
	Acanthocephala	Spiny-headed worms	600
	Entoprocta	Entoprocts	170
	Nematoda	Nematodes	12 000

Kingdom	Phyla		Estimated no. of described marine species
	Nematomorpha	Horsehair worms	<240
	Ectoprocta	Ectoprocts	5000
	Phoronida	**Phoronid worms**	**16**
	Brachiopoda	**Lamp shells**	**350**
	Mollusca	Molluscs	?75 000
	Priapulida	**Priapulids**	**8**
	Sipuncula	**Sipunculans**	**150**
	Echiura	**Echiurids**	**140**
	Annelida	Annelid worms	12 000
	Tardigrada	Water bears	few
	Chelicerata	Chelicerates	1000
	Mandibulata	Mandibulate arthropods	few
	Crustacea	Crustaceans	38,000
	Pogonophora	**Beard worms**	**120**
	Bryozoa	Bryozoans	4000
	Echinodermata	**Echinoderms**	**7000**
	Chaetognatha	**Arrow worms**	**70**
	Hemichordata	**Hemichordates**	**100**
	Urochordata	**Tunicates and ascidians**	**2000**
	Cephalochordata	**Lancelets**	**23**
	Craniata	Craniates	15 000
Fungi	3 phyla		500
Plantae	Anthophyta		50
Total			c 250 000

diverse in terms of species number than the moist tropical forests that are often taken as their terrestrial counterparts.

The apparent lower total species diversity of the marine biosphere is likely in part at least to be a result of the physical characteristics of water, particularly its very high heat capacity and its miscibility (ability to mix). Because of these, marine environments (particularly deep water ones) tend to show much less variation in time and space in their physical characteristics than terrestrial ones. This lack of physical variation seems to result in a similar lack of ecological variation over wide areas.

However, part of the difference may also be because marine environments are far less well known overall than terrestrial ones. Recent work concerned particularly with benthic faunas (outlined below) and with very small planktonic organisms (picoplankton, see above), has revealed unsuspected levels of diversity. However it seems increasingly unlikely that these new discoveries will be sufficient to overturn the dominance of terrestrial diversity at species level.

In contrast to terrestrial faunas, where one phylum – the Mandibulata – vastly outnumbers all others in terms of known species, marine species are much more evenly distributed across higher taxa. The largest marine phyla – Mollusca and Crustacea – each comprise far fewer than 100 000 known marine species, in contrast with the mandibulata, of which around one million terrestrial species have been identified to date. The only major eucaryote phyla (ie. those with 10 000 or more described species) that are believed to have comparable levels of diversity on land and in the sea are the Platyhelminthes, Nematoda, Mollusca and Craniata.

VERTEBRATE DIVERSITY IN THE SEAS

As on land, vertebrates are by far the best known group of marine organisms. Of the 50 000 or so described extant species, around 15 000 may be considered marine (Table 6.4), the overwhelming majority of which are fishes (Table 6.5) and few are tetrapods (Table 6.6).

Table 6.4
Diversity of craniates in the sea by class

Class		Number of marine species
Cyclostomata	Lampreys and hagfishes	52
Chondrichthyes	Sharks, skates and rays	821
Osteichthyes	Bony fishes	13 800
Reptilia	Reptiles	60
Aves	Birds	307
Mammalia	Mammals	110

Marine fishes

Fishes are considered a paraphyletic group (see Chapter 2), that is, all living species are thought to share a common ancestor, but are thought to share this ancestor with a major group – the tetrapods – that are not categorised as fishes. Apart from some 50 or so species

of generally parasitic lampreys and hagfishes in the superclass Agnatha, fishes are divided into two unequal-sized groups, the Chondrichthyes or cartilaginous fishes and the Osteichthyes or bony fishes. Some 60% of all known living fish species (ie. around 14 000 species) occur in marine habitats. They range in size from an 8mm-long goby *Trimmatom nanus* in the Indian Ocean to the 15m whale shark *Rhincodon typus*, respectively the smallest and largest of all fish species, and occur in virtually all habitats, from shallow inshore waters to the abyssal depths.

The Chondrichthyes comprise the sharks, skates and rays, an overwhelmingly marine group with around 850 living species in ten orders. Although far less diverse than the bony fishes, the cartilaginous fishes include many of the largest fish species, a number of which are top predators in marine ecosystems. The Osteichthyes are a remarkably diverse group, with an enormous range of morphological, physiological and behavioural adaptations. Of the 32 orders with marine representatives (Table 6.5), by far the largest and most diversified is the Perciformes. This is the largest of all vertebrate orders and dominates vertebrate life in the ocean, as well as being the dominant fish group in many tropical and subtropical freshwaters.

As with other groups of organisms, the majority of fishes in the sea are strictly marine, occurring only in salt water. A proportion, however, may also occur in inland waters, often passing a particular part of their life-cycle there. Species that spend most of their life in marine waters but ascend rivers to breed, such as many salmonids (family Salmonidae, order Salmoniformes) and sturgeons (family Acipenseridae, order Acipenseriformes), are referred to as *anadromous*. Those, such as most eels in the family Anguillidae (order Anguiliformes), that breed at sea but spend their lives otherwise in freshwater, are referred to as *catadromous*. Species with a wide salinity tolerance that may occur in marine, brackish and fresh waters (eg. some sawfishes, family Pristidae; order Rajiformes) are referred to as *euryhaline* while those with narrow tolerances, be they to marine, brackish or fresh water, are referred to as *stenohaline*.

Order	Common fishes	No. of families	No. of genera	No. of species occurring in marine habitats	Percentage of family in marine habitats
Myxiniformes	**Hagfishes**	I	6	43	100
Petromyzontiformes	Lampreys	I	6	9	22
Chimaeriformes	**Chimaeras**	3	6	31	100
Heterodontiformes	**Bullhead sharks and horn sharks**	I	1	8	100
Orectolobiformes	**Carpet sharks**	7	14	31	100
Carcharhiniformes	**Ground sharks**	7	47	207	100
Lamniformes	**Mackerel sharks**	7	10	16	100
Hexanchiformes	**Cow sharks**	2	4	5	100
Squaliformes	**Dogfishes and sleeper sharks**	4	23	74	100
Squatiniformes	**Angel sharks**	I	1	12	100
Pristiophoriformes	**Saw sharks**	I	2	5	100
Rajiformes	Rays	12	62	432	95
Coelacanthiformes	**Coelacanths**	I	1	2	100
Acipenseriformes	Sturgeons	2	6	12	46
Salmoniformes	Salmonids	I	11	21	32
Stomiiformes	**Lightfishes, hatchetfishes, barbeled dragonfishes**	4	51	321	100
Ateleopodiformes	**Jellynose fishes**	I	4	12	100
Aulopiformes	**Greeneyes, pearleyes, waryfishes, lizardfishes, barracudinas, lancetfishes**	3 I	42	219	100
Myctophiformes	**Lanternfishes**	2	35	241	100
Lampridiformes	**Oarfishes, ribbonfishes, crestfishes, opahs**	7	12	19	100
Polymixiiformes	**Beardfishes**	I	1	5	100
Ophidiiformes	Pearlfishes, cusk-eels, brotulas	5	92	350	99
Gadiformes	**Cods, hakes, rattails**	12	85	481	100
Batrachoidiformes	Toadfishes	I	19	64	93
Lophiiformes	**Anglerfishes, goosefishes, frogfishes, batfishes, seadevils**	16	65	297	100
Mugiliformes	Mullets	I	17	65	98
Atheriniformes	Silversides	8	47	139	49
Beloniformes	Needlefishes, sauries, flyingfishes, halfbeaks	5	38	140	73
Cyprinodontiformes	Rivulines, killifishes, pupfishes, four-eyed fishes, poeciliids, goodeids	8	88	13	2
Stephanoberyciformes	**Gibberfishes, pricklefishes, whalefishes, hairyfish, tapetails**	9	28	86	100

Order	Common fishes	No. of families	No. of genera	No. of species occurring in marine habitats	Percentage of family in marine habitats
Beryciformes	**Fangtooths, spinyfins, lanterneye fishes, roughies, pinecone fishes, squirrelfishes**	**7**	**28**	**123**	**100**
Zeiformes	**Dories, boarfishes, oreos, parazen**	**6**	**20**	**39**	**100**
Gasterosteiformes	Pipefishes, seahorses, sticklebacks, sandeels, seamoths, snipefishes, shrimpfishes, trumpetfishes	11	71	238	93
Synbranchiformes	Swamp-eels	3	12	3	3
Scorpaeniformes	Gurnards, scorpionfishes, velvetfishes, flatheads, sablefishes, greenlings, sculpins, oilfishes, poachers, snailfishes, lumpfishes	25	266	1219	96
Perciformes	Perches, basses, sunfishes, whitings, remoras, jacks, dolphinfishes, snappers, grunts, damselfishes, dragonfishes, wrasses, butterflyfishes etc	148	1496	7371	79
Pleuronectiformes	Plaice, flounders, soles	11	123	564	99
Tetraodontiformes	Triggerfishes, puffers, boxfishes, filefishes, molas	9	100	327	96
Total				c 14 000	

Reptiles

Present-day diversity of reptiles in the seas is low. One important reason for this appears to be that modern reptilian kidneys cannot tolerate high salinities and thus the only reptiles that have adapted to marine environments are those that have developed specialised salt-excreting glands. The most thoroughly marine reptiles are undoubtedly the sea snakes in the subfamily Hydrophiinae (family Elapidae). These spend their entirely lives in the sea, giving birth to live young there. Although largely air-breathing like other reptiles, they can also absorb some oxygen directly from seawater and are thus able to remain submerged for long periods. Around 50 species are known, widely distributed in tropical parts of the Indo-Pacific region. In addition the little file snake *Acrochordus granulatus* (family Acrochordidae), from northern Australia and south-east Asia is also entirely aquatic, but occurs in brackish estuaries as well as seawater.

Five species of sea krait in the subfamily Laticaudinae are also largely marine, feeding mainly on eels. However they return to land to breed, generally on small tropical islands. They too are confined to the Indo-Pacific region. One species of lizard, the Galápagos marine iguana *Amblyrhynchus cristatus* (family Iguanidae), feeds underwater on marine algae but spends a considerable proportion of time on land. Several other reptile

Table 6.6

Marine tetrapod diversity

Note: Birds follow the list of seabirds recognised in Croxall et al.[71] with the additional inclusion of four eider ducks and three steamer ducks in the family Anatidae and the Grey Phalarope *Phalaropus fulicaria* (Scolopacidae). Taxonomy follows Sibley and Monroe (see Chapter 2 references). Figures in parentheses indicate species that breed largely or entirely inland. Strictly marine orders shown in bold.

Class	Order	Family		Number of Marine Species
Reptilia	**Chelonia**	**Dermochelyidae**	**Leathery turtle**	1
		Cheloniidae	**Sea turtles**	6
	Squamata	Elapidae	Sea snakes and sea kraits	55
		Acrochordidae	File snakes	1
		Iguanidae	Iguanas	1
Aves	Anseriformes	Anatidae		7
	Ciconiiformes	Scolopacidae		1
		Laridae	Gulls, terns, skuas, auks, skimmers	120 (13)
		Phaethontidae	**Tropicbirds**	3
		Sulidae	**Gannets and boobies**	9
		Phalacrocoracidae	Cormorants and shags	36 (2)
		Pelecanidae	Pelicans	2
		Fregatidae	**Frigatebirds**	5
		Spheniscidae	**Penguins**	17
		Procellariidae	**Petrels, albatrosses, shearwaters**	115
Mammalia	Cetacea	**Balaenidae**	**Right whales**	3
		Balaenopteridae	**Rorquals**	6
		Eschrichtiidae	**Grey whale**	1
		Neobalaenidae	**Pygmy right whale**	1
		Delphinidae	**Dolphins**	32
		Monodontidae	**Beluga and narwhal**	2
		Phocoenidae	**Porpoises**	6
		Physereidae	**Sperm whales**	2
		Platanistidae	River dolphins	1
		Ziphiidae	**Beaked whales**	19
	Sirenia	Trichechidae	Manatees	1
		Dugongidae	**Dugong**	1
	Carnivora	Mustelidae	Otters and weasels	2
		Odobenidae	**Walrus**	1
		Otariidae	**Eared seals**	14
		Phocidae	Earless seals	17

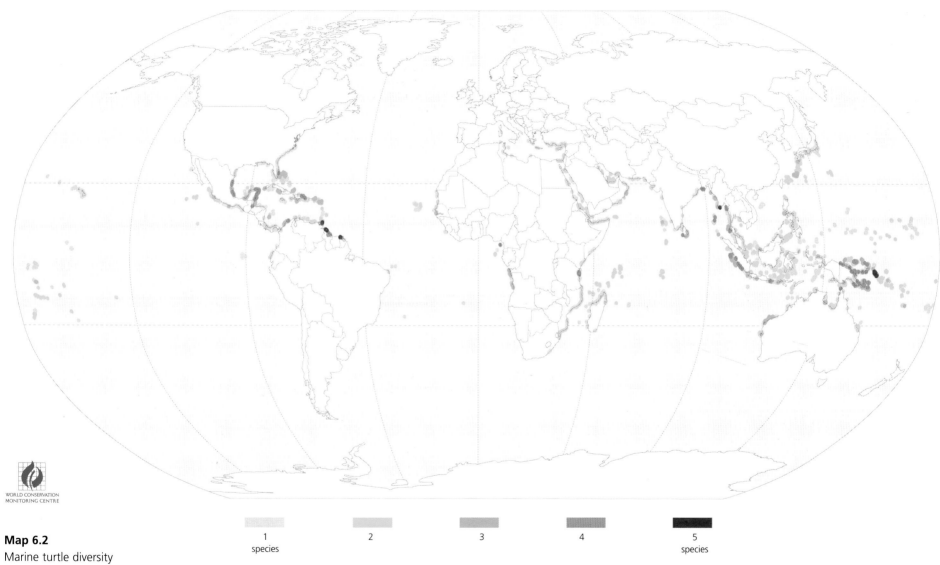

1	2	3	4	5
species				species

Map 6.2

Marine turtle diversity

The map shows marine turtle diversity, represented by the number of marine turtle species nesting in any one area. Each symbol shows the location of a nesting site or area, coloured according to the number of species present, up to a   maximum of five species in some parts of the tropics.

Source: analysis by WCMC using spatial data from GIS database of marine turtle nesting beaches.

species regularly enter sea-water, most notably a number of homalopsine mangrove snakes (family Colubridae) from the Indo-Pacific and the estuarine crocodile *Crocodylus porosus* (family Crocodylidae) from the same region.

Undoubtedly the most prominent group of marine reptiles is the sea turtles, comprising the leathery turtle *Dermochelys coriacea* in the family Dermochelyidae and six members of the family Cheloniidae. All species are large (ranging from 70 cm adult carapace length in *Lepidochelys kempii* to, exceptionally, 250 cm in *Dermochelys coriacea*) and most are widely distributed in tropical and subtropical waters (Map 6.2). Sea turtles are almost completely marine, only the females emerge to nest on land, mostly within the tropics. One species, the loggerhead *Caretta caretta*, nests largely in temperate areas of the northern hemisphere. Sea turtles typically have a long period to maturity (often up to 25 years in the case of the green turtle *Chelonia mydas*) and a long life span. Females often only nest every two or three years. They habitually return to the same nesting beaches, sometimes undergoing protracted migrations from feeding grounds. They may lay two or three clutches in a season, sometimes comprising over 100 eggs each, depending on the species. Nest, hatchling and juvenile mortality are often very high.

Mammals

Wholly aquatic mammals (those that never normally emerge onto land) are confined to two orders, the Cetacea and the Sirenia. The Cetacea comprises some 78 species, all except five marine, distributed throughout the world's seas. They include the largest living animals – the rorquals in the family Balaenopteridae. All cetaceans are carnivorous; the baleen or whalebone whales (families Balaenidae, Balaenopteridae, Eschrichtiidae and Neobalaenidae) are filter feeders, feeding on organisms several orders of magnitude smaller than they are.

Of four living members of the order Sirenia, only one – the dugon *Dugong dugon* – is exclusively marine, occurring widely in coastal waters of the Indo-Pacific. One other, the Caribbean manatee *Trichechus manatus*, is found in both marine and inland waters while the other two (the Amazonian manatee *Trichechus inunguis* and West African manatee *T. senegalensis*) enter coastal waters marginally if at all. One other species, the very large Steller's sea cow *Hydrodamalis gigas*, survived in waters around Bering and Copper Islands in the North Pacific until the early 18th century. All Sirenians are herbivores; marine populations feed mainly on sea-grasses.

The remaining marine mammals are all included in the order Carnivora. Two New World otters in the family Mustelidae, the sea otter *Enhydra lutris* from the north temperate Pacific coast and the marine otter *Lutra felina* from the south temperate Pacific coast, feed very largely or exclusively in marine waters; other otter species may frequent coastal areas but are predominantly inland water animals. Members of the three pinniped families Odobenidae (the walrus), Otariidae (eared seals) and Phocidae (earless seals) are all very largely aquatic, emerging on land to breed and rest, particularly when moulting; all are marine with the exception of one or two species of Phocidae (the Baikal seal *Phoca sibirica* and, if the Caspian is regarded as a lake rather than a sea, the Caspian seal *Phoca caspica*). One member of the family Phocidae, the Caribbean monk seal *Monachus tropicalis*, has become extinct this century. All species are carnivorous.

In contrast to most terrestrial mammal families, pinnipeds are considerably more diverse and more abundant at higher rather than lower latitudes. Of the 32 extant or recently extant species, only five occur within the tropics (two marginally). Part of the explanation for this undoubtedly lies in the greater availability of suitable habitat at higher latitudes: as noted above 70% of continental shelf waters and just over 60% of the world's marine area are found outside the tropics. However, this in itself is unlikely to account for the entire difference. It is likely that the greater productivity of shelf waters at high latitudes, discussed above, and of upwelling areas at mid-latitudes (e.g. the

Benguela current off the western coast of South Africa and the Humboldt current off Chile and Peru) plays a major part. The isolated character of many island breeding sites in temperate and sub-polar parts of the southern hemisphere may also have encouraged speciation of pinnipeds here (see below).

Birds

Defining marine birds, or seabirds, is somewhat more problematic than defining marine species in other groups. All birds breed in terrestrial habitats, but a large number (almost all of them non-passerines) obtain all or much of their food from aquatic or littoral habitats. Some of these, including all frigatebirds (Fregatidae), tropicbirds (Phaethontidae), gannets and boobies (Sulidae), penguins (Spheniscidae) and petrels, albatrosses and shearwaters (Procellariidae) are indisputably marine, in that they obtain all their food from marine habitats, almost invariably breeding along coastlines and spending most or all of their time when not breeding out at sea. Many others, however, have less clear-cut habits. Some, such as a number of cormorants and shags (Phalacrocoracidae) have both resident inland and marine populations. Others, such as a number of gulls and terns (Laridae) and some ducks and geese (Anatidae) may breed inland but spend the rest of the year living in coastal areas or out at sea. Yet others, such as sandpipers (Scolopacidae) and other waders, typically feed in littoral or intertidal habitats rather than in the sea itself; many of these species also occur inland.

Adopting a somewhat arbitrary division, and excluding all wading birds with the exception of the grey phalarope *Phalaropus fulicaria* (a truly pelagic species outside the breeding season), something over 300 species of birds can be considered wholly or largely marine.

In common with pinnipeds, seabirds show a latitudinal distribution in which diversity is much higher at higher latitudes (temperate and polar regions) than it is in the tropics. Two-thirds of all seabirds are confined as breeding species to these latitudes, compared with only 7% that are exclusively tropical. This stands in sharp contrast to the pattern found in most major terrestrial groups (see Chapter 5) and many marine groups such as sea turtles, mangroves and reef-building corals (see, respectively, Maps 6.2, 6.3, 6.4) in which species diversity increases dramatically with decreasing latitude. Diversity is also markedly higher in the southern than in the northern hemisphere, with over half of all seabird species breeding in southern temperate and polar latitudes. Dominance of this region is even more marked in the Procellariidae, the family with the greatest number of truly marine species, in which over 60% of species breed at these latitudes and half are confined to it (Table 6.7).

MAJOR MARINE COMMUNITIES

There is a fundamental distinction between the processes and patterns observed in open oceans, dominated by global winds and large-scale vertical and horizontal movement of water masses, and those observed nearer to coasts, where shelf bathymetry, coastal winds and local input of nutrients, pollutants and sediments generate a diversity of smaller-scale phenomena.

Table 6.7
Regional distribution of breeding in seabirds

Note: several species breed in more than one latitudinal band so that overall totals exceed actual number of seabirds

Breeding distribution	Ocean area (million km²)	Procellariidae	Laridae	Others	Total
Northern temperate and polar	79.5	24 (17)	50 (35)	50 (36)	124 (88)
Northern tropical	75.2	15 (5)	27 (0)	19 (0)	61 (5)
Southern tropical	78.1	25 (8)	30 (4)	26 (5)	81 (17)
Southern temperate and polar	130.1	70 (58)	34 (15)	56 (41)	160 (114)

SHALLOW WATER COMMUNITIES

Mangroves

Table 6.8

Diversity of mangroves

Note: Where two figures are given, second figure indicates no. of hybrids. *Families composed solely of mangrove species

Source: based on Duke[3] and Spalding et al.[7]

Order	Family	Species
Filicopsida	Adiantaceae	3
Plumbaginales	Plumbaginaceae	2
Theales	Pelliciceraceae*	1
Malvales	Bombacaceae	2
	Sterculiaceae	3
Ebenales	Ebenaceae	1
Primulales	Myrsinaceae	2
Fabales	Leguminosae	2
Myrtales	Combretaceae	4+1
	Lythraceae	1
	Myrtaceae	1
	Sonneratiaceae	6+3
Rhizophorales	Rhizophoraceae	17+2
Euphorbiales	Euphorbiaceae	2
Sapindales	Meliaceae	2+1
Lamiales	Avicenniaceae*	8
Scrophuliariales	Acanthaceae	2
	Bignoniaceae	1
Rubiales	Rubiaceae	1
Arecales	Palmae	1

Mangroves or mangals are truly hybrid terrestrial/marine ecosystems. They are a diverse collection of shrubs and trees (including ferns and palms) which live in or adjacent to the intertidal zone and are thus unusual amongst vascular plants in that they are adapted to having their roots at least periodically submerged in sea water. Mangrove species are generally divided into those found only in mangrove habitats and those that may also be found elsewhere but which are nevertheless an important component of mangrove habitats. Both groups come from a wide range of families. The former includes around 62 species and seven hybrids in some 22 genera (Table 6.8). The appearance of mangroves is far from uniform: they vary from closed forests 40-50 m high to widely separated clumps of stunted shrubs less than 1 m high[4].

Mangrove communities are largely restricted to the tropics between 30°N and 30°S, with extensions beyond this to the north in Bermuda (32°20'N) and Japan (31°22'N) and to the south in Australia (38°45'S) and New Zealand (38°03'S)[5] (Map 6.3). They are only able to grow on shores that are sheltered from wave action. Mangrove forests are particularly well developed in estuarine and deltaic areas. They may also extend some distance upstream along the banks of rivers, for example as far as 300 km along the Fly River in Papua New Guinea.

Mangroves occur over a larger geographical area than coral reefs (see below) and, unlike reefs, are well developed along the western coasts of the Americas and Africa. They have a more restricted distribution than coral reefs in the South Pacific. There are two main centres of diversity, termed the eastern and western groups. The eastern group occurs in the Indo-Pacific (the Indian Ocean and western part of the Pacific Ocean) and is the most species-rich[6,7]. The western group is centred around the Caribbean and includes mangrove communities along the west coast of the Americas and Africa.

Mangrove communities are unique: owing to the vertical extent of the trees, true terrestrial organisms can occupy the upper levels and true marine animals can occupy the bases[8]. A wide variety of organisms is associated with them including a number of epiphytic, parasitic and climbing plants, and large numbers of crustaceans, molluscs, fishes and birds[9].

Mangroves stabilise shorelines and decrease coastal erosion by reducing the energy of waves and currents and by holding the bottom sediment in place with their roots. They also act as windbreaks and provide protection from coastal storms. The are generally highly productive ecosystems and are important

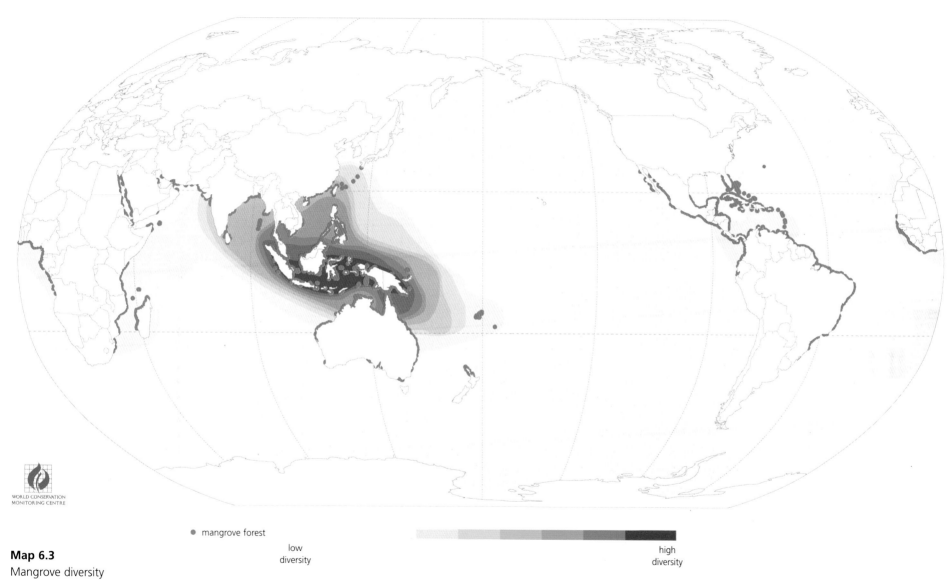

● mangrove forest

low
diversity

high
diversity

Map 6.3

Mangrove diversity

This map shows the location of existing mangrove forest, together with contours representing gradients of mangrove species richness. Note that graphic presentation at this scale enormously exaggerates actual forest area.

Source: reproduced by permission, with modification, from Spalding[76].

sources of crustaceans, shellfish and finfishes. Most of the larger commercial penaeid shrimps are mangrove-dependent; these and other species are harvested both on a subsistence basis and commercially and may provide a major source of income in some countries. As well as providing habitat for adults of many species of finfish and invertebrates, mangroves serve as spawning and nursery areas for many others, often of major economic importance.

Direct uses of mangroves are manifold. The wood provides building material, used locally in houses, as fence-poles and to build fish-traps and is also harvested on a large scale for production of pulp and particle board. In many areas mangroves are also an important source of fuel, both firewood and charcoal. The most important species for this purpose are those belonging to the genus *Rhizophora*, which has heavy, clean-burning wood[10]. Mangrove foliage may provide an important source of fodder for domestic livestock in some countries, particularly during dry seasons when other sources of greenery are in short supply.

Current mangrove cover

Table 6.9
Current mangrove cover
Source: Spalding *et al.*[7]

Region	Mangrove area (to nearest 1000 km²) and percent total
South and Southeast Asia	75 000 (42%)
Australasia	19 000 (10%)
The Americas	49 000 (27%)
West Africa	28 000 (16%)
East Africa and the Middle East	10 000 (6%)

The most recent (1996) best estimate for mangrove area is just over 180 000 km², divided regionally as shown in Table 6.9. Mangroves occur in over 100 countries (including dependent territories) but many of these only have very small areas. Four countries (Indonesia, Brazil, Australia and Nigeria) between them account for over 40% of the world's mangrove area, and Indonesia alone possesses nearly one quarter of the global mangrove area[7].

Although it is known that mangroves in most parts of the world have been extensively degraded and cleared, it is difficult to obtain reliable data on the global extent of mangrove loss over time. Mangroves by their very nature occupy highly dynamic and unstable environments so that even without human action the location and extent of mangrove cover would be constantly changing. Nevertheless it is possible to map areas were mangroves might be expected to occur under natural conditions. A global potential forest map (see Chapter 7) indicates that mangrove area might exceed 400 000 km² if all suitable areas were forested. If this figure is accepted, then current cover would represent around 45% of hypothetical original mangrove cover. This accords well with an independent assessment[11] that over 50% of the world's mangrove forest cover had been destroyed.

CORALS AND CORAL REEFS

Corals along with sea anemones comprise the class Anthozoa of the phylum Cnidaria. Corals may be categorised as hermatypic (reef-building) or ahermatypic (non-reef-building). The great majority of hermatypic corals belong to the order Scleractinia, the stony corals. They collectively deposit calcium carbonate to build colonies. The coral polyps have symbiotic algae (zooxanthellae) within their tissues which process the polyps' waste products. The zooxanthellae use the nitrates, phosphates and carbon dioxide produced by the coral, and through photosynthesis generate oxygen and organic compounds that provides much of the polyps' nutrition. The zooxanthellae give corals their colour and, because they photosynthesise, restrict the corals that contain them to the photic zone[12]. Ahermatypic, non-symbiotic corals typically do not form reefs and can exist in deeper colder waters.

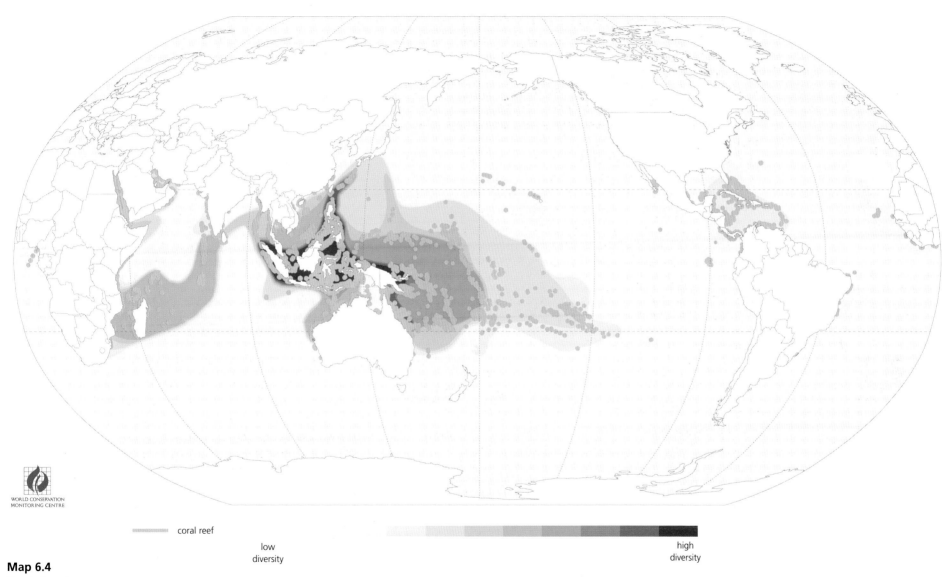

coral reef

low
diversity

high
diversity

Map 6.4

Coral diversity

This map shows the location of coral reefs, together with contours representing gradients of species richness among reef-building scleractinian coral species. Note that graphic presentation at this scale enormously exaggerates actual reef area.

Source: reproduced by permission, with modification from Veron[75].

WORLD CONSERVATION
MONITORING CENTRE

Table 6.10

Diversity of stony corals in the order Scleractinia

Family	Genera	Species
Acroporidae	4	228-300
Agaricidae	6	45-60
Anthemiphylliidae	2	5
Astrocoenidae	2	2
Caryophylliidae	71	296-336
Dendrophylliidae	21	127-173
Faviidae	27	120-153
Flabellidae	11	102
Fungiacyathidae	1	18
Fungiidae	11	43-75
Guyniidae	7	8-10
Meandriniidae	4	6-8
Merulinidae	5	13-14
Micrabaciidae	4	13
Mussidae	13	30-52
Oculinidae	11	31-46
Pocilloporidae	5	34-68
Poritidae	4	111-127
Rhizangiidae	7	42-50
Siderastreidae	7	30-39
Trachyphylliidae	1	1
Total	**224**	**1320-1678**

The term *coral reef* applies to a variety of calcium carbonate structures developed by stony corals. Coral reefs are tropical shallow water ecosystems of high biodiversity largely restricted to the seas between the latitudes of 30°N and 30°S[13]. They are most abundant in shallow, well flushed marine environments characterised by clear, warm, low-nutrient waters that are of oceanic salinity[13]. There are two basic categories: *shelf reefs*, which form on the continental shelves of large land masses, and *oceanic reefs*, which are surrounded by deeper waters and are often associated with oceanic islands. Within these two categories there are a number of reef types: *fringing reefs*, which grow close to shore; *barrier reefs* which develop along the edge of a continental shelf or through land subsidence in deeper waters and are separated from the mainland or island by a relatively deep wide lagoon; and atolls, which are roughly circular reefs around a central lagoon and are typically found in oceanic waters, probably originating from the fringing reefs of long submerged islands. Two other less clearly defined categories are *patch reefs* which form on irregularities on shallow parts of the seabed and *bank reefs*, which occur in deeper waters, both on continental shelf and in oceanic waters[9].

The diversity of stony corals

It is difficult to estimate precisely how many species of reef-building coral there are. The majority, but by no means all, such species belong to the order Scleractinia. However, not all species in the order are reef-building corals, some being found in areas where true coral reefs do not develop. Moreover, the taxonomy of much of the order is outdated and many genera have not been recently reviewed. In many families far more species have been named than are believed valid – for example of some 211 nominal species of *Montipora* (family Acroporidae), only between 60 and 80 are considered likely to be valid. Table 6.10 represents the best currently available estimates for the 21 recognized families of scleractinian coral[14]. This gives a general indication of the number of reef-building coral species and genera.

Not all reefs are constructed primarily by corals. Several genera of red algae (Phylum Rhodophyta) and green algae (Phylum Chlorophyta) in particular grow heavily calcified encrustations which bind the reef framework and in places are the main contributors to shallow reef growth.

Coral reefs are among the most productive and diverse of all natural ecosystems. As with mangroves, the centre of diversity for reef-building corals is Southeast Asia, with an estimated minimum of 450 species found associated with reefs around the Philippines, Borneo, Sulawesi and associated islands. This area is part of a single, vast, Indo-West Pacific biogeographic province that extends from the Red Sea in the west to the Pitcairn Islands in the east. Many coral genera and a significant number of species are found throughout the region although overall diversity in the province decreases as one travels away from this centre. In the east of this region, the central and eastern Pacific forms a series of somewhat distinct sub-regions, characterised by a number of genera and species (particularly in Hawaii) not found further west. This area also shares many species with the Indo-West Pacific province but overall has much lower diversity than most of the latter.

The Atlantic, including the Caribbean and the Gulf of Mexico, forms a very distinct province with few species in common with the Indo-West Pacific. It is also very depauperate compared with most of the latter (Map 6.4).

This pattern is paralleled by the distribution of coral reef fish species. As is generally the case in high diversity environments, most reef fish species are relatively rare in terms of individuals in the community. Thus, at Toliara (southwest Madagascar) only about 25% (136) of the total number of fish species present were ranked as abundant[15]. Many families of coral reef fishes have a circum-tropical distribution, although there are pronounce differences at species level; the number of reef fish species within a single zoogeographic region varies between hundreds and thousands. Most families in tropical seas include species that occur in the coral reef fauna, and some families are particularly associated with reefs, such as Chaetodontidae and Scaridae.

Within the demersal component (feeding on benthic organisms), the families Acanthuridae, Balistidae, Belennidae, Holocentridae, Chaetodontidae, Gobiidae, Ostraciodontidae, Pomacentridae and Serranidae tend to dominate. Principal pelagic families associated with reefs, other than the top predators such as Carangidae, *Sphyraena* and sharks, include Atherinidae, Pomacentridae and the Caesioninae (a sub-family of Lutjanidae)[1]. Small-sized species tend to predominate, although the range is from 1 cm or less for some *Trimmatom* species to over 5 m for some sharks.

There is a strong positive correlation between coral and fish species richness at given sites, although this is less evident on a small scale within reef zones. There may also be a positive correlation between the degree of live coral cover and species richness and abundance of reef fishes[16]. In addition, the presence of dietary specialist fish species is often related to specific coral growth forms; for example, the exclusive coral feeders in the Chaetodontidae are positively correlated with the abundance of tall-branched coral colonies[17].

Current distribution and status of coral reefs

The global extent of coral reefs is not known with certainty, although it probably exceeds 600 000 km². Near-surface reefs are estimated to cover around 255 000 km²; a regional breakdown is provided in Table 6.11. Although coral reefs occur in around 110 countries and territories, just six countries – Australia, Indonesia, Philippines, Papua New Guinea, Fiji and the Maldives – account for over half of the global total.

Table 6.11
Shallow reef area
Source: Spalding and Grenfell[74]

Region	Shallow reef area (to nearest thousand km²)
Middle East	20 000
Caribbean	20 000
Atlantic (excl. Caribbean)	3 000
Indian Ocean	36 000
Southeast Asia	68 000
Pacific	108 000

SEAGRASSES

Seagrasses are flowering plants (not true grasses) that are adapted to live submerged in seawater. There are approximately 48 species found in shallow coastal areas between the Arctic and Antarctic. Seagrasses are placed in two families: Potamogetonaceae (9 genera, 34 species) and Hydrocharitaceae (3 genera, 34 species)[18]. They occur from the littoral region to depths of 50 or 60 m but appear to be most abundant in the immediate sublittoral area. There are more species in the tropics than in the temperate zones, and of the 12 seagrass genera, 7 are confined to tropical seas and 5 to temperate seas[18]. Most seagrass species are very similar in external morphology, with long thin leaves and an extensive rhizome root system which enables them to fasten to the substrate. A variety of substrates are occupied from sand and mud to granite rock, but the most extensive beds occur on soft substrates[8].

Seagrass beds have high productivity and contribute significantly to the total primary production of inshore waters. They are a significant source of food for many organisms both by direct grazing and detritus feeding, including invertebrates, fishes, birds, the green turtle *Chelonia mydas*, the West Indian manatee *Trichechus manatus* and the dugong *Dugong dugon*. The beds also serve as nursery grounds for many commercial species such as the bay scallop *Aquipecten irradians* and shelter the inhabitants from predators and adverse environmental conditions. They serve to protect coastlines from the erosive force of wave action.

ALGAE AND KELP FORESTS

Algae lack vascular tissue (the transport system for water and nutrients) found in higher plants. They are almost exclusively aquatic; three of the four principal groups comprised of large-sized species are mainly marine in occurrence. These three, the green, brown and red algae ('seaweeds'), are all cosmopolitan in distribution and occur in a range of environments, although some constituent families have somewhat restricted ranges. There are more marine species of red algae (Rhodophyta) – around 4000 – than the greens (Chlorophyta) (ca 1000) and browns (Phaeophyta) (ca 1500) combined. As with pinnipeds and seabirds, the cold and cool temperate regions of the world appear to be surprisingly rich in species. On present incomplete information, the region around Japan (northwest Pacific), the North Atlantic, and the tropical and subtropical western Atlantic, hold the most species of marine algae. Southern Australia is not so species-rich but appears to have the highest proportion of endemics. There are few species of larger algae in regions of cold water upwelling; small isolated islands and polar regions also have few species. In contrast, coral reefs support a unique and generally diverse algal flora that includes many crustose coralline algae (more species of which are likely to be discovered). Mangrove areas also support a well-defined algal vegetation, contrasting with that of saltmarshes in the temperate zones, which are generally more species-poor. Sandy coastlines hold few species of large algae and often form barriers to seaweed dispersal.

Throughout a large part of the cold temperate regions of the world, hard subtidal substrates are occupied by very large brown algae collectively known as kelps (order Laminariales). These associations are known technically as kelp beds if the algae do not form a surface canopy[8], and kelp forests where there is a floating surface canopy. They occur primarily in the cold currents of the Atlantic and Pacific oceans and may be found within the tropics, typically in areas of upwelling and cold water surface currents. The extent of the kelp beds and forests on various coasts depends on several factors: kelps flourish with a hard substrate for attachment, moderate wave surge, cool, clear ocean water and high-nutrient waters[19].

Kelp beds and forest are extremely productive and provide the framework for an associated community including many different species of algae, invertebrates, and fishes. Despite the high productivity of the kelp, relatively few herbivores graze directly on them. It has been estimated that only 10% of the net production enters the food web through grazing and the remaining 90% enters the food chain in the form of detritus or dissolved organic matter[8]. The main causes of kelp mortality can be attributed to mechanical forces, mainly wave action, and nutrient depletion, mainly nitrogen. Adults are only occasionally destroyed by grazing herbivores[8].

OCEANIC PELAGIC COMMUNITIES

The oceanic pelagic zone is dominated by the activity of plankton in the euphotic surface waters. Plankton are by definition drifting or weakly-swimming organisms, comprising a wide range of small to microscopic animals, protoctists and bacteria. Plankton tend to be unevenly distributed – concentrated along major circulation currents (gyres), contact zones and upwellings. Species richness appears to vary with depth and latitude. For example ostracod diversity in the North Atlantic was found to peak at depths of around 1000 m, while planktonic diversity in general in the same region was found to be at a maximum in the tropics and a minimum at high latitudes[20]. At any one locality, planktonic species richness can compare with richness at terrestrial sites, but the very large scale of oceanic ecosystems means that species composition tends to be uniform over large areas, eg. the richest locality in samples from the northeast Atlantic included 81 ostracod species while the entire region has a maximum of around 120 species overall[20].

Free-swimming pelagic organisms, predominantly fishes but also cetaceans and cephalopod molluscs (squid), are collectively termed nekton. These organisms, when adult, are predators of plankton or smaller nekton. They in turn – as vertically-migrating fishes or larvae, and as dead organic material – provide food for deep sea and benthic organisms. With few exceptions, the only other food source for creatures in the aphotic zone is the 'rain' of organic matter, such as faeces moulted crustacean exoskeletons, and a variety of other organic material derived from plankton in the surface waters of the ocean.

The marked vertical gradients within the pelagic zone – of light, temperature, pressure, nutrient availability and salinity – lead to vertical structuring of pelagic species assemblages. Several zones based on changes in species composition with depth have been recognised, including *epipelagic* (usually taken as from the surface to a depth of 200-250 m and including the euphotic zone); *mesopelagic*, which underlies the epipelagic zone to a depth of 1000 m or so; and below this the *bathypelagic* which changes in a somewhat less well-defined fashion to *abyssopelagic* at around 2500-2700 m depth. These zones, however, tend to fluctuate in time and space. As well as seasonal changes in water characteristics, many components of the epipelagic and mesopelagic nekton undergo marked diel migrations (ie. on a 24 hour cycle), ascending to surface waters at night to feed and descending, sometimes over 1 km, during the day. Many species of nekton, particularly cetaceans and larger fishes are also highly migratory, ranging over enormous expanses of ocean in more or less regular and predictable patterns.

It has generally been assumed that biomass in the pelagic zone everywhere below the euphotic zone is low. However, recent studies have indicated that biomass of tropical mesoplegaic animals may be surprisingly high. Study of mesopelagic faunas has been limited to date, as it requires the use of expensive high seas research vessels; knowledge of taxonomy, distribution and biology of most of the species concerned remains very incomplete[1]. A 1980 study[21], recognised around 160 fish genera in 30 families as important components of the fauna. Most species are small <10 cm, and often bizarrely shaped. Estimates based on a variety of surveys carried out indicate that global biomass of this stock may be very large: a figure of 650 million tonnes (some 6-7 times total current marine fisheries landings) has been suggested, although this should be regarded with extreme circumspection[1].

From available data, it appears that the mesopelagic biomass is greatest in the northern Indian Ocean, and particularly in the northern Arabian sea, one of the five major upwelling zones. Surveys here indicated extremely high biomass (25-250 g m^{-2}) in the Gulf of Aden and Gulf of Oman as well as off the western coastline of Pakistan. These figures are around an order of magnitude higher than those recorded elsewhere in the tropics, indicating either great over- estimate for the northern Indian Ocean, or underestimate elsewhere, or that this region genuinely is ten times as productive as the rest of the tropical ocean system. Although this appears as yet unresolved, it is nevertheless apparent that there is substantial global mesopelagic fish biomass and that the Arabian Sea is particularly rich in these species.

DEEP-SEA COMMUNITIES

Approximately 51% of the earth's surface is covered by ocean over 3,000 m in depth. Deep-sea communities are thus prevalent over a major proportion of the planet. All deep-sea habitat is in the aphotic zone, well below the distance sunlight can penetrate. As deeper and deeper levels are reached biomass falls exponentially[22].

Despite their enormous volume, the deep oceans appeared to be relatively simple ecosystems, and to make little contribution to global species diversity, but discoveries during the past decade have shown that in some regions, species diversity in the benthic community increases with increasing depth. This was revealed by novel sampling techniques, principally the epibenthic sled[23]. The rate of discovery of new species and the proportion of species currently known from only one sample both suggest that a great number remain to be discovered[24,25]. As with arthropods in tropical moist forests, estimates for the number of unknown species vary widely, with some suggestions that there may be as many as ten million undescribed species in the deep sea[26]. Others consider that the true figure is more likely to be around 500 000[27].

In the megafauna, echinoderms of several classes are often the dominant mobile life forms on or in association with the sea bottom. Giant scavenging amphipod crustaceans, growing up to about 18 cm in length, are also characteristic in many areas. However, the high mobility of these animals means they are rarely caught in trawls and have been less well studied than less active animals. Other arthropods include a variety of sea spiders (Pycnogonida, Phylum Mandibulata) and decapod crustaceans of several families. Mobile animals of several other taxa occur, including polychaete annelid worms, hemichordates, cephalopod molluscs and fishes. Sponges, especially the glass sponges, and coelenterates, particularly anthozoans, are also well represented.

Ocean trenches

Ocean trenches are formed as a consequence of plate tectonic processes where sectors of expanding ocean floor are compressed against an unyielding continental mass or island arc, resulting in the crust buckling downwards (subducting) and being destroyed within the hot interior of the Earth. As oceanic crust ages and cools, it becomes denser and stiffer, resulting in a steeper angle of subduction and a deepening trench. Trenches along the western edge of the Pacific are deepest and oldest. Seismically, ocean trenches are highly active, as subduction is an episodic rather than continuous process. This results in an unstable and unpredictable habitat compared to the relative environmental stability of the adjacent abyssal plains[28].

Ocean trenches are typically close to land masses and tend to have high rates of sedimentation, a significant amount of which is of organic origin and an important available food source for trench communities. Several trenches also underlie highly productive cold water upwelling zones, the organic fallout from which contributes greatly to productivity there. The water within trenches generally originates from the surrounding bottom water, which is derived from cold surface water at high polar latitudes and is relatively well oxygenated[28].

Trenches tend to be isolated linear systems with high seismic activity; trench faunas are not rich in species but are often high in numbers of endemic species. There are some 25 genera restricted to the ultra-abyssal (hadal) zone, representing some 10-25% of the total number of genera present, and two known endemic hadal families; the Galatheanthemidae (Cnidaria) and Gigantapseudidae (Crustacea). The latter family contains a single species: *Gigantapseudes adactylus*. The greatest number of endemic species known from a single trench is a sample of 200 from the Kurile-Kamchatka Trench; this may be compared with 10 endemic species known from the Ryukyu and Marianas Trenches.

Hydrothermal vents

Hydrothermal vent communities were first discovered in 1977, at a depth of 2500 m on the Galápagos Rift. They are now known to be associated with almost all known areas of tectonic activity at various depths. These tectonic regions include ocean-floor spreading centres, subduction and fracture zones, and back-arc basins[29]. Cold bottom-water permeates through fissures in the ocean floor close to ocean-floor spreading centres, becomes heated at great depths in the Earth's crust, and finds its way back to the surface through hydrothermal vents. The temperature of vent water varies greatly, from around 23°C in the Galápagos vents, to around 350°C in the vents of the East Pacific Rise, and they may be rich in metalliferous brines and sulphide ions[28]. Most species live out of the main flow at temperatures of around 2°C, the ambient temperature of deep-sea water. The biomass of vent communities is usually high compared to other areas of similar depth, and dense colonies of tube-worms, clams, mussels and limpets typically constitute the major components.

Vent communities are separated by gaps of between one and 100 km, and although they may persist only for several years or decades, sites of vent activity move relatively slowly allowing dispersal of vent organisms[29]. Vent communities could be part of a unique ecosystem as old as plate tectonic activity on Earth[30].

Hydrothermal vents communities are of particular interest in that they flourish in the dark at high pressures and low temperatures[31], and unique in that they are supported by chemolithoautotrophic bacteria (predominantly *Thiomicrospira* species (Phylum Proteobacteria)) which form dense bacterial carpets in the rich hydrothermal fluid and derive their energy chiefly from oxidising hydrogen sulphide[29, 31]. Many of the vent species filter-feed on these bacteria, whilst others rely on symbiotic sulphur bacteria for energy[33].

The overall species diversity at vents is low compared with other deep-sea soft sediment areas[31], but endemism is high. More than 20 new families or sub-families, 50 new genera and nearly 160 new species have been recorded from vent environments, including brine and cold seep communities[29,32].

Other seeps

There are two further seep patterns. Cold sulphide and methane-enriched groundwater seeps occur near the base of the porous limestone of the Florida Escarpment, as well as in the Gulf of Mexico. The seeps support a dense faunal community associated with a covering or mat of bacteria on the sediment surface. These communities are strikingly similar in taxonomic composition to the hydrothermal vents of the east Pacific, a fact which points to a common origin and evolutionary history for both community types[33]. The community consists of large mussels and the vestimentiferan worm *Escarpia laminata* (Phylum Pogonophora), as well as galatheid crabs, serpulid worms, anemones, soft corals, brittle stars, gastropods and shrimps. Methane-rich seeps occur in the North Sea, the Kattegat, and elsewhere.

Tectonic subduction zone seeps are more diffuse and lower in temperature than hydrothermal vent seeps, and are rich in dissolved methane. They are known to occur off Oregon, where the fauna includes species of *Lamellibrachia* (Phylum Pogonophora) and large vesicomyid bivalve molluscs, and in the

Guaymas Basin in the Gulf of California, where thick bacterial mats cover the sulphide and hydrocarbon-coated sediment. The cold Japanese subduction zone seeps occur at a depth of 1000 m in Sagami Bay near Tokyo and in the subduction zones of the trenches off the east coast of Japan. The communities vary, but include dense benthic assemblages dominated by *Calyptogena* clams associated with a stone crab *Paralomis* sp., serpulid worms, sea anemones, galatheid crabs, swimming holothurians and amphipod crustaceans[29].

HUMAN USE OF AND IMPACT ON THE OCEANS

The seas provide many biological resources used by humans. In the form of marine fisheries they provide by far the most important source of wild protein, a source which is of particular importance to many subsistence communities around the world and which makes use of a wide range of animal species, notably fishes, molluscs and crustaceans. Marine algae are also an increasingly important foodstuff, notably in the Far East, with current annual world production of around two million tonnes. Marine organisms are also proving extremely fruitful sources of pharmaceuticals and other materials used in medicines. More minor although locally important uses include exploitation of coastal resources for building materials (eg. coral limestone, mangrove poles) and other industrial products (eg. tannins from mangroves).

Access to marine resources is not equitably distributed amongst the world's nations. Most obviously, some 39 states are landlocked, ie. have no seaboard (five of these have seaboards on the Caspian). Those that do have seaboards show great variation in length of coastline, and area of territorial waters and Exclusive Economic Zones (EEZs – see below), both absolutely and relative to their land areas. They also show great variation in their capacities to exploit marine resources, both on the high seas and within their territorial waters and EEZs.

Human activities, directly and indirectly, are now the primary cause of changes to marine biodiversity. Approximately one-third of the world's human population lives in the coastal zone (within 60 km of the sea) and indications are that this proportion will rise during the 21st century[34]. Pressures exerted by the human population on the marine biosphere are immense and increasing.

Most identified threats relate to coastal and inshore (continental shelf) areas. However, threats to the oceanic realm are undoubtedly increasing: fisheries and their attendant physical effects, such as habitat alteration owing to dredging and trawling, have entered deeper continental slope waters having previously been largely confined to the epipelagic zone, and deep water oil and gas mining is planned. Even abyssal and hadal areas are susceptible to human impact. A small, steady increase in abyssal temperature of 0.32°C in 35 years has been attributed to global climate change brought about by the activities of man. Ocean waste dumping and the potential for deep water mining and mineral extraction are also causes for concern, as are the changes in biomass and species composition in the waters above these regions[35].

The following five activities have been identified as the most important agents of present and potential change to marine biodiversity at genetic, species and ecosystem levels[35]:

- fisheries operations;
- chemical pollution and eutrophication;
- alteration of physical habitat;
- invasions of exotic species;
- global climate change.

All these factors are likely to interact with each other, making the effective, long-term management of marine resources one of the major – and most intractable – problems currently facing humankind.

Figure 6.1

Species contributing
most to global
marine fisheries in
1997

Source: FAO[37]

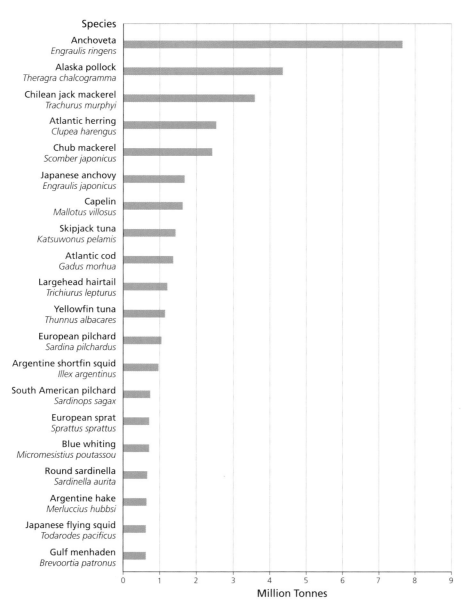

Species

Species	
Anchoveta	*Engraulis ringens*
Alaska pollock	*Theragra chalcogramma*
Chilean jack mackerel	*Trachurus murphyi*
Atlantic herring	*Clupea harengus*
Chub mackerel	*Scomber japonicus*
Japanese anchovy	*Engraulis japonicus*
Capelin	*Mallotus villosus*
Skipjack tuna	*Katsuwonus pelamis*
Atlantic cod	*Gadus morhua*
Largehead hairtail	*Trichiurus lepturus*
Yellowfin tuna	*Thunnus albacares*
European pilchard	*Sardina pilchardus*
Argentine shortfin squid	*Illex argentinus*
South American pilchard	*Sardinops sagax*
European sprat	*Sprattus sprattus*
Blue whiting	*Micromesistius poutassou*
Round sardinella	*Sardinella aurita*
Argentine hake	*Merluccius hubbsi*
Japanese flying squid	*Todarodes pacificus*
Gulf menhaden	*Brevoortia patronus*

Million Tonnes

WORLD MARINE CAPTURE FISHERIES

World marine capture fisheries have grown five-fold in the past half-century, with annual landings increasing from nearly 18 million tonnes between 1948 and 1952 to around 87 million tonnes during the period 1994-1997. The latter comprises just over 70% of current recorded global production of aquatic resources, the remainder being accounted for by inland capture fisheries (see Chapter 8) and aquaculture[36,37].

Composition of marine fisheries

Marine fisheries encompass a very wide range of organisms, including algae, invertebrate animals in various phyla and vertebrates including fishes (often termed finfishes in fisheries analysis), reptiles, mammals and birds (although by convention the last of these groups is not normally considered in fisheries analysis). FAO recognize in total just under 1000 "species items" (species, genera or families) that feature at least periodically in national catch statistics. However, globally important marine fisheries are confined to relatively few groups, with over 80% of landings by weight being finfishes and virtually all the remainder molluscs and crustaceans. Around one quarter of all landings in 1997 comprised just five species of finfish, Anchoveta *Engraulis ringens*, Alaska Pollock *Theragra chalcogramma*, Chilean jack mackerel *Trachurus murphyi*, Atlantic herring *Clupea harengus* and Chub mackerel *Scomber japonicus*, each of which accounted for over two million tonnes in that year (Figure 6.1). A further seven species had recorded catches of between one and two million tonnes. These twelve species together amounted to well over one third of marine catches. Currently by far the most important single species is the Anchoveta *Engraulis ringens*, of which just under 8 million tonnes were harvested in 1997, but whose fishery (off the west coast of South America) was nearly 13 million tonnes in 1994 and just over 13 million tonnes in 1970, constituting by far the largest single-species fishery the world has ever seen.

In terms of major species groups, by far the most important are the herrings and anchovies in the order Clupeiformes, which accounted for over 22 million tonnes, or around 25% of marine landings, followed by the jacks and mullets (some Perciformes and Mugiliformes) with 10.7 million tonnes and cod, hake and haddock (Gadiformes) with 10.2 million tonnes. The most important invertebrate group overall is cephalopod molluscs (squid, cuttlefish and octopus) of which some 3.3 million tonnes were reported landed.

Distribution of marine fisheries

The geographical distribution of marine fisheries is determined both by the distribution of harvestable fish stocks and by a range of complex socio-economic factors. The former is largely determined by variations in productivity, which, as noted above are themselves largely determined by nutrient availability, so that overall the most productive fisheries areas are on continental shelves at higher latitudes and in upwelling zones at lower latitudes. As a generalisation, the latter are associated with pelagic fish stocks and the former more with demersal or semi-demersal (deep water or bottom-dwelling) stocks, although pelagic stocks play an increasingly important rôle even here.

As might be expected purely on the basis of its size, the Pacific Ocean is by far the most important major fisheries area, accounting for over 60% of marine landings. The northwest Pacific alone – an area with very extensive continental shelf development – accounts for nearly half this total.

The various upwelling zones are not all of equal importance in fisheries. That associated with the Humboldt Current off Peru and Chile is the single most productive, while those associated with the California, Benguela and Canary Currents are of some-what lesser importance, although each is still a major fisheries area. The Arabian Sea upwelling appears to be anomalous, in that it evidently supports major populations of mesopelagic (ie. middle depth) rather than epipelagic species. Not only are the former generally considered of low value, with an identified market only as animal feed, but capture and processing requires expensive, advanced technology. They thus remain virtually unexploited at present and are considered along with the Antarctic krill stocks to be the major unexploited fisheries resources left.

Trends in marine fisheries

During the 1950s and 1960s, total landings increased steadily as new stocks were discovered, while improved fishing technology and an expansion of fishing effort enabled fuller exploitation of existing stocks of both pelagic and demersal species. Long-range fleets increased in size during this period, concentrating their efforts in the richest ocean areas, and were largely responsible for the rapid increase in world catches.

At the beginning of the 1970s, the Peruvian anchoveta fishery alone comprised some 20% of marine fisheries production. These stocks, and the fishery, collapsed between 1971 and 1973, at the same time as the important South African Pilchard fishery in the Atlantic, leading to a sharp drop in overall marine fisheries production. However, the decline in these stocks was offset by a massive rise in sardine (pilchard) stocks in the Pacific. Increasing harvest of these, and the general growth in fisheries elsewhere, meant that by the end of the 1970s global landings had once again increased above the level of the early 1970s. Landings of most demersal fish stocks remained relatively constant, however, implying that they were close to full exploitation. Long-range fleets continued to expand in importance.

The 1980s once again saw a period of continuous growth (averaging 3.8% a year) in world landings. As in the 1970s landings of demersal stocks were generally static or declining so that shoaling pelagic species provided most of the increase in fish production. In fact, just three pelagic species (Peruvian Anchoveta, South American Sardine *Sardinops sagax*, and

Figure 6.2

Marine fisheries landings,
1950-1997

Note: 1984-1997 data are for capture
fisheries only; pre-1984 data include
aquaculture (significant only for non-
cephalopod molluscs and crustaceans:
data for these included only for 1984-1997).

Source: FAO Fisheries Dept.[37]

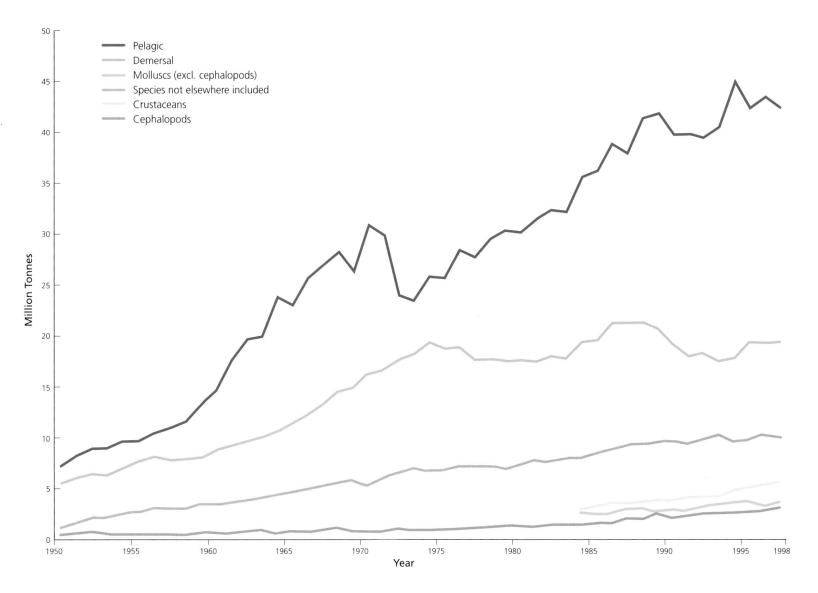

Pelagic
Demersal
Molluscs (excl. cephalopods)
Species not elsewhere included
Crustaceans
Cephalopods

Figure 6.3

Percentage of major marine fish
resources in different phases of
fishery development
Source: FAO Fisheries Dept.[37]

Japanese Sardine *Sardinops melanostictus*) and one semi-demersal species (Alaska Pollock) accounted for 50% of the increase in world landings during the 1980s[36]. Most of this increase appears to have been because of favourable climatic effects on stock sizes rather than new fishery developments or improved management practices[36].

During the 1990s there has been negligible growth in marine capture fisheries. Indeed landings decreased sharply between 1989 and 1990, although grew again to just over the 1989 total in the period 1994-1997. Four of the five most important fishes in fisheries in the late 1990s are pelagic, the exception being the Alaska Pollock. This dominance of pelagic over demersal species is reflected in overall fisheries figures, with pelagic landings well over twice demersal landings globally. This contrasts sharply with the situation in the early 1950s when pelagic landings were only some 30% greater in volume than demersal landings (Figure 6.2).

This increasing dependency on pelagic fish stocks is sympto-matic of a major crisis in global marine fisheries. In general demersal fishes are more valuable per unit weight than pelagic species so that all else being equal the former are preferentially harvested. The increased importance of the latter in the past forty years is indicative of the growing overexploitation of fisheries stocks worldwide – as valuable demersal stocks have been depleted so attention has turned to the intrinsically less valuable pelagic stocks.

Recent analysis by FAO confirms this picture of widespread overexploitation. Of the 200 major marine fishery resources (stocks of a particular species in a particular area) that between them account for nearly 80% of world marine fishery production, in 1994 35% were classified as 'senescent' (showing declining yields), around 25% were 'mature' (plateauing at a high exploitation level), around 40% were developing (showing annually increasing yields) and none were at low exploitation levels (Figure 6.3). All mature and senescent stocks were

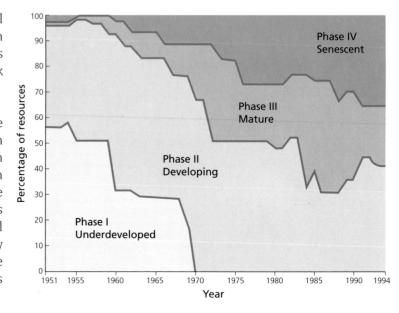

considered in urgent need of management to prevent further deterioration. A parallel study of stocks for which detailed information was available concluded that 44% were intensively to fully exploited, 16% were actively overfished, 6% depleted and 3% slowly recovering, indicating that nearly 70% of analysed stocks were in urgent need of management. The proportion of high value fisheries (eg. many demersal finfishes such as cod and hake, many crustacean fisheries) in this state is even higher.

There are three major reasons for this. First, and most funda-mental, most fisheries have traditionally been regarded as an 'open access' resource, so that, in essence, it pays any one fisher to harvest as much as possible at any given time because if they do not, somebody else will. Secondly, technological inno-vations have made fishing much more efficient. Thirdly, extreme over-capitalisation of the world's commercial fishing fleet

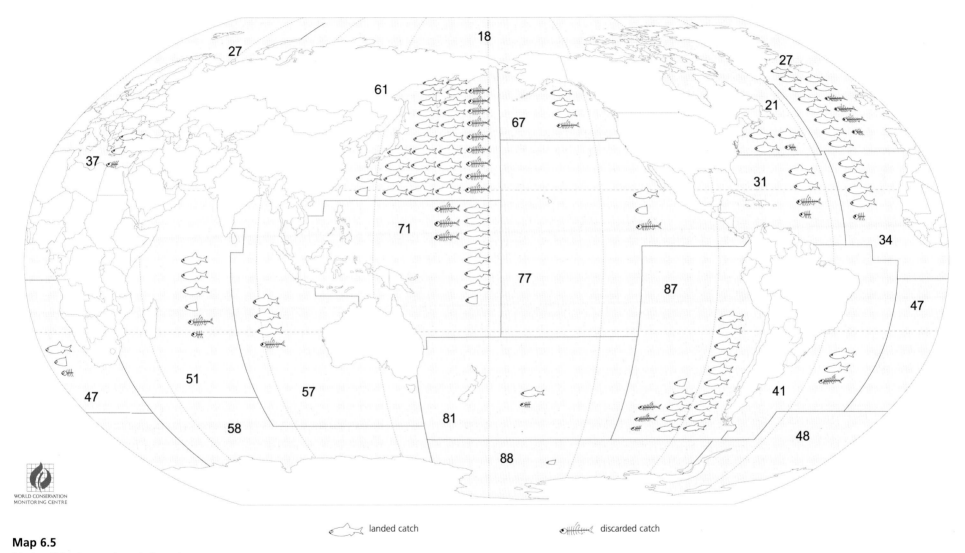

18 Arctic Sea
21 Atlantic, Northwest
27 Atlantic, Northeast
31 Atlantic, Western Central

34 Atlantic, Eastern Central
37 Mediterranean and Black Sea
41 Atlantic, Southwest
47 Atlantic, Southeast

48 Atlantic, Antarctic
51 Indian Ocean, Western
57 Indian Ocean, Eastern
58 Indian Ocean, Antarctic

61 Pacific, Northwest
67 Pacific, Northeast
71 Pacific, Western Central
77 Pacific, Eastern Central

81 Pacific, Southwest
87 Pacific, Southeast
88 Pacific, Antarctic

WORLD CONSERVATION
MONITORING CENTRE

landed catch   discarded catch

Map 6.5

Marine fisheries catch and discards

The location of the fishery areas recognised by FAO for statistical purposes is shown on this map, with symbols representing the approximate late 1990s yield from capture fisheries and the volume of discarded catch, most of which is presumed not to survive. Each symbol represents approximately 1 million tonnes.

Source: data from FAO[37].

(partly a consequence of the nature of fisheries as an open access resource but also for complex socio-economic and political reasons) has intensified fishing pressure.

Bycatches and discards

The effects of overfishing are compounded by the wastefulness of many marine capture fisheries. A 1994 FAO global assessment of bycatches and discards estimated that 18 to 40 million tonnes (mean 27 million tonnes) of marine fisheries catch were discarded annually (Map 6.5). This represented just over 25% of annual estimated total catch (ie. landings represent around 75% of actual catch). Although figures are not available, it is generally assumed that the great majority of discards die. Further losses are caused by mortality of animals which escape from fishing gear during fishery operations, but it is impossible at present to estimate the importance of this. Shrimp fishing produces the largest volume of discards (around 9 million tonnes annually).

Bycatches include non-target, often low-value or "trash" species, as well as undersized fish of target species. Non-target species may include marine mammals, reptiles (sea turtles) and seabirds, as well as finfishes and invertebrates. Of particular concern in recent years has been mortality of marine mammals, especially dolphins, in pelagic drift nets, of sea turtles in shrimp trawls and more recently, of diving seabirds, especially albatrosses, in long-line fisheries.

Discarding may be a side-effect of management systems intended to regulate fisheries (eg. non-transferable quotas may cause discarding of over-quota catch; species-specific licensing may cause discard of non-license but still commercially valuable species).

Solutions to bycatches and discards will be found essentially through improvement in the selectivity of fishing gear and fishing methods. Much of the research in this has been carried out in higher latitudes and is not readily transferable to multi-species tropical fisheries, where the tropical shrimp trawls still produce high rates of by-catch. Improved use of by-catch either as fish-meal or human food-fish is also a possibility; however, this does not address the problem of mortality of potentially threatened species (sea turtles, seabirds, cetaceans), nor the wasteful capture of immature specimens of harvestable species.

In 1994 FAO estimated that it should be possible to reduce discards by 60% by the year 2000 by: a concentrated effort to improve the selectivity of fishing gear; the development of international standards for research; greater interaction between research staff, industry and fisheries managers; and the application of technology through fisheries regulations. Where overfishing is occurring and quotas set in response, eg. in EU waters, it may be unrealistic to expect radical reduction in discarding without similar reduction in effort.

A further problem in the efficient use of marine resources is post-harvest loss. It is almost impossible to estimate this accurately, but FAO believe it to exceed 5 million tonnes per year (ie. around 5% of harvest). Most significant are physical losses of dried fish to insect infestations and loss of fresh fish through spoilage. These problems are particularly significant in developing countries.

Aquaculture

One major response to the growing crisis in marine capture fisheries has been the rapid rise in various forms of aquaculture (Figure 6.4). The latter may be defined as the rearing in water of organisms (animals, plants and algae) in a process in which at least one phase of growth is controlled or enhanced by human action. The animals used are generally finfishes, molluscs and crustaceans, although a number of other groups such as sea squirts (Tunicata), sponges (Porifera) and sea turtles are cultured in small quantities. Seaweeds of various kinds are also cultured, some in large amounts. Most of the species grown in any quantity are low in the food chain, being either primary producers, filter feeders or finfishes that in their adult stages are either herbivores or omnivores.

Figure 6.4

Marine aquaculture production,
1984-1997

Source: FAO Fisheries Dept.

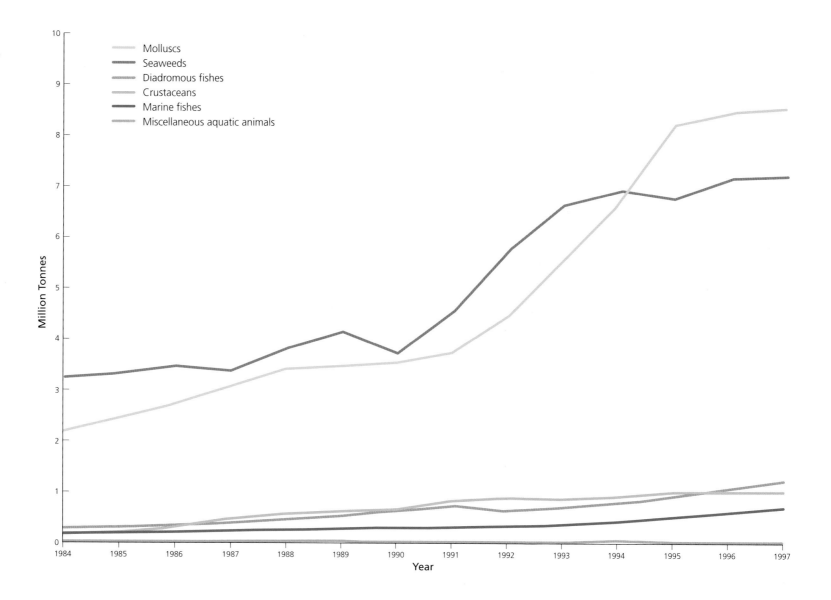

Legend:
- Molluscs
- Seaweeds
- Diadromous fishes
- Crustaceans
- Marine fishes
- Miscellaneous aquatic animals

Y-axis: Million Tonnes

X-axis: Year

FAO note that aquaculture is the world's fastest growing food production sector, annual output having increased at an average rate of around 10% in the period 1984-1996 (compared with less than 2% for capture fisheries) (see Figure 6.4). In 1996 aquaculture provided around one fifth of recorded global fisheries production and nearly 30% of food fish. Of the 26.4 million tonnes recorded in total just over 15 million tonnes originated in freshwater, 9.7 million tonnes were produced in marine environments and about 1.6 million tonnes in brackish water environments. In addition, some 7.7 million tonnes of algae and plants were produced, almost all of this seaweed, chiefly Japanese kelp *Laminaria japonica*, nori *Porphyra tenera* and wakame *Undaria pinnatifida*. The first of these was, in terms of volume, the most important of all aquaculture species, with around 4.4 million tonnes produced.

In marine and brackish (usually estuarine) environments, by far the most important animal group in terms of volume is the molluscs, whose 1997 recorded production of some 8.6 million tonnes comprised over 75% of all animal production in these environments. Around 50 mollusc species are produced in significant quantity, almost all bivalves. As with most culture systems, production is heavily skewed to a small number of species, with 65% of production comprising just three: the Pacific cupped oyster *Crassostrea gigas*, Japanese carpet shell *Ruditapes philippinarus* and Yesso scallop *Pecten yessoensis*. The Far East dominates production, with around 75% of that recorded taking place in China and most of the remainder in Japan.

Although production of marine crustaceans accounts for only 10% or so by volume of marine and brackish water animal aquaculture, it has disproportionately high economic importance, and is also the sector that has given rise to most environmental concerns. Between 1984 and 1997 annual production grew nearly six-fold, from less than 200 000 tonnes to over 1 million tonnes. The great majority of production takes place in tropical and sub-tropical Asia and is dominated by *Penaeus* species;

globally this genus produces over 90% of aquaculture crustacean supply by weight. Three species of *Penaeus* account for around three-quarters of crustacean production. The giant tiger prawn *P. monodon* is the most widely cultivated and accounts for nearly half; the whiteleg shrimp *P. vannamei* is cultured in the Americas and accounts for around 15% of estimated global supply (around 70% of this originating in Ecuador); and the fleshy prawn *P. chinensis* is cultured in China and currently accounts for around 10% of production, having declined considerably since the early 1990s when around 200 000 tonnes were produced annually. Other marine crustaceans cultivated include other *Penaeus* species, some *Metapenaeus*, and spiny lobsters *Panulirus*. These groups, however, make an insignificant contribution to global supply.

Growth in crustacean aquaculture has been fuelled by the high value of the product: the market in 1996 was estimated to be worth nearly US$7.5 billion, or around one quarter of the total value of marine and brackish water aquaculture[37]. The great majority of production takes place in low income countries – the five countries producing over 100 000 tonnes annually being China, Thailand, Indonesia, Ecuador and Bangladesh – and is aimed mainly at the export market (primarily to Europe, the USA and Japan) and to a lesser extent at the domestic luxury market. Pressure is high to produce maximum returns on investment so that increasingly intensive farming methods are used. These are widely acknowledged to be having adverse social and environmental impacts in the countries of production, as well as leading to increasing difficulties in maintaining supply, owing to the spread of major diseases. The last of these accounts for the major decline in the Chinese fleshy prawn industry during the 1990s. Impacts include:

- loss of mangrove habitat;
- abstraction of freshwater;
- introduction of pathogens and other damaging non-native species;

- escape of cultured non-native species;
- pollution;
- diversion of low quality or cheap fish food resources (may lead to more efficient use of by-catches and trash fish, but may lead to more indiscriminate catch fisheries);
- diversion of effort from other forms of aquaculture (notably milkfish *Chanos chanos*).

OTHER MAJOR IMPACTS ON THE MARINE BIOSPHERE

Alteration of physical habitat

Physical alteration of habitats through human action chiefly affects coastal and inshore areas. Impacts here can be very severe, although few attempts have been made to quantify them on a global basis. Major causes include coastal development, particularly landfilling and construction of groynes and jetties, aquaculture, dredging of channels for navigational purposes, extraction of materials such as sand and coral, stabilisation of shorelines, and destructive fishing methods such as beam- trawling, use of explosives and *muro-ami* (using rocks on ropes to drive fishes into nets). Upstream activities, such as deforestation and dam-construction can greatly alter sediment loads in rivers, affecting patterns of sediment deposition in estuarine areas.

Chemical pollution and eutrophication

Human activities have increased inputs of a huge range of organic and inorganic chemicals into marine ecosystems. Such inputs may enter by direct discharge (eg. in sewage outflow pipes), via river and stream outflow, as land runoff, through the atmosphere or from sea-going vessels. Because virtually all such input originates on land, as with most other human impacts on marine ecosystems, areas most affected are coastal and inshore regions, particularly enclosed or semi-enclosed water-bodies. In oceanic regions, mixing of the enormous volume of seawater generally ensures that inputs become rapidly diluted.

Major categories of inputs include nutrients of various kinds (eg. nitrates and nitrites, phosphates, dissolved organic matter), persistent organic pollutants (POPS), including a range of chlorinated hydrocarbons, and heavy metals such as Cadmium (Cd), Copper (Cu), Mercury (Hg), Lead (Pb), Nickel (Ni) and Zinc (Zn). Quantifying these inputs and assessing their impact is problematic, particularly because many occur naturally in sea water.

Many POPs and heavy metals can act as toxins above certain concentrations, inducing mortality or morbidity or impairing reproductive success, particularly in cases where they become increasingly concentrated towards the top of food chains. Their overall impact on marine ecosystems remains uncertain. More easily observable is the impact of eutrophication on the increased input of organic and inorganic nutrients (particularly nitrogen and phosphorus) into coastal waters, mainly through fertiliser runoff and sewage disposal. It is believed that human intervention has increased river inputs of nitrogen and phosphorus worldwide into coastal areas by more than fourfold over background levels. These inputs lead to increases in productivity in coastal waters, often in the form of algal blooms. These blooms may themselves be noxious; they also typically cause the euphotic zone to reduce in vertical extent and are implicated in the development of hypoxic (low dissolved oxygen concentration) and anoxic (zero dissolved oxygen) zones. A shallowing of the euphotic zone may cause die-off of photosynthesising benthic algae in shallow-water areas. This has occurred, for example, in the Black Sea where the euphotic zone has decreased from 50-60 m vertical extent in the early 1960s to around 35 m by 1990, leading to a decrease of up to 95% in living biomass of benthic macrophytic algae such as Phyllophora, formerly an important harvested resource.

Hypoxia and anoxia result from the activities of oxygen-respiring bacteria below the euphotic zone feeding on accumulated dead algae and other organisms and waste matter raining down from above. Hypoxia results in the emigration of mobile aerobic species and mortality of sedentary ones. This may have catastrophic impact on local fisheries. Most hypoxic zones vary in extent through the year and from year to year and some are only seasonal, disappearing when winter mixing causes re-oxygenation of bottom waters. They may be very extensive – the hypoxic zone to the west of the Mississippi Delta covered some 16 000 km² in 1997, having covered some 9000 km² in 1989. Over 50 such zones have been identified worldwide to date; some appear to be at least in part natural phenomena induced while others are believed entirely anthropogenic.

Invasions of exotic species

As on land, the breakdown of biogeographic barriers in the sea appears to be having a major, and increasing, impact on marine ecosystems. Sources of such breakdown are chiefly the construction of marine corridors between previously isolated areas and the deliberate or accidental translocation of organisms. The most notable of the former are the Panama Canal (opened in 1914), which joins the Pacific and Atlantic Oceans in central America and the Suez Canal (opened in 1869), which joins the Mediterranean Sea to the Red Sea and thereby the Indo-Pacific region.

Data on global introductions and their impacts are very incomplete. At present the best known, and almost certainly most important, phenomenon of this type is the so-called Lessepsian migration from the Red Sea into the Mediterranean basin through the Suez Canal, named after Ferdinand de Lesseps who planned the canal. The canal, which is 165 km long, is a continuous seawater channel with no gates or locks and a water level at the Red Sea end some 1.2 m higher than at the Mediterranean end, leading to a constant flow of water from the former to the latter. It is estimated that to date some 400-500 marine species have migrated northwards through the canal and established themselves in the Mediterranean while a far smaller number have moved in the other direction. New species are believed to arrive in the Mediterranean at the rate of 4-5 annually. The great majority of Lessepsian migrants are widespread Indo-Pacific species and some, such as several decapod crustaceans, have established major populations in the eastern Mediterranean[38]. Little information is available on the impact of these new arrivals on the native biota.

In contrast to the Suez Canal, the Panama Canal has a central, freshwater section some 25 m above sea-level, reached through a series of locks. This acts as a barrier to most marine organisms, so that unaided migration through the canal has been limited to date. Other forms of translocation involve deliberate stocking of harvestable species such as some molluscs, accidental escapes from aquaculture operations, transport in ballast water of ships which is flushed out far distant from where it was taken on, and release of fouling organisms that adhere to the hulls of ships and boats.

The extent of introductions of this kind has yet to be fully assessed but is certainly large. Most observations concern individual species, such as the Asian clam Potamocorbula amurensis in San Francisco Bay, USA [39], the alga *Caulerpa taxifolia* in the western Mediterranean[40] and the ctenophore or comb-jelly *Mnemiopsis leidyi* in the Black Sea, believed to have been translocated from the western North Atlantic in ship's ballast waters [41,42]. In each of these cases the introduced species appears to be having a major impact on native biota, although in general it is difficult to separate the effects of a particular species from general ecosystem deterioration.

Global climate change

The potential impact of climate change on marine ecosystems remains very incompletely understood. It has been suggested that one possible impact is an increasing frequency and severity of perturbations such as El Niño Southern Oscillation (ENSO) events, which may have disruptive effects on fish stocks and dependent fisheries and are associated with phenomena such as coral bleaching. It is also thought that increasing atmospheric temperatures may have impact on the generation of cold, oxygen-rich bottom waters beneath Arctic and Antarctic ice-sheets, with major implications for deep-sea biota and for global patterns of sea water circulation, in particular the Great Conveyor, driven by bottom water generated in the North Atlantic.

THE CURRENT STATUS OF MARINE BIODIVERSITY

Because they are usually much less readily observed, marine species are in general much more difficult to monitor and assess than terrestrial ones. Assessment is based on sampling and, in the case of harvested species, often on the basis of catch rates, although the latter may vary in response to a wide range of factors in addition to changes in population of the species concerned. An exception lies with those groups such as pinnipeds, sea turtles and seabirds that nest or breed on land; because many of these tend to be colonial species and because they tend to breed in open habitats (beaches, cliff-tops, ice-sheets), they may be easier to monitor than many other species, either terrestrial or aquatic. In the case of large, commercially valuable fish stocks, monitoring at large scale has in some cases been carried out for many years, so that estimates at the stock level are obtainable.

Threatened and extinct species

The only major marine species groups (classes or above) that have been comprehensively assessed in terms of threatened species status to date are mammals and birds. In addition, sea turtles and a number of fish families and genera (eg. the sturgeons in the order Acipenseriformes and the sea-horses

Hippocampus in the order Gasterosteiformes) have also been assessed. Other threatened marine species have been identified on more of an ad hoc basis. Data are summarised in Table 6.12.

Relatively speaking, far fewer marine species are known to have become extinct since 1600 than either terrestrial or freshwater ones. Catalogued extinctions comprise two marine mammals (the Caribbean monk seal *Monachus tropicalis* and Steller's sea cow *Hydrodamalis gigas*) and three seabirds (an unnamed petrel *Pterodroma* from Mauritius, Pallas's cormorant *Phalacrocorax perspicillatus* and the Great auk *Alca impennis*). In addition eight coastal or island duck species have disappeared at various times from the late seventeenth century onwards; however there is in most cases insufficient information to determine whether these species were predominantly marine or terrestrial.

As a gross generalisation, marine species appear to be somewhat less extinction-prone as a result of mankind's activities than freshwater or terrestrial ones. There are arguably two main reasons for this. First, because of the size of the World Ocean and the fact that people do not actually live in it, the marine biosphere remains as a whole considerably more buffered from human intervention than terrestrial and inland water areas. Second, marine species on the whole appear to be more widespread than terrestrial or inland water ones. In the open ocean, there are vast areas with apparently similar habitat conditions and there are few barriers to dispersal so that many species have circumglobal distributions. In addition, many forms that as adults are sessile (eg. sponges and corals) or sedentary (many molluscs and crustaceans) have planktonic larvae that are often widely dispersed in water currents. For this reason, many coral reef species, for example, are found in suitable habitat throughout the Indo-Pacific region. In addition, many of the most heavily exploited fish species have very high fecundity (in the case of some tunas amounting to several million eggs in a single spawning), so that they have at least potentially very high population growth rates unparalleled in terrestrial vertebrates.

Table 6.12

Taxonomic distribution and status of threatened marine animals.

Note: none of the major groups listed has been comprehensively assessed for species at risk.

Source: WCMC database, data compiled in part for IUCN Red List[72].

Phylum and class or order	Common name	Critically Endangered	Endangered	Vulnerable
Cnidaria				
Anthozoa	Stony corals			2
Mollusca				
Bivalvia	Bivalves			4
Gastropoda	Gastropods	1		4
Craniata – Pisces				
Hexanchiformes	Cow sharks			1
Lamniformes	Mackerel sharks			4
Carcharhiniformes	Ground sharks	1	1	2
Squaliformes	Dogfishes and sleeper sharks			1
Rajiformes	Rays	1	3	
Coelacanthiformes	Coelacanths		1	
Acipenseriformes	Sturgeons	1	8	4
Clupeiformes	Herrings and anchovies		1	
Siluriformes	Catfishes	1		
Salmoniformes	Salmonids		2	1
Gadiformes	Cods, hakes, rattails	1		2
Ophidiiformes	Pearlfishes, cusk-eels, brotulas			1
Batrachoidiformes	Toadfishes			5
Lophiiformes	Anglerfishes etc	1		
Gasterosteiformes	Pipefishes, seahorses etc	1		37
Scorpaeniformes	Gurnards, scorpionfishes etc	1	1	1
Perciformes	Perches etc	6	4	33
Pleuronectiformes	Plaice, flounders, soles		1	1

There are of course significant exceptions to all these. Coastal regions in many parts of the world, and enclosed or semi-enclosed marine areas such as the Baltic, Black and Yellow Seas are often under intense pressure from a range of human activities. A number of marine species do appear to have restricted ranges (eg. the Hawaiian coral reefs are relatively rich in species found nowhere else while many southern hemisphere seabirds are apparently confined to a small number of breeding sites) and significant numbers have low or very low reproductive rates (many chondrichthyine fishes, marine mammals and seabirds).

Until recently by far the most important human activity affecting marine species was uncontrolled exploitation. Where species are either easily exploitable or are highly sought-after (ie. have high unit value), or both, they may suffer catastrophic declines. This is the case with sea turtles and a number of marine mammals and birds that are or have been harvested principally at their terrestrial breeding sites (which are often colonial), as well as with the great whales and the dugong, which although strictly

Phylum and class or order	Family	Common name	Critically Endangered	Endangered	Vulnerable
Tetraodontiformes		Triggerfishes etc		3	
Craniata – Reptilia					
Squamata	Iguanidae	Iguanas			1
Testudines	Dermochelyidae	Leathery Turtle		1	
	Cheloniidae	Sea turtles	2	3	1
Craniata – Aves					
Anseriformes	Anatidae	Ducks			2
Ciconiiformes	Laridae	Gulls, terns, skuas, auks, skimmers	1	1	9
	Sulidae	Gannets and boobies			1
	Phalacrocoracidae	Cormorants and shags			8
	Fregatidae	Frigatebirds	1		1
	Spheniscidae	Penguins			5
	Procellariidae	Petrels, albatrosses, shearwaters	11	6	15
Craniata – Mammalia					
Cetacea	Balaenidae	Right whales		1	
	Balaenopteridae	Rorquals	3	1	
	Delphinidae	Dolphins		1	
	Monodontidae	Beluga		1	
	Phocoenidae	Porpoises	1	1	
	Physeteridae	Sperm whales		1	
Sirenia	Dugongidae	Dugong		1	
	Trichechidae	Manatees		1	
Carnivora	Otariidae	Eared seals	1	5	
	Phocidae	Earless seals	1	1	

marine are air-breathing and therefore spend much time at the sea surface. Most of these species have relatively low reproductive rates, so that even if they are ultimately afforded protection population recovery rates may be very slow.

Land-breeding species may also be susceptible to other threats, such as predation, coastal development and pollution. It is noteworthy in this context that the family Procellariidae contains nearly three times as many threatened species (32 out of 115 species, or 28%) as the average bird family, in which 11% of species are threatened, and nearly six times as many critically endangered species as would be expected at random. It is almost certainly the tendency of these birds to nest on islands, whose biotas have in general suffered enormously more from mankind's influence in the past few centuries (see Chapter 5), rather than their sea-going habits, that has resulted in this.

For truly marine species (chiefly finfishes and invertebrate animals) the situation appears somewhat different. Even when these have been exploited to the point of stock collapse, as has occurred for example with the Cod *Gadus morhua* stocks off Newfoundland in the North Atlantic, the species concerned do not appear to have become imminently threatened with biological extinction. This is in part because once stocks are reduced below a certain level it is often no longer economically viable to continue harvesting them. Generally, the residual population at this stage is still large enough to allow recovery if harvesting ceases, particularly in the case of species with high fecundity and therefore high potential intrinsic rates of increase. Exceptions to this are species that have low fecundity, particularly if they also have a long period to maturity, with limited ranges and which may either have high unit value or be caught as by-catches. In the case of by-catches, because the fishery is not directed at the species concerned, its intensity will not decrease as population levels decrease so that it may theoretically be possible at least locally to extirpate species, particularly if they are habitually caught before they reach maturity. Examples include several sawfish species (family Pristidae). These are large, slow-growing predominantly inshore species that give birth to relatively small numbers of live young. Population densities appear to be naturally low and animals are widely caught as bycatch in inshore fisheries before they are large enough to reproduce. As a result three species are classified as Endangered and one as Critically Endangered. In addition, it is possible that trophic shifts may occur when populations of some species are severely reduced, inhibiting recovery of these populations when exploitation ceases. This has been suggested in the case of some great whale populations that have not apparently recovered as rapidly as projected following the cessation of their harvest.

Fig 6.5

Marine living planet index
Note: see Chapter 5 for outline methodology
Source: Loh *et al.*[73]

The marine living planet index

An overall impression of the current status of marine ecosystems can be gained by applying the Living Planet Index approach developed by WCMC for WWF described in Chapter 5. Time-series data on some 102 species or populations of species of marine reptiles, birds, fishes and mammals have been amassed, with a wide geographic spread. Most of the species included are fishes. The index – which is a measure of the population status of an average species – has declined by around 35% in the period 1970-1995[73]. Self-evidently, the index is dominated by those stocks and species that humans have an interest in monitoring, most of these being of commercial importance as a fisheries resource. These should also be stocks that humans have an interest in managing as well as possible. That the index has declined in every five-year interval since 1970 is ample evidence that such management is failing, as confirmed by the picture painted above of global marine capture fisheries.

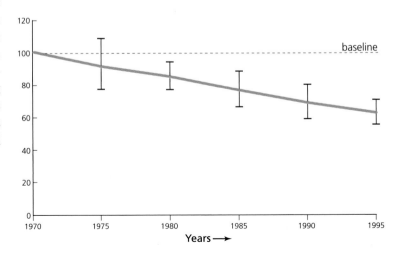

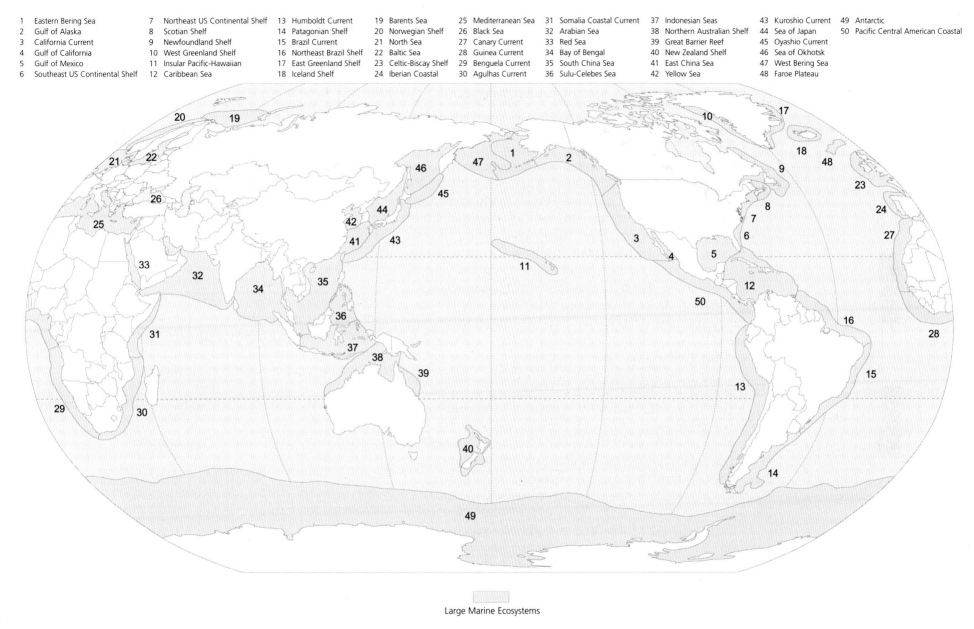

1	Eastern Bering Sea	7	Northeast US Continental Shelf	13	Humboldt Current	19	Barents Sea	25	Mediterranean Sea	31	Somalia Coastal Current	37	Indonesian Seas	43	Kuroshio Current	49	Antarctic
2	Gulf of Alaska	8	Scotian Shelf	14	Patagonian Shelf	20	Norwegian Shelf	26	Black Sea	32	Arabian Sea	38	Northern Australian Shelf	44	Sea of Japan	50	Pacific Central American Coastal
3	California Current	9	Newfoundland Shelf	15	Brazil Current	21	North Sea	27	Canary Current	33	Red Sea	39	Great Barrier Reef	45	Oyashio Current		
4	Gulf of California	10	West Greenland Shelf	16	Northeast Brazil Shelf	22	Baltic Sea	28	Guinea Current	34	Bay of Bengal	40	New Zealand Shelf	46	Sea of Okhotsk		
5	Gulf of Mexico	11	Insular Pacific-Hawaiian	17	East Greenland Shelf	23	Celtic-Biscay Shelf	29	Benguela Current	35	South China Sea	41	East China Sea	47	West Bering Sea		
6	Southeast US Continental Shelf	12	Caribbean Sea	18	Iceland Shelf	24	Iberian Coastal	30	Agulhas Current	36	Sulu-Celebes Sea	42	Yellow Sea	48	Faroe Plateau		

Large Marine Ecosystems

Map 6.6

Large Marine Ecosystems

The general location of the 50 Large Marine Ecosystems (LMEs) currently identified is shown in this map.

Source: modified from data at the LME website[70], and see[43-48].

Assessing the status of marine and coastal ecosystems

Threatened species inventories and the marine living planet index can give a very general overall impression of the status of marine biodiversity. Assessing marine ecosystem 'health' is much more problematic. However, snapshots can be obtained from examining particular ecosystems, such as mangroves and coral reefs[12]. In the former, an overall assessment can be made on the basis of the area destroyed or severely degraded. In the latter areal measures are much more problematic, in part because reef extent is much more difficult to measure than mangrove extent and, more importantly, because the vast majority of a reef is composed of non-living calcareous deposits. Measures of the change in extent of these give little insight into the state of the living component of the reef. For this reason other measures, such as estimates of incidence of coral disease, may be feasible.

MANAGEMENT OF MARINE ECOSYSTEMS

Although the particular nature of the marine biosphere has to some extent buffered it from the impacts of humans, it also imposes its own set of difficulties and constraints for rational management. In the first instance, because almost all of it is generally out of sight, impacts are not immediately apparent, so that extreme deterioration may take place before anyone is aware of the fact; even when deterioration become apparent, there is usually less incentive to take action than with terrestrial ecosystems where such deterioration has a direct impact on people. Second, with some exceptions (such as communal property resources on reefs and other inshore areas in parts of the South Pacific), living marine resources have been widely considered open-access resources, particularly those outside territorial waters (usually up to 12 nautical miles from shore). There is thus, quite simply, an incentive for any given individual to exploit a resource as fast and as intensively as possible before someone else does. Third, the ability of water to transport large amounts of dissolved and suspended materials, including living organisms, means that it is extremely difficult to manage limited areas of marine habitat in isolation.

Some changes have taken place in this. With the introduction of the Exclusive Economic Zone (EEZ) under the United Nations Convention of the Law of the Sea (UNCLOS), which allows nations control over resources (including living resources) in an area up to 200 nm offshore, a far greater proportion of the world's seas now come within the control of individual nations. At present 99% of world fisheries catch is taken within EEZs. Although this should theoretically allow more rational management of marine resources, and more effective enforcement of management measures, in practice progress in both has been limited to date, as evinced by the increasing proportion of world's fisheries that are overexploited. This is because fisheries management regimes are frequently subject to political pressure, so that in many countries quotas are habitually set higher than those recommended by fisheries biologists, and also because active enforcement of regulations is difficult and expensive. Many countries lack the resources or the political will, or both, to enforce such regulations adequately.

It is also increasingly apparent that individual marine resources cannot be effectively managed in isolation from each other. Complex interactions between populations of different organisms, when combined with perturbations in the environment and variations in human impact (eg. changes in fisheries technology or fishing effort) create responses that may be far from intuitively predictable. Recognising that the understanding of such responses will require modelling and management of large-scale ecosystem processes, a number of large ecosystem units have been identified[43-48], based on the world's coastal and continental shelf waters, which are regarded as central to such analysis. These Large Marine Ecosystems (LMEs) are:

Map 6.7

Marine protected areas

This map shows the location of protected areas in IUCN categories I-VI that are entirely or in part marine, with map symbols graded according to protected area size (including any land present).

Source: WCMC Protected Areas database.

| <100 km² | >100 km² | >1000 km² | >5000 km² |

"Regions of ocean space encompassing near-coastal areas from river basins and estuaries out to the seaward boundary of continental shelves and the seaward margins of coastal current systems. They are relatively large regions of the order of 200 000 km² or larger, characterised by distinct bathymetry, hydrography, productivity, and trophically dependent populations."

Over 95% of the usable annual global biomass yield of fishes and other living marine resources is produced within 50 identified LMEs which lie within and immediately adjacent to the boundaries of EEZs of coastal nations [43] (Map 6.6, see page 175).

Management of LMEs

Many LMEs include the coastal waters of more than one state. In these cases, it will be effectively impossible for individual nations to assess whether their use of marine resources is sustainable in isolation from neighbouring nations. Co-ordination between states in monitoring and resource management will thus become increasingly necessary as the pressures placed on these areas increase. At present no single international body is in a position to co-ordinate action and reconcile the needs of individual nations operating within particular LMEs.

A critical need in monitoring marine ecosystems is the development of a consistent long-term database for understanding between-year changes and multi-year trends in biomass yields. For example, during the late 1960s when there was intense fishing within the Northeast US Continental Shelf LME, marked alterations in fish abundances were observed. The biomass of economically important finfish species (eg. cod, haddock, flounders) declined by approximately 50%. This was followed by increases in the biomass of small elasmobranchs (dogfish and skates), and led to the conclusion that the overall carrying capacity of the ecosystem for finfishes did not change [49,50]. However, the excessive fishing effort on the more highly valued species allowed for the lower-valued elasmobranchs to increase in abundance. Management of marine fisheries will need to take these kinds of species dominance shifts into account in developing strategies for long-term, economic sustainability of the fisheries [49,51]. The theory, measurement, and modelling relevant to monitoring the changing states of LMEs have received considerable attention[44-48,52-54].

Among the LMEs being assessed, monitored, and managed from a more holistic ecosystem perspective are the Yellow Sea [55]; the Benguela Current [56]; the Great Barrier Reef [57]; the Northwest Australia Continental Shelf [58]; and the Antarctic marine ecosystem [59]. Movement toward ecosystem-level assessment, monitoring and management is emerging also for the North Sea [60]; the Barents Sea [61]; and the Black Sea [62]. The driving forces of variability in biodiversity and biomass yields are currently being examined in several LMEs.

Programmes intended to enhance sustainability and ecosystem health are being implemented for several LMEs adjacent to developing nations. The programmes are being supported by international agencies as part of an effort by the World Bank, the Global Environment Facility, UNDP, and UNEP to assist countries in: (1) conducting transboundary diagnostic analayses that would identify priority transboundary productivity, and environmental concerns; (2) developing a Strategic Action Programme (SAP) to address high priority assessment, monitoring, and management actions; and (3) implementing use of science-based technologies and innovations in these activities.

Marine Protected Areas

The long term management of Large Marine Ecosystems is a highly complex undertaking, politically, economically and scientifically. Clearly there is enormous scope – and indeed an urgent need – for smaller-scale and more immediate approaches. As with terrestrial ecosystems, the establishment of protected areas in marine ecosystems has been viewed as a major contribution to maintenance of biodiversity. A 1995 review [63] identified just over 1300 marine protected areas in existence at that time, ranging in size from one hectare to 34.4 million hectares (the Great Barrier Reef Marine Park) (Map 6.7, see page 177). Effective management and control of marine protected areas is problematic, particularly if, as is often the case, they are in areas of intensive and potentially conflicting resource use. As noted above, marine ecosystems are also in general more difficult to protect from allochthonous inputs (ie. those originating elsewhere) than terrestrial ones. Although a no-catch regime can be effective in small marine reserves, in general it has been found that large, carefully zoned multiple-use areas are more practical and effective than small reserves. Sanctuaries or strict reserves may still be required for critical habitat areas such as nutrient sources, areas of high biological diversity, nesting sites of threatened species or to protect breeding stocks of important fishes [64,65].

Reference

1 Longhurst, A.R. and Pauly, D. 1987. *Ecology of tropical oceans*. Academic Press Inc., California, USA. ICLARM Contribution No 389.
2 Vaulot, D. 1995. Marine biodiversity at the micron scale. *International Marine Science Newsletter* No 75/76. UNESCO.
3 Duke, N.C. 1992. Mangrove floristics and biogeography. In: Robertson, A.I. and Alongi, D.M. (eds). *Tropical mangrove ecosystems. Coastal and Estuarine Studies* 41. Pp 329. American Geophysical Union, Washington D.C.
4 Finlayson, M. and Moser, M. (eds). 1991. *Wetlands*. Facts on File Limited, Oxford.
5 Woodroffe, C.D. and Grindrod, J. 1991. Mangrove biogeography: the role of Quaternary environmental sea-level change. *Journal of Biogeography* 18:479-492.
6 Tomlinson, P.B. 1986. *The botany of mangroves*. Cambridge University Press.
7 Spalding, M.D., Blasco, F. and Field, C.D. (eds). 1997. *World Mangrove Atlas*. The International Society for Mangrove Ecosystems, Okinawa, Japan.
8 Nybakken, J. 1993. *Marine biology: an ecological approach*. 3rd edition. HarperCollins College Publishers, New York.
9 World Conservation Monitoring Centre. 1992. Groombridge, B. (ed). *Global Biodiversity: status of the Earth's living resources*. Chapman & Hall, London.
10 Saenger, P., Hegerl, E.J. and Davie, J.D.S. (eds). 1983. *Global Status of Mangrove Ecosystems*. Commission on Ecology Papers, No. 3. IUCN, Gland, Switzerland.
11 Lasserre, P. 1995. Coastal and marine biodiversity. *International Marine Science Newsletter*. No 75/76. Pp 13-14.
12 IUCN. 1993. *Reefs at Risk*. A programme of action. Pp24.
13 Wilkinson, C. R. and Buddemeier, R. W. 1994. *Global climate change and coral reefs: implications for people and reefs*. Report of the UNEP-IOC-ASPEI-IUCN Global Task Team on the Implications of Climate Change on Coral Reefs. Pp 124.
14 World Conservation Monitoring Centre. 1999. Checklist of fish and invertebrates listed in the CITES Appendices and in EC Regulation 338/97. 4th edition. JNCC Reports, No. 292.
15 Harmelin-Vivien, M.L. 1989. Reef fish community structure: an Indo-Pacific comparison. In, Harmelin-Vivien, M.L. and Bourlière, F. (eds). *Vertebrates in complex tropical systems*. Springer-Verlag, New York.
16 Bell, J.D. and Galzin, R. 1984. The influence of live coral cover on coral reef fish communities. *Marine Ecology Progress Series* 15(3): 265-274.
17 Bouchon-Navarro, Y., Bouchon, C. and Harmelin-Vivien, M.L. 1985. Impact of coral degradation on a chaetodontid fish assemblage (Moorea, French Polynesia). *Proceedings of 5th International Coral Reef Symposium*. 5: 427-432.
18 Phillips, R.C. and Meñez, E.G. 1988. Seagrasses. *Smithsonian contributions to the marine sciences*, number 34. Pp 104.
19 Dybas, C. L. 1993. Sequoia forest beneath the sea. *Wildlife Conservation* 96 (4):24-35.
20 Angel, M.V. 1993. Biodiversity of the pelagic ocean. *Conservation Biology* 7(4):760-772.
21 Gjøsaeter, J. and Kawaguchi, K. 1980. A review of the world resources of mesopelagic fish. FAO Fisheries Technical Paper 193:1-151.

22 Rowe, G.T. 1983. Biomass and production of the deep-sea macrobenthos. In, Rowe, G.T. (ed). *Deep-Sea Biology.* Vol 8. *The Sea.* Pp 453-472. John Wiley and Sons, New York.

23 Hessler, R.R. and Sanders, H.L. 1967. Faunal diversity in the deep-sea. *Deep-Sea Research* 14:65-78.

24 Angel, M.V. 1991. *Biodiversity in the deep ocean.* A working document for ODA. Unpublished MS.

25 Grassle, J.F. 1991. Deep-sea benthic biodiversity. *Bioscience* 41(7).

26 Grassle, J.F. and Maciolek, N.J. 1992. Deep-sea species richness: Regional and local estimates from quantitative bottom samples. *Amer. Nat.* 139:313-341.

27 May, R.M. 1992. Bottoms up for the oceans. *Nature* 357:278-279.

28 Angel, M.V. 1982. *Ocean trench conservation.* Commission on Ecology papers No. 1. IUCN.

29 Gage, J.G. and Tyler, P.A. 1991. *Deep-sea biology: a natural history of organisms at the deep-sea floor.* Cambridge University Press.

30 Grassle, J.F. 1985. Hydrothermal vent animals: distribution and biology. *Science* Vol 229.

31 Grassle, J.F. 1986. The ecology of deep-sea hydrothermal vent communities. *Advances in Marine Biology* Vol. 23. Academic Press.

32 Grassle, J.F. 1989. Species diversity in deep-sea communities. TREE (*Trends in Ecology and Evolution*) 4(1).

33 Hecker, B. 1985. Fauna from a cold sulphur-seep in the Gulf of Mexico: comparison with hydrothermal vent communities and evolutionary implications. *Bulletin of the Biological Society of Washington* 6:465-473.

34 UNCED. 1992. *The global partnership for environment and development.* A guide to Agenda 21. UNCED, Geneva.

35 Committee on Biological Diversity in Marine Systems. 1995. *Understanding marine biodiversity: a research agenda for the nation.* National Academy Press, Washington D.C.

36 FAO. 1990. *Review of the state of world fishery resources.* FAO Marine Resources Service, Rome.

37 FAO 1999. *The State of World Fisheries and Aquaculture* 1998. FAO, Rome, Italy. Fisheries data from FISHSTAT, the FAO Fisheries database.

38 Goldschmid, A. 1999. Essay about the phenomenon of Lessepsian Migration. Colloquial Meeting of Marine Biology I, Salzburg 1999. Http://www.sbg.ac.at/ipk/avstudio/pierofun/lm/Lesseps.htm

39 Carlton, J.T., Thompson, J.K., Schemel, L.E. and Nichols, F.H. 1990. Remarkable invasion of San Francisco Bay (California USA) by the Asian Clam *Potamocorbula amurensis.* I. Introduction and dispersal. Marine Ecology – Progress Series 66: 81-94.

40 UNEP. 1998. Report of the workshop on invasive *Caulerpa* species in the Mediterranean, Heraklion, Crete, Greece, March 18-20th 1998. UNEP(OCA)/MED WG. 139/4

41 Griffin, M. 1993. It's collapsing completely. *Ceres – the FAO Review* 26(4): 28-31.

42 Mee, L.D. 1992. The Black Sea in crisis: the need for concerted international action. Ambio 21(4): 278-286.

43 Sherman, K. and Busch. D.A. 1995. Assessment and monitoring of large marine ecosystems. In: Rapport D.J., Guadet, C.L. and Calow, P. (eds). Evaluating and monitoring the health of large-scale ecosystems. Springer-Verlag, Berlin. (Published in cooperation with NATO Scientific Affairs Division). NATO Advanced Science Institutes Series. Series 1: *Global Environmental Change,* Vol 28. Pp 385-430.

44 American Association for the Advancement of Sciences. 1986. *Variability and management of large marine ecosystems.* AAAS Selected Symp. 99. Pp 319. Westview Press Inc, Boulder.

45 American Association for the Advancement of Sciences. 1989. *Biomass yields and geography of large marine ecosystems.* AAAS Selected Symp. 111. Pp 493. Westview Press Inc, Boulder.

46 American Association for the Advancement of Sciences. 1990. *Large marine ecosystems: patterns, processes and yields.* Pp 242. AAAS Press, Washington DC.

47 American Association for the Advancement of Sciences. 1991. *Food chains, yields, models, and management of large marine ecosystems.* Pp 320. Westview Press Inc, Boulder.

48 American Association for the Advancement of Sciences. 1993. *Large marine ecosystems: stress, mitigation, and sustainability.* Pp 376. AAAS Press, Washington DC.

49 Anthony, V.C. 1993. The state of groundfish resources off the northeastern United States. *Fisheries* 18(3):12-17.

50 Collie, J.S. 1991. Adaptive strategies for management of fisheries resources in large marine ecosystems. In: Sherman, K., Alexander, L.M., and Gold, G.D. (eds). *Food chains, yields, models, and management of large marine ecosystems.* Pp 225-242. Westview Press Inc, Boulder.

51 Murowski, S.A. 1996. Can we manage our multispecies fisheries? In, *The northeast shelf ecosystem: assessment, sustainability, and management.* Pp 491-510. Blackwell Science.

52 Beddington, J.R. 1984. The response of multispecies systems to perturbations. In, May, R.M. (ed). *Exploitation of marine communities.* Pp 209-205. Springer-Verlag, Berlin.

53 Levin, S.A. 1993. Approaches to forecasting biomass yields in large marine ecosystems. In, Sherman, K., Alexander, L.M. and Gold, B.D. (eds). *Large marine ecosystems: patters, processes and yields.* Pp 179-187. AAA Press, Washington DC.

54 Mangel, M. 1991. Empirical and theoretical aspects of fisheries yield models for large marine ecosystems. In, Sherman, K., Alexander, L.M. and Gold, B.D. (eds). *Food chains, yields, models, and management of large marine ecosystems.* Pp 243-261. Westview Press, Boulder.

55 Tang, Q. 1989. Changes in the biomass of the Yellow Sea ecosystems. In, Sherman, K. and Alexander, L.M. (eds). *Biomass yields and geography of large marine ecosystems.* AAAS Selected Symp. 111. Pp 7-35. Westview Press Inc, Boulder.

56 Crawford, R.J.M., Shannon, L.V. and Shelton, P.A. 1989. Characteristics and management of the Benguela as a large marine ecosystem. In, Sherman, K. and Alexander, L.M. (eds). *Biomass yields and geography of large marine ecosystems.* AAAS Selected Symp. 111. Pp 169-219. Westview Press Inc, Boulder.

57 Kelleher, G. 1993. Sustainable development of the Great Barrier Reef as large marine ecosystem. In, Sherman, K., Alexander, L.M. and Gold, B.D. (eds). *Large marine ecosystems: stress, mitigation, and sustainability.* Pp 272-279. AAAS Press, Washington DC.

58 Sainsbury, K.J. 1988. The ecological basis of multispecies fisheries, and management of a demersal fishery in tropical Australia. In: Gulland, J.A. (ed). *Fish population dynamics.*2nd edition.. Pp 349-382. John Wiles & Sons, New York.

59 Scully, R.T., Brown, W.Y. and Manheim, B.S. 1986. *The convention for the conservation of antarctic marine living resources: a model for large marine ecosystem management. Variability and management of large marine ecosystems.* AAAS Selected Symp. 99. Pp 281-286. Westview Press Inc, Boulder

60 North Sea Quality Status Report. 1993. Oslo and Paris Commissions, London. Pp 132 + vi pp. Olsen and Olsen, Fredensborg, Denmark.

61 Eikeland, P.O. 1992. *Multispecies management of the Barents Sea large marine ecosystem: a framework for discussing future challenges.* The Fridtjof Nansen Institute, Polhogda, Norway.

62 Hey, E. and Mee, L.D. 1993. Black Sea. The ministerial declaration: an important step. *Environmental Policy and Law* 2315:215-217, 235-236.

63 Kelleher, G., Bleakley, C and Wells, S. (eds). 1995. A *Global Representative System of Marine Protected Areas Vols* 1-4. IBRD/World Bank, Washington, USA.

64 Salm, R.V. and Clark, J.R. 1984. *Marine and coastal protected areas: a guide for planners and managers.* IUCN, Gland, Switzerland.

65 Wells, S.M. (ed). 1988. Coral Reefs of the World. Volume 1: Atlantic and Eastern Pacific. UNEP Regional Seas Directories and Bibliographies. IUCN, Gland, Switzerland and Cambridge, UK/UNEP, Nairobi, Kenya.

66 Angel, M.V. 1997. What is the Deep Sea? In: Randall, D.J. and Farrell, A.P. (eds) *Deep Sea Fishes.* Academic Press, San Diego, California.

67 Malakoff, D. 1998. Death by suffocation in the Gulf of Mexico. Science 281: 190-192.

68 Diaz, R.J. and Rosenberg, R. 1995. Marine benthic hypoxia: a review of its ecological effects and the behavioural responses of benthic macrofauna. *Oceanography and Marine Biology* 33:245-303.

69 Rana, K. and Immink. A. 1998. Trends in Global Aquaculture Production: 1984-1996. Fishery Information, Data and Statistics Service. FAO, Rome.

71 Nelson, J.S. 1994. *Fishes of the world.* John Wiley & Sons., Inc., NY.

72 IUCN. 1996. 1996 IUCN *Red List of threatened animals.* IUCN, Gland, Switzerland.

73 Loh, J., Randers, J., MacGillivray, A., Kapos, V., Jenkins, M., Groombridge, B., Cox, N. and Warren, B. 1999. *Living planet report* 1999. WWF International, Gland, Switzerland.

74 Spalding, M.D. and Grenfell, A.M. 1997. New estimates of global and regional coral reef areas. *Coral Reefs* 16:225-230.

75 Veron, J.E.N. 1995. *Corals in space and time: the biogeography and evolution of the Scleractinia.* University of New South Wales Press. 321pp.

76 Spalding, M.D. 1998. Biodiversity patterns in coral reefs and mangrove forests: global and local scales. Unpublished PhD thesis, University of Cambridge.

Suggested introductory sources

Summerhayes, C.P. and Thorpe, S.A. 1996. *Oceanography.* Manson Publishing, Southampton.

Longhurst, A.R. and Pauly, D. 1987. *Ecology of tropical oceans.* Academic Press Inc., California, USA. ICLARM Contribution No 389.

7

TERRESTRIAL BIODIVERSITY

Terrestrial ecosystems extend over little more than one quarter of the Earth's surface; they are far more accessible than aquatic habitats and so are far better known. The land supports fewer phyla than the oceans but most of the global diversity of species, and is characterised above all by an extensive cover of vascular plants, with associated animals and soil systems.

Types of forest and woodland form the predominant natural land cover over most of the Earth's surface. These systems generate around half the terrestrial net primary production, and forests in the tropics are believed to hold most of the world's species. Approximately half the area of forest developed in post-glacial times has since been cleared or degraded by humans, and the amount of old growth forest continues to decline.

Grassland and open shrubland, and deserts, between them cover most of the unforested land surface below the high latitudes with frozen subsoil where tundra predominates. These areas tend to have lower species diversity than most forests, with the notable exception of Mediterranean-type shrublands, which support some of the richest floras on earth.

Humans have extensively altered most grassland and shrubland areas, usually through conversion to agriculture, burning and introduction of domestic livestock. They have had less immediate impact on tundra and true desert regions although these remain vulnerable to global climate change.

THE TERRESTRIAL BIOSPHERE

About 71% of the earth's surface is covered by ocean; the land remaining extends over nearly 150 million km² or about 29% of the total surface of the planet.

With the continents in their present position, more than two-thirds of the land surface is in the northern hemisphere, and the area of land situated north of the Tropic of Cancer slightly exceeds that in the rest of the world put together (Table 7.1). A very small proportion of the total land area is occupied by inland water ecosystems: lakes and rivers cover around 2% and swamps and marshland a similar amount.

About half of the land surface, approximately 52%, is below 500m in elevation, and the mean elevation is 840m. A minor but significant proportion of the land surface is mountainous in nature, with alpine landscapes tending to replace trees above 1000m at higher latitudes and above 3 500m in the tropics.

Features of the terrestrial environment

The distinction between terrestrial and aquatic habitats is profound, and essentially reflects the difference between air and water as media for living organisms.

The most far-reaching difference is that most terrestrial organisms occupy an environment where water loss is a constant threat. The anatomical and physiological solutions to this problem are many and varied, including epidermal and/or cuticular layers with reduced permeability to water, water storage tissues, and metabolic processes that conserve water. However, the greater the risk of dehydration, the more carbon and energy are needed for water conservation mechanisms, leaving less available for other adaptations and diversification. Thus, life remains very sparse and low in diversity in hyperarid environments, whether exceptionally hot or exceptionally cold.

It takes far less energy (about 500 times less) to raise the temperature of a given mass of air by one degree Centigrade than to heat the same mass of water by the same amount, and water conducts heat far more rapidly. This means that while aquatic organisms are buffered against rapid fluctuation in their surroundings, terrestrial organisms can be subjected to wide extremes in temperature, corresponding to daily and seasonal variation in insolation, and also that change between extremes can be very rapid.

The air surrounding terrestrial organisms is much less dense (about 800 times less) than water, and so land organisms must support themselves against the full effects of gravity, but are not subject to the large forces exerted on aquatic organisms by moving water.

Oxygen is freely and uniformly available in the atmosphere, but in water is far less concentrated and much more variable in time and space. On the other hand, mineral nutrients dissolve readily in water and it is possible for aquatic organisms to extract nutrients directly from their immediate surroundings; whereas on land mineral nutrients occur in soil where they are available to microorganisms and plant roots, but not directly to other organisms.

Land cover

The particular conditions faced by terrestrial organisms vary greatly over the surface of the earth, reflecting differences in climate, elevation, terrain, geology, and soil, but in attempting to define broad zones of biological significance, climate and the

Table 7.1

Global distribution of land area, by latitude bands

Region		Land area (million km2)
Northern hemisphere	North of Tropic of Cancer	74
	Equator north to T. Cancer	26
Southern hemisphere	Equator south to T. Capricorn	23.5
	South of Tropic of Capricorn (incl. Antarctica)	23.5

predominant form of vegetation are often given greatest weight.

In the broadest possible simplification, most of the world land surface could be assigned to one of two categories: areas with extensive or significant tree cover, and areas with few or no trees. In terrestrial conditions, primary production (ie. plant growth) is favoured by high soil water availability and relatively elevated temperature during all or some part of the year. Trees tend to be the main plant growth form in such conditions, and forest the main vegetation. Conversely, primary production is limited by a shortage of soil water. Grasses and low shrubs tend to be the main plant growth forms in such dryland regions, and where vegetation exists, it consists mainly of grassland, savanna or scrub.

In some parts of the world, water is present, but temperatures are so low for all or part of the year that tree growth remains insignificant. For example, in polar regions, there is an abundance of water, but because of permanent low temperature it is mainly in solid form and so unavailable to living organisms. In tundra areas at high latitudes, the subsurface is permanently frozen and plant growth is restricted to the few summer months when thawing of the superficial layers makes liquid water available. In upper alpine regions, liquid water is often present in seasonal abundance, but temperatures are too low over the year as a whole to allow tree growth.

Where very low gradients exist between different water and temperature regimes, vegetation types tend to intergrade imperceptibly, and it is impossible without arbitrary definitions to distinguish grassland with a few trees from open woodland with grass ground cover.

Classifying ecosystems

Classification of the immense range of environments at the Earth's surface into a manageable system is a major problem in biology, one not merely of theoretical interest, but of considerable importance in the management and conservation of the biosphere. A wide variety of terms has been coined to describe the units of classification (community, habitat, ecosystem, biome), none of which has a precise and universally accepted definition. Some of these can be seen as forming a loose and ill-defined hierarchy partly analogous with the taxonomic system developed for classifying organisms (Chapter 2), but the classification of the natural environment is far more problematic. Indeed there are good theoretical grounds for questioning the basis of such a classification, because any such system is based on the assumption that the natural environment can be divided into a series of discrete bounded units, whereas the world in reality appears to form a highly variable continuum.

Attempts to classify ecological units are often based on identification of the species that occur in them, along with a description of the physical characteristics of the area. Most terrestrial ecosystems, for example, are generally identified on the basis of plant communities, that is areas with similar plant species composition and structure. A problem with this approach is that the more precisely a community is defined the more site-specific it becomes and hence the more limited its use in higher level analysis and planning. At the other extreme, very general habitat classifications ('forests', 'grasslands', 'wetlands') are based on the physical characteristics and appearance of an area independent of species composition. These terms cover a wide range of possible conditions and so cannot be used alone for precise classification: the term 'forest' applies both to highly diverse lowland tropical rainforest and coniferous monoculture, systems that may have no species in common. Defining such terms entails setting arbitrary limits, eg. for the density of tree cover necessary before an area can be called a woodland (see below). Similarly, it is impossible to determine for how long and how intensely an area must be flooded before it can be classified as a wetland rather than a terrestrial ecosystem. This naturally makes any mapping of habitats a problematic task.

Most global habitat classification systems have attempted to steer a middle course between the complexities of community ecology and the oversimplified general ecosystem labels. Generally these systems will use a more or less elaborate combination of a general definition of habitat type with a climatic descriptor ('tropical moist forest', 'temperate grassland', 'warm deserts and semi-deserts'). Simplification of this kind is valid for general global purposes, although it would be inadequate for any attempt to map land cover or assess change in ecosystems at finer resolution. The task of assessing modification or loss of habitat remains problematic. In large part this is because habitat alteration covers a wide spectrum of change, from short-term, slight and reversible disturbance to complete, and perhaps irreversible, destruction.

Vegetation and chlorophyll

Technical advances in gathering and interpreting data gathered by satellite, particularly by Advanced High Resolution Radiometer (AVHRR) equipment, have made it possible to map the amount of actively photosynthesising vegetation on the world's land surface at a resolution of around 1km. The satellite sensors actually measure the reflectance of vegetation, primarily of the green photosynthetic pigment chlorophyll, in the visible and the near infrared part of the spectrum. On land this can be interpreted as broadly equivalent to the density and vigour of green plant growth, and is commonly represented by the Normalized Difference Vegetation Index (NVDI). Mapped versions of the vegetation index are one useful way to visualise the global distribution of actively growing vegetation without preconceptions or criteria relating to structure, physiognomy or species composition.

The world map shown in Map 7.1 is a plot of observations of chlorophyll density aggregated over one year. This reveals a clear distinction between areas rich in standing growth of plants, whether cropland or natural vegetation, and areas where standing plant growth is sparse or absent - essentially the drylands and rangelands of conventional land use classifications.

Spatial variation in chlorophyll density is only indirectly related to variation in primary production levels, NPP depends further on soil and climate conditions and community dynamics within ecosystems (see Chapter 1). So, while Map 7.1 shows high chlorophyll in both tropical and high latitudes, net primary production is far higher in the former than the latter, where it is restricted mainly by seasonally unfavourable climatic conditions (compare with Map 7.2).

FORESTS

Forests and woodlands once covered about half of global land area and now cover about one quarter. They provide habitat for half or more of the world's species. They are responsible for just under half of the global terrestrial annual net primary production (NPP, Table 1.1), and they and their soils house about 50% of the world's terrestrial carbon stocks. In addition to carbon storage, forests perform many other important ecosystem services, such as regulating local hydrological and nutrient cycles, and stabilising soils and watersheds, especially in areas of unstable topography.

Forests also provide products, including food and fuel, medicines, construction materials and paper, which are important both for human subsistence and for economic activity. Wood products are one of the most economically important natural resources. In the region of 3.4 billion m³ of wood is extracted from forests and other habitats annually, the equivalent of several hundred million trees. Just over half of this volume is used as fuelwood and charcoal, of which developing countries consume 90%. The remainder is industrial roundwood which is processed into various wood products, of which developed countries consume 70% (FAO, 1999). Forests are frequently important culturally and play a significant rôle in the spiritual life of communities worldwide.

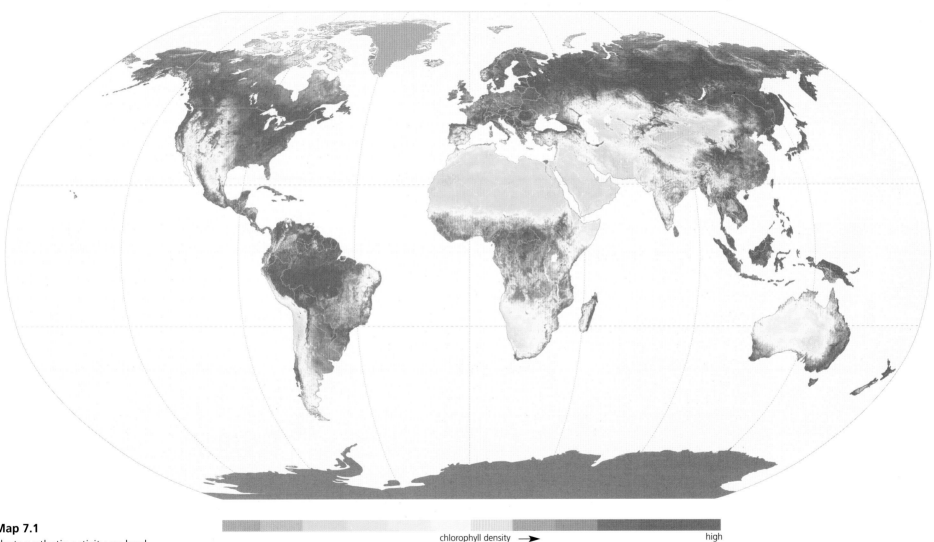

Map 7.1

Photosynthetic activity on land.

chlorophyll density ⟶   high

This map is a plot of observations of Normalized Difference Vegetation Index (NDVI) calculated from AVHRR satellite data and aggregated over the year 1998. NDVI values vary with absorption of red light by plant chlorophyll and the reflection of infrared radiation by water-filled leaf cells, and are therefore correlated with photosynthesis. The map reveals a clear distinction between areas rich in standing growth of plants, whether cropland or natural vegetation, and areas where standing plant growth is sparse or absent.

Source: modified from image provided by the SeaWiFS Project, NASA/Goddard Space Flight Centre and ORBIMAGE.

Table 7.2

Different definitions of forest cover

UNESCO[1]	Closed Forest	trees >=5m with crowns interlocking
	Woodland	trees >=5m tall with crowns not usually touching but with canopy cover >=40%
US classification standards[2]	Closed Tree Canopy	trees with crowns interlocking, with crowns forming 60–100% cover
	Open Tree Canopy	trees with crowns not usually touching forming 10–60 or 25–60% cover
FAO[3,4]	Forest (Developing Countries)	10% crown cover of trees and/or bamboos
	Forest (Developed Countries)	tree crown cover (stand density) of more than 20% of the area
	Closed forest (tropical countries)	tree crown cover greater than 40%

What is a forest?

Despite their importance in a number of different human contexts and the large amount of research focused on forest ecosystems, a precise definition of 'forest' remains elusive. Although it is generally accepted that the term indicates an ecosystem in which trees are the predominant life form, the problem arises because of the broad range of systems in which trees occur. For example, tree species may dominate at high altitude, but be barely recognisable as trees because of their spreading prostrate forms; savannas may have a significant presence of trees, but it is problematic to define where trees are

Table 7.3

Sample effects on forest area estimates of different forest definitions

Country	Forest definition	Forest area (km²)
Australia	Tree canopy cover >20%	384 115
	Tree canopy cover >70%	30 729
Senegal	Tree canopy cover ≥ 10% (includes dry woodland)	78 689
	'Closed' forest (canopy cover >40%)	3 934

predominant. There is surprisingly little consensus on the definition of forest among the many groups that evaluate and monitor natural resources (Table 7.2).

Using different degrees of canopy closure in defining forest can make an appreciable difference to the estimated total area of forest cover for any given location (Table 7.3). The data presented in this volume are based on a threshold of 30% canopy cover for closed forest and between 10 and 30% canopy cover for sparse trees and parkland (or open forest).

Where are forests?

Forests and woodlands were originally distributed throughout the temperate and tropical latitudes of the Earth, except for areas of desert climate, extreme high altitude or latitude, as well as some areas of prairie and steppe. The factors determining their distribution are largely climatic: tree establishment and growth requires a minimum number of days in the year with adequate climatic conditions for active growth. Substrate characteristics are also important: trees require access to enough soil for nutrient and water supply. Other non-anthropogenic factors that limit the distribution of forests include flooding, the incidence of wildfire and the presence of toxic minerals in the substrate.

Forest types

As the difficulty in defining forest implies, there is great variation in the forms and types of forest throughout the world. Information about this variation and the distribution of forest vegetation classes is crucial to understanding the different rôles of forests in supporting biodiversity, in carbon and hydrological cycles and other ecosystem processes, and in supplying wood and non-wood forest products. However, if the definition of forest is problematic, arriving at consensus on how to classify forests is an extremely difficult task.

A number of global classification systems have been

Key to forest types:

Temperate and Boreal

1 Evergreen needleleaf
2 Deciduous needleleaf
3 Mixed broadleaf/needleleaf
4 Broadleaf evergreen

5 Deciduous broadleaf
6 Freshwater swamp forest
7 Sclerophyllous dry forest
8 Disturbed natural forest

9 Sparse trees and parkland
10 Exotic species plantation
11 Native species plantation

Tropical

12 Lowland evergreen broadleaf rainforest
13 Lower montane forest
14 Upper montane forest
15 Fresh water swamp

16 Semi-evergreen moist broadleaf
17 Mixed needleleaf and broadleaf
18 Needleleaf
19 Mangrove

20 Disturbed natural forest
21 Deciduous/semideciduous broadleaf
22 Sclerophyllous

23 Thorn
24 Sparse trees and parkland
25 Exotic species plantation
26 Native species plantation

1 2 3 4 5 6 7 8 9 10 11

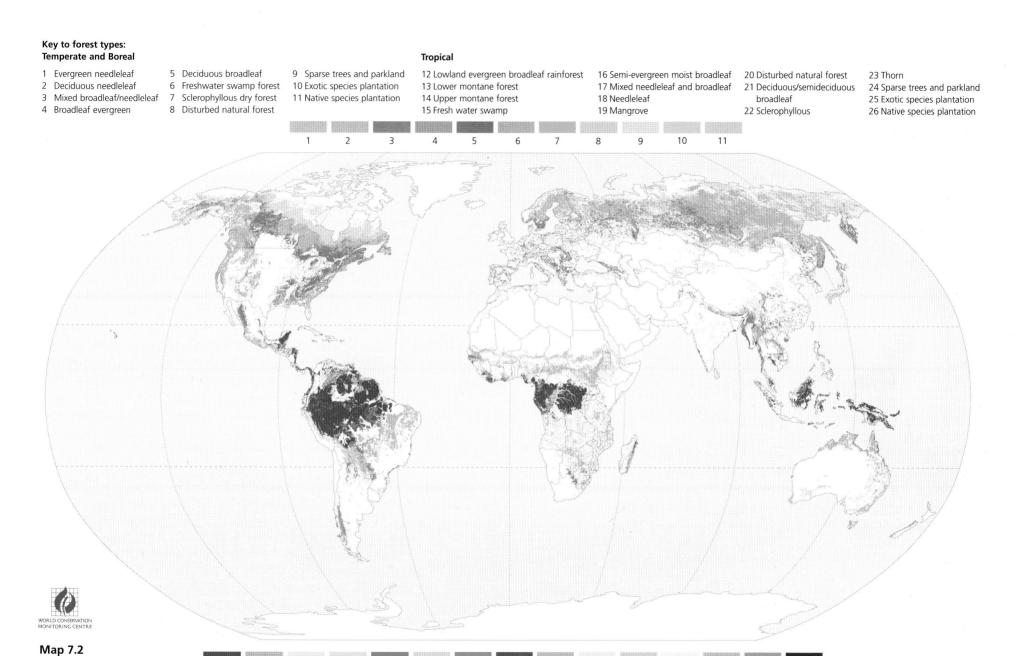

WORLD CONSERVATION
MONITORING CENTRE

Map 7.2

Current forest cover

12 13 14 15 16 17 18 19 20 21 22 23 24 25 26

This map shows the world's forest cover derived from a number of different national and international sources. The dates vary between sources, and this synthesis can be considered to show global cover in approximately 1995. The forest classification used was designed to reflect characteristics of forests that are relevant to conservation and to facilitate harmonisation between various national and international classification systems.

Source: detailed information on data sources is available from WCMC.

proposed, but as yet none has gained universal acceptance. The UNESCO system proposed by Ellenberg and Mueller-Dombois[1] is one such system. It includes nearly 100 forest and woodland 'subformations' and allows for yet finer subdivisions, but many of the characteristics that separate categories can only be determined in the field. Other classifications, such as the EROS Data Centre seasonal landcover regions, with nearly a thousand classes, reflect more strongly the nature of land cover data obtained from Earth orbiting satellites and the processes involved in their classification[6]. This complex system has been translated into a much less complex one in the International Geosphere-Biosphere Programme (IGBP) classification, which includes seven forest and woodland types that reflect phenology and canopy closure worldwide, but provides little other information on forest physiognomy, composition or environment within the class names[6].

A simplification of the more complex systems, which retains a certain amount of basic information on forest physiognomy

and phenology, divides the world's forests into 25 major types, which reflect climatic zones as well as the principal types of trees (Table 7.4, Map 7.2). Of course, each of these major types comprises a great range of forest ecosystems.

TEMPERATE NEEDLELEAF FORESTS

Distribution, types and characteristic taxa

Temperate needleleaf forests cover a larger area of the world than other forest types. They mostly occupy the higher latitude regions of the northern hemisphere, as well as high altitude zones and some warm temperate areas, especially on nutrient-poor or otherwise unfavourable soils. These forests are composed entirely, or nearly so, of coniferous species (Coniferophyta). In the northern hemisphere, pines Pinus, spruces Picea, larches Larix, silver firs Abies, Douglas firs Pseudotsuga and hemlocks Tsuga, make up the canopy, but other taxa are also important. In the southern hemisphere most coniferous trees, members of the Araucariaceae and Podocarpaceae, occur in mixtures with broadleaf species in systems that are classed as broadleaf and mixed forests.

Structure and ecology

The structure of temperate needleleaf forests is often comparatively simple, as conifer canopies are efficient light absorbers, reducing the possibilities for development of lower strata in the canopy. The tallest of these forests, the giant redwood forests of the west coast of the USA, may reach 100 m tall, but most are much shorter, and indeed some pine forests at high altitude or in arid environments are quite stunted. The distribution of temperate needleleaf forest is limited at the northern and high altitude extremes of the range by lack of enough days with temperatures suitable for growth, and at the southern lower altitude extremes, by competition with broadleaf species. In about a quarter of the area of temperate needleleaf forest, deciduous conifers of the

Table 7.4

Areas covered globally by 22 main forest types

Note: plantations excluded. See Map 7.2.

Forest Type	Area (km²)	Forest Type	Area (km²)
Temperate needleleaf		Fresh water swamp	516 142
Evergreen needleleaf	8 894 690	Semi-evergreen moist broadleaf	1 991 013
Deciduous needleleaf	3 616 372	Mixed needleleaf and broadleaf	17 848
		Needleleaf	61 648
Temperate broadleaf & mixed		Mangrove	121 648
Mixed broadleaf/needleleaf	1 803 222	disturbed	842 269
Broadleaf evergreen	342 892		
Deciduous broadleaf	3 738 323	**Tropical dry**	
Freshwater swamp forest	126 963	Deciduous/semideciduous broadleaf	3 034 038
Sclerophyllous dry forest	485 093	Sclerophyllous	405 553
disturbed	60 533	Thorn	262 292
Tropical moist		**Sparse trees & parkland**	
Lowland evergreen broadleaf rainforest	6 464 455	Temperate	2 407 735
Lower montane forest	620 014	Tropical	2 340 959
Upper montane forest	730 635	**TOTAL**	**38 884 337**

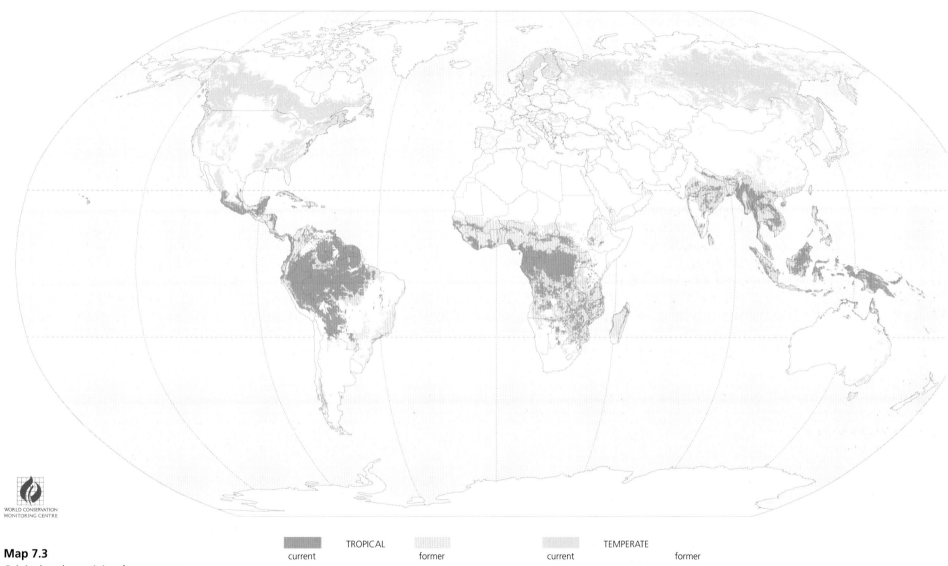

Map 7.3

Original and remaining forest cover

This maps shows the approximate original distribution of temperate and tropical forest cover under current (post-glacial) climatic conditions and before significant human impacts, and the distribution of remaining forest. Approximately half the world's original forest cover has disappeared, and more original forest has been replaced by secondary tree cover, which cannot be distinguished on this map.

Source: detailed information on data sources is available from WCMC.

TROPICAL

current former

TEMPERATE

current former

WORLD CONSERVATION
MONITORING CENTRE

genus *Larix* replace the evergreen species. This is especially true in far northern continental areas with extreme winter temperatures.

In many, if not all, temperate needleleaf forests, wildfire is an important factor affecting the dynamics and maintenance of the forest ecosystem. Many coniferous species produce resins that increase flammability, and many are characterised by thick bark that increases the resistance of adult trees to fire-induced mortality. A number of tree species, such as the jack pine *Pinus banksiana*, have serotinous cones, which depend on the high temperatures of forest fires to open and release their seeds. Non-coniferous species are generally less resistant to fire than conifers, so periodic fires are an important factor in maintaining the composition and extent of these forests.

In general, increasing the frequency of fire decreases its intensity by decreasing the standing amount of fuel. It also tends to increase the diversity of the herb layer by severely affecting the shrub layer. Fires are caused by natural events such as lightning strikes and by human activities. Changing forest management has significantly altered fire regimes in coniferous forests. The 1988 fires that affected over 5000 km[2] in and around Yellowstone National Park, USA, were attributed to the accumulation of fuel in the forests resulting from a long-term policy of fire suppression. Climatic variation plays an important part in determining fire occurrence and severity. The Yellowstone fires, Canadian fires in 1989, and the 1987 Black Dragon Fire in the boreal forest region of China and Siberia 1987, were all ascribed to unusual drought conditions. This last fire burned over 70 000 km[2], and qualified as the largest forest fire in recorded history[7]. There is concern that global climate change may increase the frequency and impact of fires in boreal coniferous forests.

Biodiversity

Although tree species richness is low in most temperate needleleaf forests, those species are in some cases of great conservation concern. A well-known example occurs in the giant redwood forests of northern California, where the redwood *Sequoiadendron giganteum* is considered vulnerable to extinction[8]. However, most of the other 355 globally threatened conifer taxa are characteristic of mixed forests, especially in the Southern Hemisphere. Old growth conifer stands, several centuries in age, represent an irreplaceable gene pool.

Total species richness in temperate needleleaf forests is commonly increased by a relatively high diversity of mosses and lichens, which grow both on the ground and on tree trunks and branches. For example, there are at least 100 species of moss growing in the coniferous forests between 1300 and 2000 m altitude on Baektu Mountain, Korea[9]. Mosses and lichens are important sources of food for many coniferous forest animals.

Vertebrate richness is generally lower in boreal needleleaf forests than in broadleaf temperate and tropical forests. Many of the species are wide-ranging generalists, often with a holarctic distribution, eg. wolf *Canis lupus*, brown bear *Ursus arctos*.

There are a number of animals of conservation concern that are dependent on temperate needleleaf forests. For example, the northern spotted owl *Strix occidentalis caurina* requires large expanses of old growth coniferous forest in the northwest USA to provide nesting habitat and adequate food resources; Kirtland's Warbler *Dendroica kirtlandii*, needs young regrowing Jack Pine as a nesting habitat. Fire suppression programmes have reduced the available habitat for this species to critical levels. While there is relatively little information available on the conservation status of invertebrates, many common old growth species are known to become much rarer in modern managed forests, often through the loss of essential microhabitats[10].

Rôle in carbon cycle

Temperate needleleaf forests are significant in the global carbon balance, accounting for more than a third of the carbon stored in forest ecosystems (Table 7.6) and about 8% of global annual net primary production (Table 1.1). Furthermore, the soils under these forests store large amounts of carbon (up to 250 tonnes per ha), some of which may be liberated by increasing decomposition rates related to climate change. Of particular note in this context are the giant conifer forests of the Pacific Northwest of the USA. These forests may store more than twice as much carbon per hectare as tropical rain forests.

Use by humans

Global industrial roundwood production is dominated by coniferous species. Pines *Pinus*, spruces *Picea*, larches *Larix*, silver firs *Abies*, Douglas firs *Pseudotsuga* and hemlocks *Tsuga* from the needleleaf forests of the northern hemisphere are the major sources of softwood. A few species from the southern hemisphere and the tropics are also excellent timbers. Large-scale exploitation of natural coniferous or mixed forests is taking place around the Pacific Rim, notably in North America, Russia, Chile and Australia. Temperate needleleaf forests are also the principal source of pulpwood for paper production.

Other ecosystem services

Like other forest types, temperate needleleaf forests stabilise soils on sloping topography, especially in mountainous regions. The recognition of this function in early 20th century Switzerland was the basis for a new programme of forest planting to control avalanches in the Alps. Particularly in northern Europe, coniferous forests have high recreational and cultural values.

TEMPERATE BROADLEAF AND MIXED FORESTS

Distribution, types and characteristic taxa

Temperate broadleaf and mixed forests, which include a substantial component of trees in the Anthophyta, cover over 6.5 million km^2 of the Earth's surface (Table 7.4). They are generally characteristic of the warmer temperate latitudes, but extend to cool temperate ones, particularly in the southern hemisphere. They include such forest types as the mixed deciduous forests of the USA and their counterparts in China and Japan, the broadleaf evergreen rain forests of Japan, Chile and Tasmania, the sclerophyllous forests of Australia, the Mediterranean and California, and the southern beech *Nothofagus* forests of Chile and New Zealand.

Especially in the southern hemisphere, trees of the Anthophyta and the Coniferophyta grow in mixtures in many of these forests. For example, the Valdivian and Magellanic rainforests of Chile include mixtures of *Nothofagus* and *Weinmannia* with *Podocarpus*[11]. Much of the forest of New Zealand was originally a mixture of the conifers *Podocarpus*, *Phyllocladus*, *Dacryocarpus* and *Dacrydium* with such Anthophyta as *Metrosideros*, *Elaeocarpus* and *Weinmannia*. In the various North American mixed forests, pines *Pinus*, hemlocks *Tsuga* and cypress *Taxodium*, among other conifers, mix in various proportions with oaks *Quercus*, maples *Acer*, ashes *Fraxinus*, hickories *Carya*, beeches *Fagus*, and other hardwoods. The same hardwood genera are important in the predominantly broadleaf forests of the USA. The beech family (Fagaceae) is generally important in temperate broadleaf forests (Table 7.5), with such genera as *Castanopsis* and *Cyclobalanopsis* playing an important rôle in Japan and China, and many different *Quercus* species being important elements of hardwood forests in North America, Asia and Europe and of sclerophyllous forests in California and the Mediterranean. Sclerophyllous forests in Australia are largely made up of *Eucalyptus* species, as are the wet forest of Tasmania, which may also include *Nothofagus*[13].

Structure and ecology

Depending on the precise forest type, these forests tend to be structurally more complex than the coniferous forests, having more layers in the canopy. It is not uncommon for an upper

Family	Genus	Common name	Northeast America	Europe	East Asia	South America
Fagaceae	Quercus	oak	37	18	66	
	Lithocarpus			1	47	
	Castanopsis				45	
	Cyclobalanopsis	asian oak			30	
	Castanea	chestnut	4	1	7	
	Fagus	beech	1	2	7	
	Nothofagus	southern beech				10
Aceraceae	Acer	maple	10	9	66	
	Dipteronia				1	
Betualceae	Betulus	birch	6	4	36	
	Alnus	alder	5	4	14	
Salicaceae	Salix	willow	13	35	97	
Juglandaceae	Carya	hickory	11		4	
	Juglans	walnut	5	1	4	
	Platycarya				2	
Leguminosae	Cercis	redbud	1	1	2	
	Gleditsia	honey-locust	2		7	1
	Gymnocladus		1		3	
	Maackia				3	
	Robinia	locust	1		1	
	Acacia	acacia				1
	Albizia				10	
Magnoliaceae	Liriodendron	tulip tree	1		1	
	Magnolia	magnolia	8		50	
Oleaceae	Fraxinus	ash	4	3	20	
	Osmanthus		2		10	
Rosaceae	Malus	apple	1	1	8	
	Prunus	cherry	3	4	59	
	Pyrus	pear	1	1	5	
	Sorbus	mountain ash	3	5	18	
	Kageneckia					2
	Quillaia					1
Tilliaceae	Tilia	basswood, lime, linden	4	3	20	
Lauraceae	Phoebe				16	
	Sassafras	sassafrass	1		2	
	Beilschmidia					1
	Cryptocarya					1
	Persea					1
Myrtaceae	Amomyrtus					1
	Myrceungenella					2
	Myrceugenia					8
	Nothomyrcia					1
Ulmaceae	Celtis	hackberry	2	1	14	
	Ulmus	elm	4	3	30	
	Zelkova					3
Carpinaceae	Carpinus	hornbeam	2	2	25	
	Ostrya		1	1	3	

Table 7.5

Important families and genera, and numbers of species, in four areas of temperate broadleaf deciduous forest
Source: after Röhrig[12].

canopy layer to have as many as six distinct subcanopy and understorey layers below it. The tallest of these forests, the mixed forests of southern Chile and some *Eucalyptus* forests of Australia, reach over 50 m in height. On the other hand, some sclerophyllous forests barely reach 5 m in height. The deciduous forests may support rich herb layers, which depend on the increased penetration of sunlight early in the growing season.

As in temperate needleleaf forests, fire plays an important rôle in many types of temperate broadleaf and mixed forest. Especially in the *Eucalyptus* forests of Australia and the sclerophyllous forests of the Mediterranean and California, both natural and anthropogenic fires are important ecological factors affecting the maintenance of forest structure and composition. Spatial variation in forest structure and composition is due to this and other kinds of natural and anthropogenic disturbance. When canopy trees die, the resulting gaps in the canopy increase light levels locally, and such areas may be colonised by a different subset of the forest flora. These gap dynamic processes are important in maintaining stand diversity and are distinct from the secondary successional processes that occur after disturbance by humans. Relatively few old growth broadleaf and mixed forests remain in the temperate zones because of historical exploitation of these forests by both indigenous and immigrant human populations.

Biodiversity

As might be expected from their structural diversity, temperate broadleaf and mixed forests are also richer in species than coniferous forests. Southern mixed hardwood forests in the USA are commonly composed of as many as 20 canopy and subcanopy tree species and may include as many as 30 overstorey species, in a relatively large area[14]. European forests are less species rich, while the deciduous forests of East Asia may be the richest of all[15] (Table 7.5).

In 1996, a draft list of temperate dicotyledonous trees of

conservation concern[16] included 369 taxa. These are nearly all characteristic of temperate broadleaf and mixed forests. Japan alone has 43 threatened endemic tree species, which are mostly characteristic of its temperate broadleaf forests[17]. *Fitzroya cupressoides*, an endangered large conifer once important for local and international timber trade, is characteristic of mixed forests in southern Chile. It has been overexploited and proves to be highly dependent for its regeneration on large-scale natural disturbances, such as landslides and lightning strikes. *F. cupressoides* is now listed in CITES Appendix I and international trade is accordingly prohibited.

While temperate broadleaf forests generally have a high faunal diversity, vertebrate richness is often lower than in comparable tropical habitats. There is considerable geographical variation in richness in the northern hemisphere; the forests of East Asia being generally the most species-rich[18]. As in the boreal needleleaf forests, many of the mammal and bird families of northern broadleaf forests, are holarctic in distribution and can be found in other habitat types.

In contrast to the northern forests the temperate forests of southern South America, Australia and New Zealand contain several restricted-range mammals and birds. Analysis[19] of habitat requirements of Australian mammals indicates that the forests of southeast Australia and Tasmania are of particular importance for wildlife conservation.

Examples of temperate broadleaf forest species of special conservation concern include the huemul deer *Hippocamelus bisulcus* of the southern Andes, threatened by both habitat loss and hunting; a number of New Zealand forest birds including the kakapo *Strigops habroptilus* and some kiwi species *Apteryx* spp., which are threatened mainly by introduced predators; Leadbeater's possum *Gymnobelideus leadbeateri*, which is threatened by the loss of specific habitat within the montane ash forests of Victoria (Australia); the Amami rabbit *Pentalagus furnessi* of Amami Island (Japan), threatened by habitat loss and introduced predators; and the European bison *Bison bonasus* of central and eastern Europe, at risk from disease and low genetic diversity. Several temperate broadleaf invertebrates associated with dead wood (saproxylic) have been identified as of conservation concern, eg. the hermit beetle *Osmoderma eremita*.

Rôle in carbon cycle

Aboveground biomass of temperate deciduous forests in Europe, the USA and the former USSR ranges from 140 to 500 tonnes per hectare depending on stand age and altitude, among other factors[20]. Soil carbon storage in these systems is lower than in the needleleaf forests, due both to climatically favourable conditions for decomposition and to the inherent greater decomposability of leaf litter from broadleaf trees. One survey[21] suggests that average soil carbon storage in these forests is somewhere between 135 and 160 tonnes per ha. Thus, global carbon storage across the forest types included in this broad category may total as much as 231 petagrammes (Table 7.6).

Use by humans

Temperate broadleaf and mixed forests have provided large amounts of timber over the centuries, but they have been largely replaced by tropical forests in the bulk supply of hardwood timber. The forests remaining in most active production are those of southern Chile and Australia. Hardwood production continues elsewhere in the temperate zones, but principally for furniture and finishing wood, rather than bulk construction materials. Some non-timber products of temperate mixed forests include camphor from *Cinnamomum camphorum* in Japan (though this has now largely been replaced by synthetics), and sweet chestnuts *Castanea sativa* from southern Europe. Mushroom production is also a major income source in some parts of Europe and Asia.

Table 7.6

Biomass and carbon storage in the world's major forest types

Note: The carbon storage figures are over-, rather than underestimates as they incorporate no weighting for anthropogenic disturbance or for the variation in biomass and area among different forest classes within the broad types.

Source: after Adams[23] and Huston[23].

Forest type	Above ground biomass per area (t/ha)	Estimated total carbon storage including soil and roots (t/ha)	Estimated global total carbon storage (petagrammes =10^{15}g)
Temperate Needleleaf	200-1500 (-3400)	~300 (700 giant conifer)	375
Temperate Broadleaf and Mixed Forest	150-300 (-500)	~350	231
Tropical Moist Forest	195-500 (-1000)	~300	339
Tropical Dry Forest	98-320	~250	92

Other ecosystem services

Services provided by temperate broadleaf and mixed forests include soil and watershed protection. This is especially important in the southern Andes[11], and other areas where steep topography is responsible for a high incidence of landslides, but it is also important in Mediterranean regions where soils are prone to degradation. Natural temperate forests are important reservoirs of genetic material of trees such as eucalypts that are now commonly grown as plantation species. These forests, perhaps more than any other class, have significant recognised recreational and aesthetic values throughout much of their range.

TROPICAL MOIST FORESTS

Distribution, types and characteristic taxa

Tropical moist forests cover more than 11 million km² of the humid tropics (Table 7.4 and Map 7.2) and include many different forest types. The best known and most extensive are the lowland evergreen broadleaf rainforests, which make up over half the total area and include, for example: the seasonally inundated *varzea* and *igapó* forests and the *terra firme* forests of the Amazon basin; the peat forests and moist dipterocarp forests of Southeast Asia; and the high forests of the Congo basin. The forests of tropical mountains are also included in this broad category, generally divided into upper and lower montane formations on the basis of their physiognomy, which varies with altitude. The montane forests include cloud forest, those forests at middle to high altitude which derive a significant part of their water supply from cloud, and support a rich abundance of vascular and nonvascular epiphytes. Mangrove forests also fall within this broad category, as do most of the tropical coniferous forests of Central America.

The high diversity of many tropical forests (see below) makes it difficult to characterise them taxonomically. It is not possible to generalise about characteristic species, or even genera. However, some plant families are more prevalent than others. In neotropical moist forests, the legumes Leguminosae are both abundant and, usually, the most species-rich family. Other families that are generally among the ten most rich in tree species in lowland neotropical moist forests are: Moraceae, Lauraceae, Annonaceae, Sapotaceae, Myristicaceae, Meliaceae, Euphorbiaceae and Palmae[24]. In Southeast Asian lowland moist forests the dominant family is the Dipterocarpaceae, and the Myrtaceae is also very speciose[25]. In African rain forests, legumes are important, and the ten richest families usually include the Olacaceae, Sterculiaceae, Dichapetalaceae, Apocynaceae, Sapindaceae and Ebencaeae. The remainder of the top ten families in both Africa and Asia are often the same as those in the Americas[24].

Structure and ecology

Many tropical moist forests have canopies 40 to 50 m tall, and some have emergent trees that rise above the main canopy to heights of 60 m or more. Such large-stature forests are characteristic of lowland forests and some lower montane forests on relatively nutrient-rich soils. Another characteristic of these forests is a relatively high frequency of woody lianas and, especially in the neotropics, palms[24]. Moist tropical forests are also known for a very high abundance and diversity of vascular

Region	Number of tree species (≥ 10 cm diameter) per hectare
Africa	56 - 92
Southeast Asia	108 - 240
Americas	56 - 285

Table 7.7

Tree species richness in tropical moist forests
Source: after Phillips et al.[29].

epiphytes, which take advantage of the higher light levels found in the canopy and can survive because of abundant rainfall and high atmospheric moisture. On more nutrient-poor soils and at higher altitudes forest stature decreases substantially; forest communities such as those on white sands (*bana* and *campina*) and in upper montane environments (elfin forests) may be no more than a few metres tall. With increasing altitude, decreasing forest stature is accompanied by a reduction in the frequency of lianas and palms, and an increase in tree ferns (Cyatheaceae, Blechnaceae) and non-vascular, as well as vascular epiphytes[26]. Mangrove ecosystems are still more different from lowland rain forests in many aspects of their structure and function (see Chapter 6).

Unlike the other forest types discussed here, tropical moist forests have relatively little seasonal limitation to their growth, though seasonal drought may be a limiting factor, particularly in the semi-evergreen formations. However, the tropical moist forest environment is an intensely competitive one. Though solar energy inputs are high, canopy closure and complexity are also high, resulting in efficient capture of incident radiation and understorey light levels frequently much less than 2% of those above the canopy. This in turn limits the growth of understorey species and regenerating trees. Some species can tolerate low light levels, while others grow or regenerate only in gaps in the canopy. Such gaps are formed by the death of one or more canopy trees, and represent a significant contribution to overall environmental heterogeneity, which is an important contributor to high diversity within tropical forests[23]. Infrequently, catastrophic disturbances such as blowdowns caused by hurricanes or convective storms may create large areas of regenerating forest[27,28], and perhaps alter the long term forest composition, as can logging and other forms of forest clearance by man (see below).

Soil nutrients are another limiting resource, whose availability is regulated by competition in most tropical moist forests. Soils in the humid regions of the tropics are notoriously poor in nutrients, due to loss of nutrients through leaching by the high annual rainfall, and to the retention of nutrients in the high standing biomass, among other factors. The formation of gaps in the canopy allowing regeneration of canopy species is important for increasing the heterogeneity of availability of nutrients as well as solar energy. Forests such as the Amazonian *várzea*, which are seasonally inundated by sediment-bearing rivers are an exception to this nutrient limitation.

Biodiversity

In numerical terms, global terrestrial species diversity is concentrated in tropical rain forests. Many theories have been proposed to explain this phenomenon (and see Chapter 5)[23]. Generally speaking, the wet tropical forests of Africa have a lower tree species richness than those of Asia and America (Table 7.7). However, there is great local variation in species richness. Within the Amazon Basin, tree species richness ranges from 87 species per hectare in the eastern Amazon Basin[30] to 285 species in central Amazonia[31] and nearly 300 species in the western Amazon[32].

The high diversity of tree species in lowland evergreen rain forests is mirrored in the diversity of epiphytes and lianas, which is also much higher in neotropical forests than in other regions[33]. Fifty-three families in the Anthophyta and at least nine pteridophyte (Filicinophyta and allies) families include epiphytes. Of nearly 25,000 species of vascular epiphytes, around 15,000 belong to the Orchidaceae. Nearly a thousand others are members of the pineapple family Bromeliaceae, which is essentially neotropical. Other groups having a high diversity of epiphytic species are: the cactus family Cactaceae, the aroids Araceae, the pepper family Piperaceae and the African violet family Gesneriaceae.

Not all tropical moist forests have high species richness. Mangrove ecosystems have a low diversity of tree species despite their sometimes high productivity and high animal diversity (see chapter 6). Extremely nutrient-poor soils, such as white sands, lead to the development of low diversity forests including *bana* and *campina*[34]. As climate becomes more seasonal the extreme tree species richness of tropical rain forests decreases (see dry forests, below). Increasing altitude also tends to reduce species richness, with montane forests typically having fewer tree species than lowland ones[36]. In a few cases, low diversity forests occur adjacent to high diversity ones, with no obvious variation in soil fertility or other environmental factors to account for the difference. These forests include those in Central Africa dominated by the legume *Gilbertiodendron*, East African *Cynometra* forest, and *Mora excelsa* forest in Trinidad. One hypothesis[35] is that their existence is due partly to the large size of the dominant species and their ability to regenerate in their own shade, as well as a low frequency of disturbance.

Regionally, the forests of Asia and South America are considered to be especially rich in animal species, those of Africa much less so[36]. Many moist forest animals are largely confined to these forests, eg. the okapi *Okapia johnstoni*, but some are widespread outside, such as African elephant *Loxodonta africana* and leopard *Panthera pardus*[36]. In Africa, the Guineo-Congolean forest block contains 84 percent of African primate species, 68 percent of African passerine birds, and 66 percent of African butterflies[36]. About half of the 1100 South American reptile species are found in moist forests, with around 300 of these endemic to the habitat[37]. Amphibian species are particularly diverse in tropical moist forests[38]; ninety percent of 225 species identified in the Amazon basin forests are endemic. The importance and diversity of the fish communities in forest streams and rivers is often overlooked; the Amazon basin has the richest fish fauna known, with at least 2500 species, many important as major seed predators and dispersal agents[39]. Local species richness of insect and other arthropod groups in tropical forest canopies is much higher than in temperate forests[40].

Around one third of the animal biomass of the Amazon terra firme rain forest is comprised of ants and termites, and each hectare of soil is estimated to contain more than 8 million ants and 1 million termites[41].

Numerous tropical moist forest species are of conservation concern; notable animals include the Sumatran rhino *Dicerorhinus sumatrensis* of Southeast Asia, endangered by habitat fragmentation and hunting; the bonobo *Pan paniscus* of the Congo D.R., is generally threatened by habitat destruction, and some hunting for food; the Philippine eagle *Pithecophaga jefferyi* of the Philippines, has been reduced to small fragmented populations through habitat loss and hunting; the indri *Indri indri* of eastern moist forest on Madagascar is threatened by habitat destruction, as is the recently rediscovered Edward's pheasant *Lophura edwardsi* of Viet Nam. Relatively few tropical forest herpetofauna have had their conservation status globally assessed; however a number of frogs from the forests of Queensland, Australia (eg *Litoria lorica*, *Taudactylus acutirostris*) are recognised to be globally threatened.

Rôle in carbon cycle

Lowland evergreen broadleaf rain forests can have very high above ground biomass (Table 7.6), though not as high as some giant conifer forests. Soil carbon, however is relatively low in most tropical moist forests, with the exception of the peat forests of Southeast Asia and some swamp forests. On this basis it can be calculated that the remaining tropical moist forests store over 300 petagrammes of organic carbon, or about one fifth of global terrestrial organic carbon. They account for nearly a third of global terrestrial annual net primary production (chapter 1). Thus they are key to the global carbon cycle and potentially significant in regulating global climate.

Use by humans

Tropical hardwood species contribute almost a fifth of world industrial roundwood production and almost a third of timber export value[42]. From several thousand species (in over 200 families in the phylum Anthophyta) that show commercial potential, a few hundred may be found in international trade. Important families include Dipterocarpaceae, with species of meranti and balau *Shorea*, and keruing *Dipterocarpus*; Meliaceae, with mahogany *Swietenia* and *Khaya*, and cedar *Cedrela* and *Toona*; and Leguminosae with rosewood *Dalbergia* and *Pterocarpus*.

The exploitation of tropical hardwood from moist forests has been the subject of much publicity in the past two decades, coinciding as it has with increased rates of deforestation and forest degradation. Timber supplies from various countries are now widely exhausted, generating openings for different countries to take over as suppliers. A few major producers, however, continue to dominate supply. Indonesia, Malaysia, Brazil and India accounted for 80% of tropical log production in International Timber Trade Organisation (ITTO) countries in 1997-8, and timber from Indonesia and Malaysia makes up a large part of the export volume[42].

Important non-timber products from tropical moist forests include rattans, which are the second most important source of export earnings from tropical forests. A few other craft products and some medicinal products, such as the bark of *Prunus africana*, are significant in international trade. Brazil Nuts *Bertholettia excelsa* and native rubber *Hevea brasiliensis* are other extractive products from natural tropical moist forests that are important in international markets, and many tropical moist forests provide fruit, bushmeat and other products for local markets.

Other ecosystem services

Like other forest types, tropical moist forests often play an important rôle in soil and watershed protection. The high rainfall regimes of the humid tropics mean that exposed soil is particularly liable to leaching of mineral nutrients and to erosion. Montane forests, and especially cloud forests, serve to intercept and store water, thus regulating local and regional hydrological cycles[43]. Lowland forests are also important in hydrological cycles; it has been estimated that about half the rainfall in the Amazon Basin is derived from water recycled by forest transpiration[44].

Regenerating forest, or forest fallow, is an important part of the cycle of shifting cultivation, which is vital for restoring fertility to areas that have been previously cultivated. This regeneration can only take place if nearby forest cover is adequate to provide a source of propagules.

TROPICAL DRY FORESTS
Distribution, types and characteristic taxa

Tropical dry forests are characteristic of areas in the tropics affected by seasonal drought. Such seasonal climates characterise much of the tropics, but less than 4 million km² of tropical dry forests remain. The seasonality of rainfall is usually reflected in the deciduousness of the forest canopy, with most trees being leafless for several months of the year. However, under some conditions, eg. less fertile soils or less predictable drought regimes, the proportion of evergreen species increases and the forests are characterised as 'sclerophyllous'. Thorn forest, a dense forest of low stature with a high frequency of thorny or spiny species, is found where drought is prolonged, and especially where grazing animals are plentiful. On very poor soils, and especially where fire is a recurrent phenomenon, woody savannas develop[45,46] (see 'sparse trees and parkland' below).

Perhaps the best known tropical dry forest tree species is teak, *Tectona grandis* (Verbenaceae), a deciduous hardwood characteristic of the seasonal forests of South and Southeast Asia, widely exploited for furniture and other uses, and now an important plantation species. In Southeast Asia the dipterocarps, *Shorea* and *Dipterocarpus*, and *Lagerostroemia* (Lythraceae), and a number of species of legumes (Leguminosae) are also important components of seasonally dry forests[47]. In

Africa, dry forest occurs both north and south of the equatorial rainforests. In the north they are characterised by *Afraegele* (Rutaceae), *Diospyros* (Ebenaceae), *Kigelia* (Bignoniaceae) and *Monodora* (Annonaceae), among other taxa, while in the south the characteristic genera are *Entandophragma* (Meliaceae), *Brachystegia* (Leguminosae), *Diospyros*, *Parinari* (Chrysobalanaceae), *Syzigium* (Myrtaceae) and *Cryptosepalum* (Leguminosae)[48].

In the neotropics, tropical dry forests occur along the Pacific side of Central America, on the leeward sides of Caribbean islands, on the Caribbean coasts of Venezuela and Colombia, in northeast Brazil and in the Chaco region of Bolivia and Paraguay (see Map 7.2). The neotropical dry forests are quite rich in species and include Leguminosae, Bignoniaceae, Rubiaceae, Sapindaceae, Euphorbiaceae, Falcourtiaceae, and Capparidaceae as the families with the largest numbers of species. Important genera include *Tabebuia* (Bignoniaceae), *Trichilia* (Meliaceae), *Erythroxylum* (Erythroxylaceae), *Randia* (Rubiaceae), *Capparis* (Capparidaceae), *Bursera* (Burseraceae), *Acacia* (Leguminosae), and *Coccoloba* (Polygonaceae)[49].

Structure and ecology

Tropical dry forests are generally of lower stature than moist forests, with canopy heights ranging from only a few metres to 30 or occasionally 40 m[63]. The taller forests have multi-layered canopies. Dry forests tend to have more small trees than moist forests and a lower above ground biomass. The trees have a greater proportion of their total biomass below ground, as more extensive root systems help the trees to obtain water from the soil and avoid the most severe drought stress. Dry forests have a much lower incidence of epiphytes than wet forest, and tend to have both higher frequencies and higher diversity of vines and lianas[49].

Plants with specialised mechanisms for avoiding drought stress or conserving water are an important feature of these forests. The proportion of deciduous tree species is thought to increase steadily with decreasing annual rainfall, but factors such as the substrate and the inter-annual variation of seasonal rainfall patterns are also important. Many species have water storage tissues such as succulent stems or tubers, and specialised photosynthetic mechanisms that conserve water are especially common among the epiphytes. Most tropical dry forest trees tend to flower and sometimes to re-leaf before the end of the dry season, and stored water within the plant is essential to this pattern.

As in tropical moist forests, environmental heterogeneity is linked with increased species diversity. Gallery areas along water courses are one source of such variation, and they serve as refuge for animals during the dry season[51]. Termite mounds provide an important source of environmental variation in African dry forests, adding to local topography and supplying high nutrient microenvironments to the system, increasing tree species richness by 40-100%[48].

Biodiversity

Though of lower species richness than tropical moist forests, dry forests still have appreciably more tree species than most temperate forests, with a global range of 33-90 tree species in 1-3 ha[63]. The richest neotropical dry forests, which are not the wettest ones but those in western Mexico and in the Chaco of Southeast Bolivia, have around 90 woody species per 0.1 ha sample[49].

A full assessment has yet to be made, but dry forests are thought to have high rates of plant species endemism relative to wet forests in the tropics[49]. Sixteen percent of the plant species of the Chamela dry forest in Western Mexico are local endemics, and 19% of the flora of Capeira, Ecuador is endemic to western Ecuador[52]. Many of the dipterocarps in Thailand's seasonal forests are national endemics and distinct from the species in the country's moist forests[47].

Vertebrate species diversity is lower in dry forests than in moist forests, but many dry forests have high rates of endemism among mammals, especially among groups such as insectivores and rodents, characterised by low body weights, low mobility and short generation times[51]. Among neotropical dry forests, those of Mexico and the Chaco have the highest numbers of mammal endemics (26 and 22 respectively[51]). Remaining areas of dry forest are often important refuges for once widespread species. The Gir forest of Gujarat (India) contains the only population of Asiatic lion *Panthera leo persica*, once found throughout much of southwest Asia; the dry forests of western Madagascar are inhabited by around 40% of the island endemic lemurs, some, such as red-tailed sportive lemur *Lepilemur ruficaudatus*, are almost entirely confined to this habitat. Invertebrate species richness tends to be poorly known, but in groups such as lepidoptera and hymenoptera, richness in some dry forest areas may be comparable to adjacent wet forest[53].

Because of their high degree of endemism and because degradation and conversion of tropical dry forests has progressed further than in wet forests, their biota are often highly threatened. Hunting, especially for the wildlife trade, and habitat conversion, are important pressures on dry forest animals species. Threatened dry forest species include Spix's macaw *Cyanospitta spixii* which is nearly extinct globally through trapping and habitat loss in Brazil's northeastern caatinga region; the Chacoan peccary *Catagonus wagneri* rediscovered in the Gran Chaco of central South America during the 1970s is threatened by overhunting, habitat loss and disease; Verreaux's sifaka *Propithecus verreauxi* of western Madagascar is at risk from loss of spiny and gallery forest habitat; the Madagascar flat-tailed tortoise *Pyxis planicauda* is restricted to the western Andranomena forest of Madagascar and believed to be declining through habitat destruction.

Rôle in carbon cycle

Their lower biomass means that tropical dry forests represent a smaller reservoir of stored carbon per unit area than the well developed forms of the other forest types discussed so far. With a total biomass ranging from 98 to 320 tonnes per ha[50] and soil carbon storage in the region of 100 tonnes per ha[21], it is unlikely that relatively undisturbed tropical dry forests store more than 250 tonnes of carbon per ha (Table 7.6). This, combined with the fact that little intact tropical dry forest remains worldwide, suggests that the total contribution of seasonally dry tropical forests to global carbon storage is far less than that of other forest types.

Use by humans

Notable among the economically important species of seasonally dry tropical forests is teak *Tectona grandis*, which accounts for about 1 % of reported global tropical timber exports[54]. More than a dozen species of Thailand's seasonal forests are important for the timber trade[47], and important timber species such as mahogany *Swietenia* and several species of *Tabebuia* (Bignoniaceae) are characteristic of neotropical dry forests. The southern dry forests of Africa also contain important timber species such as *Entandrophragma*, which has become an important plantation species. Tropical dry forests are also important sources of fuel wood for local populations. A number of food plants are native to tropical dry forests and medicinal uses have been reported for many dry forest plant species[55]. Craft products are also important.

Other ecosystem services

Protection of relatively fragile soils is an important ecosystem service provided by tropical dry forests. Rains may be intense during the wet season, and erosion can be a severe problem in tropical dry forest areas, where soils are often thin and soil formation processes slow[56]. Tropical dry forests may also be important resources for native pollinators as well as nectar sources for domestic bees. Many dry forest trees produce conspicuous flowers with specialist pollination mechanisms. Their mass flowering provides a major nectar resource for pollinating insects at the end of the dry season when other such resources may well be limited[49,57]. Honey production is one of the livelihoods being promoted for local communities in dry forest areas in Mexico and elsewhere.

SPARSE TREES AND PARKLAND

Sparse trees and parkland are forests with open canopies of 10-30% crown cover. They occur principally in areas of transition from forested to non-forested landscapes. The two major zones in which these ecosystems occur are in the boreal region and in the seasonally dry tropics.

At high latitudes, north of the main zone of boreal forest or taiga, growing conditions are not adequate to maintain a continuous closed forest cover, so tree cover is both sparse and discontinuous. This vegetation is variously called open taiga, open lichen woodland, and forest tundra[58]. It is species-poor, has high bryophyte cover, and is frequently affected by fire. It is very important for the livelihoods of a number of groups of indigenous people, including the Saami and some groups of Inuit.

In the seasonally dry tropics, decreasing soil fertility and increasing fire frequency are related to the transition from closed dry forest through open woodland to savanna. The open woodland ecosystems include the more open Brachystegia and Isoberlinia woodlands of dry tropical Africa and parts of both the caatinga and cerrado vegetations of Brazil[48]. Open woodlands in Africa are more species rich than either closed dry forest or savanna. The cerrado supports a very high diversity of woody plants, though many of them are of shrubby habit.

Animal diversity is generally low in forest tundra; few species are restricted to this habitat, many also occurring in boreal forest or tundra proper. The sparsely wooded tropical savannas are generally more species-rich than temperate forests or grasslands[59]. Wooded savannas vary greatly in richness; those of America are relatively species poor while African savanna sometimes attains a richness not far below rain forest in the same continent. Sparsely wooded areas in Australia are amongst the richest wildlife habitats on the continent, sometimes more so than adjacent wet forests[59,60]. The large savanna vertebrates present in such high diversity in Africa are largely absent from other continents[62] (see Chapter 4). The density and biomass of tropical savanna soil invertebrates (mostly earthworms, ants and termites) is generally lower than temperate grasslands, but greater (at least in biomass) than tropical rain forests[61].

Species of conservation concern include the black rhinoceros Diceros bicornis, threatened primarily by hunting for the horn and the golden-shouldered parrot Psephotus chrysopterygius of northern Queensland (Australia), threatened by the burning of seeding grasses during the breeding season and predation by feral cats.

FOREST PLANTATIONS

Forest plantations, generally intended for the production of timber and pulpwood, increase the total area of forest worldwide. FAO[9] estimates that total plantation area in developed countries is about 600 000 km^2 and in developing countries is about 550 000 km^2. Commonly monospecific and/or composed of introduced tree species, these ecosystems are not generally important as habitat for native biodiversity. However, they can be managed in ways that enhance their biodiversity protection functions and they are important providers of ecosystem services, such as maintaining nutrient capital and protecting watersheds and soil structure as well as storing carbon. They may also alleviate pressure on natural forests for timber and fuelwood production.

There are no global figures for the current output of timber from forest plantations, but it is only a fraction of that from natural forests. In some countries, however, wood from plantation sources makes up a significant portion of the industrial wood supply. In 1997, New Zealand obtained 99% of its industrial roundwood from plantations, Chile obtained 84%, Brazil 62% and Zambia 50%[9].

CHANGES IN FOREST COVER

About half of the forest that was present under modern (i.e. post-Pleistocene) climatic conditions, and before the spread of human influence, has disappeared (Map 7.3), largely through the impact of man's activities. The spread of agriculture and animal husbandry, the harvesting of forests for timber and fuel, and the expansion of populated areas have all taken their toll on forests. The causes and timing of forest loss differ between regions and forest types, as do the current trends in change in forest cover.

The temperate forests of Western Europe have diminished by far more than the 50% estimated for forests globally, but much of this deforestation occurred between 7000 and 5000 years ago as Neolithic agriculture expanded[70]. The expansion of human populations and increasing demand for fuel during classical times and the middle ages put further pressure on European forests. Between Neolithic times and the late 11th century, forest cover in what is now the United Kingdom decreased by 80%. As European forests dwindled they became an increasingly valuable resource that was more carefully managed. Forest cover stabilised during the 19th century in Western Europe in response to both improved management and reduced demand for forest products (due to the increasing use of fossil fuels and changes in construction materials). Since the early 20th century, forest cover in Europe has expanded, often through the establishment of conifer plantations.

In Eastern and Central Europe and in Russia, forest clearance accelerated in the 16th and 17th centuries as sedentary agriculture expanded. One estimate[71] suggests that around one million square kilometres of forest had been cleared in the former USSR up to 1980. Timber exploitation continues to drive forest clearance in the coniferous forests of Siberia.

In North America, indigenous groups had impacts on the forests from at least 12 000 years ago, but most forest clearance evidently took place after European settlement. Eastern North American forest cover reached its minimum around 1860, but then increased with the westward movement of the agricultural frontier and subsequent urbanisation and industrialisation. Forests west of the Appalachians suffered the most severe impacts in the late 19th and early 20th centuries, but are still under pressure from demand for timber and pulp.

In Oceania, as in North America, indigenous groups had significant impacts on the forests before the arrival of Europeans. This was especially true of Aboriginal use of fire in Australia, but European colonisation greatly increased the rate of forest conversion. Over 230 000 km² of forest and 120 000 km² of woodland in Oceania are estimated to have been converted to cropland between 1860 and 1980[71].

In tropical Asia, Africa and Latin America, large-scale deforestation was precipitated by European colonial activities, including agriculture and timber exploitation. Probably more than one million km² of forest and a similar amount of woodland in tropical developing countries were converted to cropland between 1860 and 1980[71]. The bulk of this conversion was in South and Southeast Asia, where forest area declined by 39% from 1880 to 1980.

Globally, tropical dry forest has lost the greatest proportion of its original area of the four major types of closed forest, nearly 70%. About 60% of the original area of temperate broadleaf and mixed forests has disappeared, and tropical moist and temperate needleleaf forests have lost about 45 and 30% of their original area respectively.

Current trends in change in forest cover, which are shown in

Table 7.8, reveal that the rates of deforestation continue to be high in the developing countries of the tropics, in both absolute and proportional terms. In contrast, temperate countries are losing forests at lower rates, or indeed showing an increase in forest area[5].

PRESSURES ON FOREST BIODIVERSITY

The principal pressures on forests and their biodiversity are conversion to other land uses, principally forms of agriculture, and logging. Conversion of forest to agriculture is the main cause of tropical moist forest loss. This is largely due to expanding populations and the use of shifting cultivation at an intensity that does not permit adequate fallow periods. Government resettlement programmes that have moved large numbers of poor farmers have increased the rate of land colonisation and clearance in parts of Southeast Asia and Latin America. In some areas, land has been converted to ranching principally as a means of gaining title in order to permit speculation in land values. Thus, population growth, poverty and inequitable land tenure are among the causes underlying deforestation by conversion to agriculture.

Timber extraction is an important pressure on biodiversity in both tropical and temperate forests. Global consumption of industrial roundwood was nearly 1.5 million m^3 in 1996 and is projected to rise by a further 25% by 2010[5]. Although most timber species are naturally abundant, a factor which usually ensures their survival from commercial exploitation, many have suffered extensive and irreversible population and genetic losses. Furthermore, rare species that are indistinguishable in the field from their commercially important relatives are in danger of extinction through exploitation. This particular problem exists, for example, among the dipterocarp groups *meranti*, *balau* and *keruing*. A few hundred species may be traded under these names, and a significant proportion of these are geographically and ecologically restricted and so at high risk of extinction.

Region	Annual change (km2)	Annual change rate (%)
Africa – tropical	–36 950	–0.7
Africa – non-tropical	–530	–0.3
Asia – tropical	–30 550	–1.1
Asia – non-tropical	1 54	0
Oceania – tropical	–1 510	–0.4
Oceania – temperate	600	0.1
Europe	5 190	0
North America	7 630	0.2
Central America, Mexico and Caribbean	–10 370	–1.3
South America – tropical	–46 550	–0.6
South America – temperate	–1 190	–0.3

Furthermore, logging operations create access to forest areas that may otherwise have remained isolated. This improved access facilitates hunting and other activities that exert pressure on forest biodiversity, and may ultimately lead to colonisation and conversion of the land to agricultural use. There is also strong evidence that logging can increase the probability of wildfire in temperate forests and even in tropical moist forests not usually subject to burning[65].

Particularly in the drier areas of the tropics, fuelwood extraction can have serious impacts on forests and open woodlands. Fuelwood and charcoal consumption more than doubled between 1961 and 1991, and is projected to rise by another 30% to 2395 million m^3 by 2010[4]. About 90% of the consumption is in developing countries, but wood fuel may play an increasing role in some developed countries, increasing demand for wood still further[5].

In addition to loss of area, forest conversion and logging lead to changes in the condition or quality of the remaining forest. These can include fragmentation of large areas of continuous forest. Tropical forest fragments are distinct from continuous forests in both ecology and composition[66]. There are physical and biotic gradients associated with fragment edges, and forest structure undergoes radical change near edges as a result of the impacts of wind and increased tree mortality. Some animal species are 'edge-avoiders' and decline in abundance in forest fragments, while others become more abundant, and some non-forest and even non-native species of plants and animals successfully invade forest fragments but not continuous forest. In addition to directly affecting canopy composition, removal of large timber trees may also affect the availability of seed for regeneration and may affect animal species that depend on the timber species.

Other factors that affect forests and their biodiversity include acid rain and global climate change. So far, most of the effects of acid precipitation, which is caused by industrial air pollutants, have been documented in temperate needleleaf forests and associated waterways of Europe and eastern North America. The likely impacts of global climate change on forests are still being debated, but there seems to be general consensus that the boreal coniferous forests are particularly vulnerable to both range restrictions and increasing fire frequency[67]. Another forest type that has been shown to be vulnerable to climate change is tropical montane cloud forest, which depends upon clouds to supply it with atmospheric moisture. Research has shown that the mean cloud base is moving upwards on tropical mountains as a result of climatic shifts. The forest species are not able to migrate at a comparable rate, and in any case, range shifts will be limited by the land area existing at higher elevations. Local extinctions in cloud forest amphibians, including the golden toad *Bufo periglenes* assessed as critically endangered, have been attributed to climatic fluctuations that may be linked to long-term global climate change[72].

FOREST PROTECTION AND MANAGEMENT

The effects of these many pressures on forests and their biodiversity are to some degree being mitigated by forest

Forest Type	Global Protection (km²)	%
Temperate needleleaf	**675 470**	**5.4**
Evergreen needleleaf	645 406	7.3
Deciduous needleleaf	30 064	0.8
Temperate broadleaf & mixed	**457 535**	**7.0**
Mixed broadleaf/needleleaf	134 872	7.5
Broadleaf evergreen	80 249	23.4
Deciduous broadleaf	205 236	5.5
Freshwater swamp forest	3 319	2.6
Sclerophyllous dry forest	30 394	6.3
disturbed	3 465	5.7
Tropical moist	**1 382 004**	**12.2**
Lowland evergreen broadleaf rainforest	931 775	14.4
Lower montane forest	60 385	9.7
Upper montane forest	127 297	17.4
Fresh water swamp	28 135	5.5
Semi-evergreen moist broadleaf	183 106	9.2
Mixed needleleaf and broadleaf	900	5.0
Needleleaf	6 367	10.3
Mangrove	6 405	13.5
disturbed	27 634	3.3
Tropical dry	**413 524**	**11.2**
Deciduous/semideciduous broadleaf	385 481	12.7
Sclerophyllous	13 508	3.3
Thorn	14 535	5.5
Sparse trees & parkland	**274 401**	**5.8**
Temperate	146 214	6.1
Tropical	128 187	5.5
TOTAL	**3 202 556**	**8.3**

protection and management. Just over eight percent of the world's forests are included in protected areas falling in IUCN management categories I to VI (Table 7.9), with the largest areas in management category VI. Less than 4% of the world's forests are protected in categories I and II, the strictest forms of protection[69].

Temperate needleleaf forests are the least protected, probably reflecting their key role in global wood supply and their perceived lack of importance for biodiversity. Deciduous needleleaf forests, thorn forests and freshwater swamp forests in both tropical and temperate regions are especially poorly protected.

The importance of forest management outside official protected areas is increasingly being recognised. The rôles of private landowners and forest products companies are particularly important. In some regions, private nature reserves may play a significant role in the effective protection of forests[64]. The improved management of forest concessions and private production forests is another important avenue for conserving forests and their biodiversity. Reduced impact logging methods are being widely adopted as part of a suite of efforts to achieve sustainable forest management.

Clear procedures now exist for certifying individual forest management units as being sustainably managed[68]. The availability of wood products from certified sources is increasing rapidly, and market demand for them is also increasing in developed countries. The most advanced certification scheme is that of the Forest Stewardship Council (FSC), but a number of other national and international forest management certification programmes are in development.

The importance of forests for biodiversity protection and other ecosystem services are recognised in a number of international policy processes such as the Convention on Biological Diversity, the Intergovernmental Forum on Forests, and the Framework Convention on Climate Change. The evolution of these processes, in combination with changing markets and public priorities, should help to create the impetus necessary for establishing improved forest protection and management worldwide.

NON-FOREST ECOSYSTEMS

The parts of the earth that are too cold, too dry or too severely affected by fire and/or grazing do not support forest or woodland ecosystems. However, as can be seen from map 7.1, many of them do support active plant growth. The natural non-forest ecosystems with actively photosynthesising vegetation include tundra (both arctic and montane), grasslands and savannas, and shrublands. Less productive, but with unique elements of biodiversity, are the deserts and semi-deserts (Map 7.4).

TUNDRA

Tundra is the vegetation found at high latitudes beyond the low temperature limits of forest growth; the same term is sometimes used for outwardly similar vegetation at high elevation at lower latitudes. These broad categories may be distinguished as 'polar tundra' and 'alpine tundra', respectively. In the arctic, polar tundra occurs north of the northern tree line, which is determined by a number of climatic factors including the summer position of arctic air masses[73] and the depth of permafrost (permanently frozen sub-surface soil). Similarly, alpine tundra occurs above climatic tree line on mountains, and its elevation varies in a complex fashion with latitude, continental or oceanic climate, and the maximum elevation and overall size of the massif[74].

The characteristics of polar and alpine tundra environments differ in many respects. In high latitude polar tundra systems, temperatures are low and daylight hours are few for much of the year, while permafrost limits both drainage and root extension, and the growing season may last for as little as six to ten weeks. Rainfall is low, usually less than 200mm per year, and at extreme latitudes, may be so low that the environment is described as polar desert. At high elevation in temperate regions temperatures may be similar, though permafrost is rare. However, at high altitudes in the tropics, although low temperatures occur every night, high insolation causes warming during the day so that adequate temperatures for active plant growth occur throughout the year and the diurnal temperature range is large.

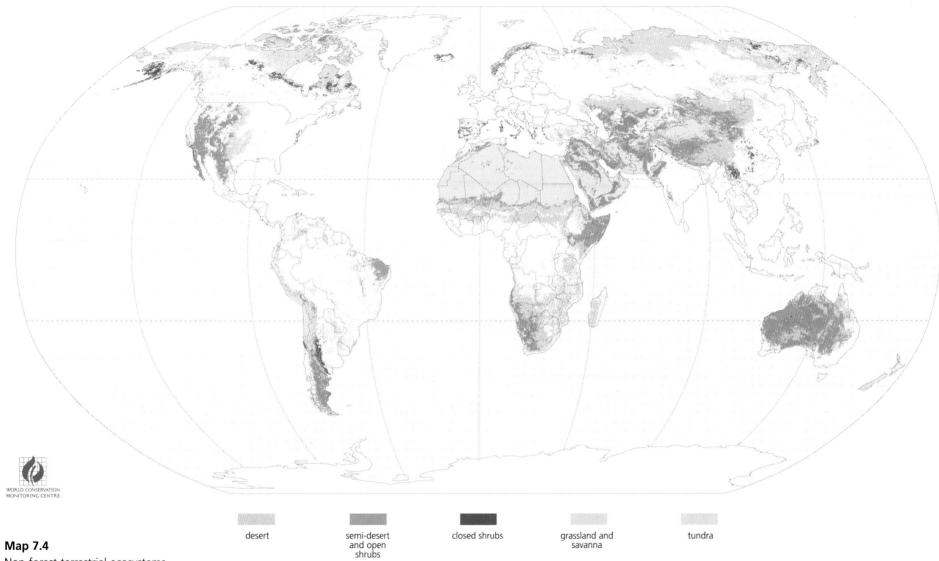

desert   semi-desert and open shrubs   closed shrubs   grassland and savanna   tundra

WORLD CONSERVATION
MONITORING CENTRE

Map 7.4

Non-forest terrestrial ecosystems

On this map, satellite data on landcover have been combined to show the distribution of natural and semi-natural terrestrial ecosystems that are not forest. These are: tundra, closed shrubland, desert and semi-desert including open arid shrublands, and grasslands and savannas.

Source: from USGS EROS Data Center Global Landcover Characteristics Database, version 1.2.

Despite these environmental differences, polar and alpine tundra vegetation have some features in common. Both lack trees, but contain woody species growing in dwarf or prostrate forms, especially in locations with less extreme climates. As latitude or altitude increases, grasses, sedges, bryophytes and lichens increase in importance while shrubs decrease. Many plants have tussock or cushion growth forms. At extremes of latitude or elevation a high proportion of bare ground is characteristic.

In the arctic tundra, plants cover 80-100% of the ground[73], and cover decreases along the climatic gradient to polar desert . The important woody plants are birches *Betula*, willows *Salix*, and alders *Alnus*. These are all genera that occur as trees in temperate regions, but as spreading, prostrate, or dwarf forms, sometimes less than 20 cm tall, in tundra. Other shrub species are also important, including *Dryas*, *Vaccinium* and *Empetrum*. Interspersed with the shrubs and of increasing importance in more extreme sites are the sedges *Carex* and the cotton grasses *Eriophorum*, among other graminoids, and a high diversity of mosses and lichens. Biomass is often much lower above than below ground, and annual production is low. This combination means that the potential of these ecosystems for recovery following disturbance is limited[75].

Temperate alpine systems are characterised by many of the same taxa as the arctic tundra, but at lower latitudes other groups become important. At high altitude in the tropics, giant rosette plants are a notable feature of alpine communities. These distinctive plants, which have a number of morphological and physiological adaptations to the high insolation, large temperature fluctuations and desiccating conditions on tropical mountains, include *Espeletia* and *Puya* in the páramos of the Andes, and *Senecio* and *Lobelia* in the high mountains of Africa[76].

In comparison with forested ecosystems, both polar and alpine tundra are relatively species-poor. It is estimated that species richness declines by a factor of between 3 and 4 between the boreal and arctic zones[75], and species richness in the polar desert is one fifth that of the tundra. The entire North American arctic has a vascular flora of about 600 species[73]. Bryophytes and lichens may add more than 300 additional species to the circumpolar flora. The tropical alpine systems are richer; Venezuelan páramos include over 400 angiosperm species[76], but this is still much lower than in surrounding forests.

Because most arctic plants species have wide geographic distributions, few are of significant conservation concern, but many alpine areas are isolated by lowlands with contrasting climates, so increasing endemism in many alpine floras. The floras of isolated mountains in North America have significant rates of endemism[77], about 15% of the flora above the timberline in the Alps are locally endemic species[78], and about 80% of the flora at high altitude in East Africa and Ethiopia is endemic[79]. Local distribution tends to make species more vulnerable to extinction, while the harsh environment of alpine regions adds to the risk.

Animal species richness tends to be low; groups represented by several species in boreal forest are often reduced in diversity by up to one-third in tundra habitat[75]. In contrast, a few groups, particularly waterbirds and waders, are able to exploit the large numbers of invertebrates found in tundra soil and can be both diverse and abundant[75]. Although the species richness of the most common invertebrate groups (Collembola and oribatid mites) decreases with increasing latitude, their total abundance may increase.

There are relatively few globally threatened species that are completely dependent on tundra. An exception is one of the world's most severely endangered species, the once abundant Eskimo curlew *Numenius borealis* of the Americas, which nests almost exclusively within this habitat.

Although the tundra accounts for less than 2% of global annual net primary production (Table 1.1), the high below-ground biomass and soil carbon in arctic tundra means that it makes an important contribution to global carbon stocks. Total biomass in tundra communities of the Russian arctic falls in the range 7-30 tonnes per ha, of which some 60-70% is below ground[80]. Sedge-moss communities in the North American arctic may have 15-30 times as much biomass below ground as above ground[73], and   3-8 times the amount of dead as live material may accumulate. Tundra soils may store around 200 tonnes of carbon per ha[21].

Because of its inhospitable climate, tundra is not much subject to pressure for conversion for other land uses. However, its lack of ecological resilience means that disturbances, eg. associated with settlements or long-distance pipelines, tend to have long lasting effects. It is anticipated that the effects of global warming on the arctic tundra will be significant, as relatively large temperature increases are predicted for this zone[73]. These will cause changes in the permafrost regime and decomposition of accumulated soil organic matter, which in turn will release additional carbon dioxide into the atmosphere. The period of active plant growth has lengthened in parts of North America. There is also evidence that species are already migrating northwards in response to climate change, so that arctic tundra is likely to be compressed into a much smaller area of remaining appropriate climate.

GRASSLANDS AND SAVANNAS

Grassland ecosystems may be loosely defined as areas dominated by grasses (members of the family Gramineae excluding bamboos) or grass-like plants, with few woody plants[40]. Natural grassland ecosystems are typically characteristic of areas with three main features: periodic drought; fire; and grazing by large herbivores[81]. In addition, they are often associated with soils of low fertility[40]. Savannas are tropical ecosystems characterised by dominance at the ground layer of grasses and grass-like plants. They form a continuum from treeless plains through open woodlands to virtually closed-canopy woodland with a grassy understorey. Some savanna areas therefore meet general definitions of grassland (fewer than 10-15 woody plants per hectare[82]) while others meet the definition of woodland. Some polar and alpine tundra communities may also meet the definition of grassland.

Around 20% of the Earth's land surface (excluding Antarctica) supports grasslands of varying degrees of naturalness[40]; temperate grasslands make up approximately one fourth of this area, and savannas the remainder (Map 7.4). The most extensive areas of temperate grasslands are the prairies of North America, the *pampas* and *campos* of southern South America, and the steppes of Central Europe, southwest and Central Asia, and Russia. Temperate grasslands are sometimes divided into tall-grass, mixed and short-grass formations, which differ both floristically and ecologically; short-grass communities are usually associated with drier climatic regimes[83]. Tropical grasslands and savannas include the llanos of the Orinoco basin in Venezuela and Colombia, the *cerrado* of central Brazil, and the savannas of tropical and subtropical Africa. In addition to many species of grasses, the sedges (Cyperaceae), and many different groups of dicotyledonous herbs are also important.

Key ecological factors in grasslands are grazing pressure and the effects of fire. All natural grasslands had, at one time, large populations of native grazing mammals. These have been replaced to a great extent by domesticated ungulates, which also exert a significant degree of grazing pressure (the magnitude depending on stocking densities). Grazing tends to increase abundance of less palatable species and to increase species richness in productive areas, or decrease it in less productive areas. At intermediate frequencies, fire tends to increase diversity and suppress invasion by woody species. Frequent, eg. annual, fires favour grasses, which usually recover easily, and low fire frequency allows the importance of woody species to increase.

These factors have important consequences for vegetation

structure in grassland and savanna systems. A high proportion of plant biomass, in the form of roots and rhizomes, and metabolic activity is located underground; there is a high turnover of those parts of the plant above ground, and the persistent or perennating parts of the plant are generally located near the soil level[84]. One important consequence of this is that grassland soils, especially in more humid environments, are often rich in organic matter and are thus particularly vulnerable to conversion to cropland, with replacement of native grasses by their domestic derivatives (cereals) and other plants[82].

At very fine spatial scales natural grasslands can be among the most species-rich habitats on Earth. For example, a square metre of 'meadow steppe' in the former USSR may have 40-50 plant species[85], a tall grass prairie remnant of less than 2 ha may contain 100 species, and 250 ha may contain 250-300 plant species[82]. However, grassland communities tend to be similar over large areas, and structurally simple, so that at the landscape scale diversity is relatively low compared with tropical moist forest or Mediterranean-type ecosystems. Because they once occupied extensive contiguous areas, grasslands tend to have low rates of endemism, however, the climatic and soil gradients within them have led to substantial ecotypic variation and high genetic diversity[82].

The world's grasslands and savannas support very distinctive plant and animal communities. Although diversity tends to increase towards the tropics, it tends to be moderate or low at landscape scale and above. Little more than 5% of the world's mammal and bird species are primarily dependent on grasslands habitats[40]. All these systems hold, or formerly held, an array of native herbivores, and these in turn support a number of high profile mammalian and avian predators. The savanna communities of East Africa are typified by large herds of ungulate herbivores, including a remarkable diversity – more than 70 species – of antelope and other medium to large sized bovids. The biomass of ungulate herbivores here may rise to 30 tonnes per km², which is the highest recorded in any terrestrial environment[23]. Most grassland invertebrate biomass is found within the soil (most commonly nematodes, enchytraeid worms, and mites) and may be in order of 100 to 1000 times as great as vertebrate biomass; soil invertebrate biomass above the soil is often dominated by Orthoptera[84].

Many grassland birds require large areas of habitat to take full advantage of the sparsely distributed food resources, and fragmentation of natural grasslands has made conservation of these wide-ranging species difficult[40]. While the number of grassland species now assessed as globally threatened is more modest than might be expected, eg. only some 6% of threatened birds, this may be because some species, such as the steppe marmot *Marmota bobac* in eastern Europe (once highly threatened by hunting and loss of shortgrass steppe) have been able to occupy other agricultural habitats or use them temporarily. On the other hand, local areas of high grassland biodiversity, eg. in the state of Victoria (Australia) have been severely impacted and have a large proportion of the native plants and animals ranked as threatened. Among notable threatened mammal species in grassland are the greater one-horned rhinoceros *Rhinoceros unicornis*, associated with tall riverine grassland in northern India and southern Nepal, where it is threatened by poaching and further loss of its restricted habitat; the saiga antelope *Saiga tatrica*, almost entirely restricted to the steppe grasslands of central Asia where it remains abundant in places but generally in decline; the vicuña *Vicugna vicugna* inhabits the arid montane grasslands and plains of the Andes, and although conservation measures have allowed numbers to increase, the species remains threatened by poaching. Among birds, the short grassland habitat of the plains-wanderer *Pedionomus torquatus* of southern Australia continues to be lost to cultivation; Rudd's lark *Heteromirafra ruddi* inhabits the montane grassland plateaus of eastern South Africa, and is threatened by habitat degradation; the lesser florican *Sypheotides indica*, of western India and Nepal, is already Critically Endangered and still declining through loss of remaining patches of suitable grassland habitat.

Grassland biomass ranges from around 2 to over 80 tonnes per hectare[84.] As a rule, more than 50% of this is below the soil surface, and the ratio of root to shoot biomass ranges from below 5 in warm humid grasslands to as much as 30 in the desert grassland of Mongolia[84]. Usually more than half of the below-ground biomass is in the upper 10 cm of the soil, and soil carbon stocks may be as high as 250 tonnes per hectare[22]. Annual production reaches 30-50 tonnes per hectare per year in some warm humid grasslands. It is estimated that grasslands and savannas together account for about a third of global terrestrial net primary production (Table 1.2), so despite their relatively low biomass grasslands play an important role in the global carbon balance.

Because the high below-ground biomass and its decomposition tend to increase the fertility of their soils, grasslands have been subject to high rates of conversion to agriculture. Less than half of North American grasslands remain in natural or semi-natural states[40], the steppes of the former USSR have been extensively irrigated and converted to agriculture, and much of the *pampas* has been converted to agriculture or grazing land. Some anthropogenic grassland consists of short-term monospecific sown pasture, while some areas support old species-rich semi-natural grassland created over centuries by pastoralists in conjunction with livestock grazing.

Domestic livestock grazing is the most extensive human use of unconverted or anthropogenic grassland ecosystems, and of most arid or semi-arid ecosystems. Livestock have an impact on ecosystems through trampling, removal of plant biomass, alteration of plant species composition through selective grazing, and competition with native species. The impact of this on the biological diversity of these ecosystems has been variable. In some areas where the native vegetation is well adapted as a result of evolution, the impact on plant species diversity has been relatively small. In other areas, where vegetation has not evolved in the presence of hooved herbivores, the changes have been great. Sometimes, particularly in tropical and semi-tropical grasslands, the dominant component of vegetation has shifted from grass to woody plants. Overgrazing can lead to reduction in plant cover, loss and degradation of soil, and invasion by non-native plant species. In almost all cases, wild animal diversity has been greatly affected (mostly through competition and hunting, but also through spread of pathogens), so that the biomass of domestic livestock greatly exceeds that of native wild herbivores. In some areas, feral species (eg. rabbits, camels, donkeys, horses, goats) may also have a very marked impact on natural or semi-natural ecosystems.

SHRUBLANDS

Shrub communities, where woody plants, usually adapted to fire, form a continuous cover, occur in all parts of the world with 200-1000 mm of rainfall[23]. In more arid areas including some semi-desert systems, shrubs are the dominant life form, but cover is discontinuous. The most distinctive and best-known shrublands are those of Mediterranean climate regions.

Mediterranean climates are typified by cool, wet winters and warm or hot dry summers. However, no single climatic or bioclimatic definition of a Mediterranean ecosystem has yet been established, so that these areas remain somewhat loosely defined. Mediterranean ecosystems encompass a wide range of habitat types including forest, woodland and grassland, but are typified by a low, woody, fire-adapted sclerophyllous shrubland (*maquis*, *chaparral*, *fynbos*, *mallee*) on relatively nutrient-poor soils. These systems occur in five distinct parts of the world: the Mediterranean basin; California (USA); central Chile; Cape Province (South Africa); and southwestern and south Australia.

Each of these regions occurs on the west side of a continent and to the east of a cold ocean current that generates winter rainfall. They cover around 2.5 million km² in total, or between 1% and 2% of the Earth's surface (according to definition). More than three-quarters of the total Mediterranean-type ecosystem area is within the Mediterranean basin.

The characteristic shrublands occur within areas of Mediterranean climate where soil fertility is too low to support grass or trees. The shrubs are typically sclerophyllous, ie. have small leathery leaves, which are efficient at conserving water and last a relatively long time for a given investment of mineral nutrients. Differences in vegetation structure between regions are in part a consequence of differences in the annual distribution of rainfall. In South Africa the sclerophyllous fynbos community contains an abundance of ericaceous species as an understorey to low broader-leafed shrubs including Proteaceae and Myrtaceae[86]. The Australian heaths are structurally similar, with Epacridaceae replacing Ericaceae. Californian shrublands, known as chaparral, are characterised by *Adenostoma* (Rosaceae) and a high richness of *Arctostaphylos* and other Ericaceae. The shrublands, or matorral, of Chile include many of the same genera as those of California, while in the Mediterranean basin itself Ericaceae, Cistaceae, Leguminosae and Oleaceae are all important.

Species richness in Mediterranean-type ecosystems, particularly among plants, is generally high – approaching values for moist tropical forest areas – and endemism is also very high. Among the five Mediterranean-type ecosystems, richness appears highest on the poorer soils of South Africa and southwest Australia (Table 7.10), and lower on the richer soils of California, Chile and the Mediterranean proper[23]. Countries around the Mediterranean basin hold some 25 000 vascular species (about 10% of all vascular plants) of which around 60% are endemic to the Mediterranean region. The other four Mediterranean-type ecosystem regions are all considered to hold a disproportionate amount of global biodiversity in relation to their area[87].

At fine scale, mean plant richness in the fynbos of South Africa is moderate, ie. around 16 species per m² quadrat, but many species have small ranges, and there is a uniquely high turnover in the species composition of plant communities along ecological and geographical gradients. At landscape scale, richness accordingly rises to very high values, 2256 species in 471 km² on the Cape peninsula, and the entire Cape floristic region (including some non-fynbos vegetation) holds some 8550 species, about 70% of which are endemic.

The Mediterranean-type ecosystems in general have a relatively high proportion of their species categorised as threatened. The Cape flora, largely within a Mediterranean-type ecosystem, occupies only 4% of the land area of southern Africa, but accounts for nearly 70% of the region's threatened plant species. About one third of the natural vegetation has been transformed by human activity; the remaining natural vegetation is at risk from a number of invasive introduced woody plants, and the effects of an introduced ant (that suppresses native seed-storing ants and thus renders seed liable to destruction by rodents or fire). Around 10% of the California flora is considered threatened (equivalent to approximately one quarter of all threatened plants in the USA). In Australia, heath habitats, primarily in the southwest Mediterranean-type ecosystem

Table 7.10

Estimated plant species richness in the five regions of Mediterranean-type climate

Source: UNEP[81]

Region	Approximate area (millions of km2)	Total number of plant species
Cape Province, South Africa	0.09	8 550
Southwestern Australia	0.31	c 8 000
California	0.32	5 050
Chile	0.14	c 2 100
Mediterranean Basin	1.87	25 000

region, rank third after 'woodland' and 'scrub' in numbers of Endangered category plants. Given their much smaller extent, this indicates that a far higher proportion of their flora is threatened than in either woodland or scrub habitats.

Although invertebrates are poorly known, they are suspected to attain high diversity in parallel with the plants they are often associated with, but levels of diversity in vertebrate animals tend to be lower than in plants, eg. in the Cape Mediterranean-type ecosystem, reptile diversity is only moderate while bird and mammal diversity is relatively low. The absence of large mammals in California and the Mediterranean basin may be linked to over-hunting by humans during the late Pleistocene[88].

Several threatened animal species rely on shrubby or scrub habitat. The Iberian lynx *Lynx pardinus*, found in the light woodland and marquis of Spain and Portugal, is possibly the most threatened cat species; habitat fragmentation and hunting have led to its decline. The riverine rabbit *Bunolagus monticularis* of South Africa is restricted to a small area of riverine bush habitat in the central Karoo, where it is threatened by further loss of this habitat to agriculture. Among birds, the island cisticola *Cisticola haesitatus* is endemic to the island of Socotra in the western Indian Ocean, where it is threatened by loss of light scrub and grassland habitat, possibly through overgrazing by goats.

Mediterranean-type shrublands are not notably high in either biomass or net primary production. Biomass at mature fynbos sites is typically 15-16 tonnes per ha, and in chaparral may be twice that[89]. Combined with their relatively low rates of primary production, related to both climate and soil fertility factors, the incidence of fire tends to reduce the accumulation of carbon in these ecosystems and their soils. Total carbon storage is probably between 100 and 150 tonnes per ha in Mediterranean-type shrublands[22].

The Mediterranean basin itself has for many centuries been subject to intense human activities, including forest clearance and grazing, such that little climax vegetation remains. It has been suggested that the plant diversity is locally high because of the number of species that have evolved as components of successional vegetation in response to frequent disturbance. In other Mediterranean-type shrublands, expanding human populations and conversion of land to agricultural or residential use are important pressures. These changes are often accompanied by changing fire and grazing regimes, and both these changes tend to facilitate invasion by non-native plant and animal species, which threaten native species populations, especially in California and South Africa[90].

DESERTS AND SEMI-DESERTS

Nearly 10 million km^2 of the earth's land area is *hyperarid*, or true desert, where rainfall is extremely low and unpredictable in space and time, so in some years none falls at all. These areas have a ratio of rainfall to potential evapotranspiration (P/PET) of less than 0.05[91]. The Sahara desert alone makes up nearly 70% of the world hyperarid zone. Other extensive areas are found in the Arabian Peninsula and central Asia with smaller areas in southwest Africa, the Horn of Africa, western South America and western North America. In semi-deserts, areas with less arid climates, the vegetation is usually more substantial than in deserts, but covers no more than 80% of the ground. Temperate deserts and semi-deserts cover nearly six million km^2 in Eurasia, and North and South America[92]. Polar regions and some high mountain areas with a permanently cold, dry climate also meet the definition of desert, but have completely different ecological characteristics from true drylands and are not usually considered with them.

True desert species show a wide range of adaptations to the extreme environment. Characteristic plants include the Cactaceae in the Americas and the succulent Euphorbiaceae in Africa. Semi-desert species include salt bush *Atriplex*, and creosote bush *Larrea*. Amongst animals, groups that are intrinsically adapted to very low moisture environments include reptiles and many arthropods, although species in a wide range of other groups have also evolved to cope with these conditions. Strategies for survival amongst both plants and animals often include long periods of dormancy (as seeds, in the case of many plants) punctuated with brief periods of high activity and productivity coinciding with rare rainfall events. In the particular conditions prevailing in the so-called fog deserts (notably the Atacama desert and the Skeleton Coast desert) different strategies have evolved. Here plants and animals make use of the regular moisture-laden fogs, which roll in from the cold offshore currents, to obtain a very low level but predictable supply of water.

Biodiversity, assessed in terms of species number, tends to be moderate in semi-arid areas and to decline to low or very low levels as aridity increases. In contrast to this general rule, diversity in some groups – scorpions and other predatory arthropods, tenebrionid beetles, ants, termites, snakes and lizards, annual plants – tends at first to increase as aridity increases[93]. Desert animals are often wide-ranging but occur at low population densities because of the low primary productivity of these areas.

The often overlooked inland water habitats of deserts may contain a particularly high proportion of locally endemic species; the pools of Cuatro Ciénegas, Mexico, for example, contain numerous mollusc and fish species found nowhere else.

Many of the larger desert and sub-desert species are threatened; the openness of these arid areas means that species such as antelopes and other bovid, are more conspicuous than forest species and thus more vulnerable to over hunting. Threatened vertebrates include the wild bactrian camel *Camelus bactrianus* with a few remnant populations in the Gobi desert of Mongolia and China, and Przewalski's gazelle *Procapra przewalskii* of China's subdesert steppes, now restricted by over hunting and habitat loss to a few small areas surrounding Lake Quinghai. The Mexican prairie dog *Cynomys mexicanus* is confined to prairies and intermontane basins with herbs and grasses where it is threatened by persecution and continuing habitat loss. The Addax antelope *Addax nasomaculatus* originally occurred from the western Sahara to Egypt and Sudan, it is now extinct throughout much of its range as a result of uncontrolled hunting, although small populations may remain in the western Sahel.

In drylands, most adverse impacts that lead to some form of land degradation can be classified as desertification. Under the UN Convention to Combat Desertification, the latter term is defined explicitly as "land degradation in arid, semi-arid and dry sub-humid areas resulting from various factors, including climatic variations and human activities".

By the above definition, hyperarid lands (true deserts) are not susceptible to desertification, as their productivity is already so low that it cannot be seriously decreased by human action. The effects of desertification on arid and semi-arid areas promote poverty among rural people, and by placing greater stress on natural resources, poverty tends to reinforce any existing trend toward desertification.

THE STATUS OF TERRESTRIAL BIODIVERSITY

Information on the causes and assessment of risk to species, and the broad patterns of taxonomic and geographic distribution of threatened species, has been presented in chapter 5. Maps 5.6 and 5.7 illustrate the actual and potential geographic distribution of species in selected groups at risk of extinction; these may serve to indicate geographic variation in levels of risk in other groups. Tables 5.3 and 5.5 include numerical data on the occurrence of threatened species by biome type and at country level, respectively.

WCMC[95] has applied the basic methodology of the WWF Living Planet Index[94] (and see p. 116) to two subsets of terrestrial species in order to provide an interim and partial view of current trends in terrestrial ecosystems.

Fig. 7.2
Population trends in northern grassland birds
Source: WCMC[25], Living Planet Index methodology.

Figure 7.1 shows an index to a series of population estimates relating to a combined group of 170 forest-occurring birds in North America and Europe. In contrast to all other indices shown in this book, the prevailing trend is for a small net increase over the period represented, rather than decrease. In other words, the 'average' forest bird in these regions is faring a little better toward the end of the century than it had done in the 1970s. This could be interpreted as reflecting a phase of relative stability in these forests during recent decades, following centuries of decline in area, and may also be correlated with local increase in forest area through spread of plantations. Management in some forest areas may also have exerted a positive effect. Perhaps most significantly, the separate North America and Europe samples (123 and 47 species, respectively) show extremely similar trends.

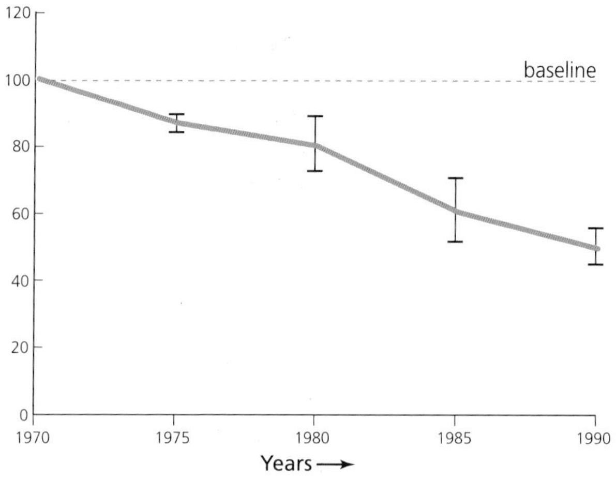

Figure 7.2 illustrates trends over a slightly shorter period in a sample of 29 grassland birds from North America (25 species) and Europe (4). The clear trend here is downward, and as with forest birds, the separate continent samples show a very similar overall pattern. This graph can readily be interpreted as indicative of declining grassland area and quality.

Fig. 7.1
Population trends in north temperate forest birds
Source: WCMC[25], Living Planet Index methodology.

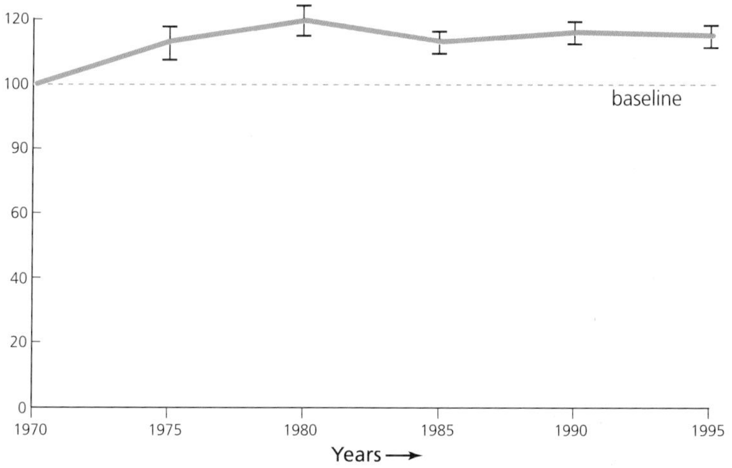

References

1   UNESCO 1973. *International classification and mapping of vegetation.* United Nations Educational, Scientific and Cultural Organisation, Paris.

2   FGDC. 1995. FGDC Vegetation Classification Standards. Federal Geographic Data Committee. Reston, VA. Unpublished.

3   FAO 1993. *Forest resources assessment 1990: Tropical countries.* FAO Forestry Paper 112. Food and Agriculture Organization of the United Nations, Rome.

4   FAO 1995. *Forest resources assessment 1990: Global Synthesis.* FAO Forestry Paper 124. Food and Agriculture Organization of the United Nations, Rome.

5   FAO. 1999. *State of the World's forests.* 1999. Food and Agriculture Organization of the United Nations, Rome

6   Loveland, T.R., Zhu, Z., Ohlen, D.O., Brown, J.F., Reed, B.C. and Yang, L. 1999. An analysis of the IGBP Global land-cover characterization process. *Photogrammetric Engineering and Remote Sensing* 65:1021-1032.

7   Salisbury, H. E. 1989. *The great black dragon fire: a Chinese inferno.* Little, Brown & Co., Boston, USA.

8   Farjon, A. and Page, C.N. 1999. *Status survey and conservation action plan: Conifers.* IUCN, Gland, Switzerland.

9   Hoang Ho-dzung 1987. The moss flora of the Baektu mountain area. Pp. 29-31. In, Yang, H., Wang, Z., Jeffers, J.N.R. and Ward, P.A. (eds). *The temperate forest ecosystem.* Institute of Terrestrial Ecology Symposium 20, NERC, U.K.

10 Väisänen, R., Biström, O. and Heliövaara, K. 1993. Sub-cortical Coleoptera in dead pines and spruces: is primeval species composition maintained in managed forests? *Biodiversity and Conservation* 2(2): 95-113.

11 Veblen, T.T., Schlegel, F.M. and Oltremari, J.V. 1983. Temperate broad-leaved evergreen forests of South America. Pp. 5-32. In, Ovington, J.D. (ed). *Ecosystems of the World 10: temperate broad-leaved evergreen forests.* Elsevier, Amsterdam.

12 Röhrig, E. 1991a. Floral composition and its evolutionary development. Pp. 17-24. In, Röhrig, E. and Ulrich, B. (eds). *Ecosystems of the World 7: temperate deciduous forests.* Elsevier, Amsterdam.

13 Ovington, J.D. and Pryor, L.D. 1983. Temperate broad-leaved evergreen forests of Australia. Pp. 72-102. In, Ovington, J.D. (ed). *Ecosystems of the World 10: temperate broad-leaved evergreen forests.* Elsevier, Amsterdam.

14 Barnes, B.V. 1991. Deciduous forests of North America. Pp. 219-344. In, Röhrig, E. and Ulrich, B. (eds). *Ecosystems of the World 7: temperate deciduous forests.* Elsevier, Amsterdam.

15 Ching, K.K. 1991. Temperate deciduous forests in East Asia. Pp. 539-556. In, Röhrig, E. and Ulrich, B. (eds). *Ecosystems of the World 7: temperate deciduous forests.* Elsevier, Amsterdam.

16 Lear , M. and Hunt, D. 1996. Updating the threatened temperate tree list. Pp. 161-171. In, Hunt, D. (ed). *Temperate trees under threat.* International Dendrology Society, U.K.

17 Ohba, H. 1996. A brief overview of the woody vegetation of Japan and its conservation status. Pp. 81-88. In, Hunt, D. (ed). *Temperate trees under threat.* International Dendrology Society, U.K.

18 Schaefer, M. 1991. The animal community: diversity and resources. Pp. 51-120. In, Röhrig, E. and Ulrich, B. (eds). *Ecosystems of the world 7: temperate deciduous forests.* Elsevier, Amsterdam.

19 Ovington, J.D. and Pryor, L.D. 1981. Temperate broad-leaved evergreen forests of Australia. Pp. 73-99. In, Ovington, J.D. (ed.). *Ecosystems of the world 10: temperate broad-leaved evergreen forests.* Elsevier, Amsterdam.

20 Röhrig, E. 1991b. Biomass and productivity. Pp. 165-174. In, Röhrig, E. and Ulrich, B. (eds). *Ecosystems of the World 7: temperate deciduous forests.* Elsevier, Amsterdam.

21 Zinke, P.J., Stangenburger, A.G., Post, W.M., Emmanuel, W.R. and Olson, J.S. 1984. *Worldwide organic soil carbon and nitrogen data.* Environmental Sciences Division, publication No. 2212. Oake Ridge National Lab, U.S. department of Energy.

22 Adams, J. 1997. *Estimates of preanthropogenic carbon storage in global ecosystem types.* http://www.esd.ornl.gov/projects/qen/carbon3

23 Huston, M.A. 1994. *Biological Diversity: the coexistence of species on changing landscapes.* Cambridge University Press, Cambridge, U.K.

24 Gentry, A.H. 1988a. Changes in plant community diversity and floristic composition on environmental and geographical gradients. *Annals of the Missouri Botanical Garden* 75: 1-34.

25 Whitmore, T.C. 1984. *Tropical rain forests of the Far East.* Clarendon Press, Oxford, U.K.

26 Grubb, P.J. 1977. Control of forest growth on wet tropical mountains. *Annual Review of Ecology and Systematics* 8:83-107.

27 Nelson, B., Kapos, V., Adams, J.B., Oliveira, W.J., Braun, O.P. and do Amaral, I. 1994. Forest disturbance by large blowdowns in the Brazilian Amazon. *Ecology* 75: 853-858.

28 Tanner, E.V.J., Kapos, V. and Healey, J.R. 1991. Hurricane effects on forest ecosystems in the Caribbean. *Biotropica* 23: 513-521.

29 Phillips, O.L., Hall, P., Gentry, A.H., Sawyer, S.A. and Vásquez, R. 1994. Dynamics and species richness of tropical rain forests. *Proceedings of the National Academy of Sciences* 91: 2805-2809.

30 Pires, J.M. 1957. Noçoes sobre ecologia e fitogeografia da Amazônia. *Norte Agronômico* 3:37-53.

31 Oliveira, A.A. de and Mori, S.A. 1999. A central Amazonian terra firme forest. I. High tree species richness on poor soils. *Biodiversity and Conservation* 8:1219-1244.

32 Gentry, A.H. 1988. Tree species richness of upper Amazonian forests. *Proceedings of the National Academy of Sciences of the USA* 85:156-159.

33 Benzing, D.H. 1989. Vascular epiphytism in America. Pp. 133-154. In, Lieth, H. and Werger, M.J.A. (eds). *Ecosystems of the World 14B: tropical rain forest ecosystems: biogeographical and ecological studies.* Elsevier, Amsterdam.

34 Prance, G.T. 1989. American tropical forests. Pp. 99-132. In, Lieth, H. and Werger, M.J.A. (eds). *Ecosystems of the World 14B: tropical rain forest ecosystems: biogeographical and ecological studies.* Elsevier, Amsterdam.

35 Hart, T.B., Hart, J.A. and Murphy, P.G. 1989. Monodominant and species-rich forests of the humid tropics: causes for their co-occurrence. *American Naturalist* 133: 613-633.

36 Jenkins, M. 1992. Biological diversity. Pp. 26-32. In, Sayer, J.A., Harcourt, C.S. and Collins, N.M.H. 1992. *The conservation atlas of tropical forests: Africa.* Macmillan Publishers Ltd., London.

37 Harcourt, C.S. and Sayer, J.A. (eds), 1996. *The conservation atlas of tropical forests: the Americas.* Simon and Schuster, New York. Pp. 335.

38 Lynch, J.D. 1979. The amphibians of the lowland tropical forests. Pp. 189-215. In, Duellman, W.E. (ed), 1979. *The South American herpetofauna: its origin, evolution, and dispersal.* Monograph No. 7 of the Museum Museum of Natural History Kansas, Lawrence, Kansas.

39 Goulding, M. Leal Carvalho, M. and Ferreira, E.G. 1988. *Rio Negro: rich life in poor waters: Amazonian diversity and foodchain ecology as seen through fish communities.* SPB Academic Publishing, the Hague, Netherlands.

40 World Conservation Monitoring Centre. 1992. Groombridge, B. (ed). *Global biodiversity: status of the Earth's living resources.* Chapman & Hall, London.

41 Hölldobler, B. and Wilson, E.O. 1990. *The Ants.* Cambridge, MA: Harvard Univ. Press.

42 ITTO. 1999. *Annual review and assessment of the world timber situation.* 1998. International Tropical Timber Organization (ITTO)

43 Hamilton, L.S., Juvik, J.O. and Scatena, F.N. 1993. *Tropical montane cloud forests.* East-West Center, Honolulu, HI. USA.

44 Salati, E. and Vose, P.B. 1984. Amazon basin: a system in equilibrium. *Science* 225: 129-138.

45 Mooney, H.A., Bullock, S.H. and Medina, E. 1995. Introduction. Pp. 1-8. In, Bullock, S.H., Mooney, H.A. and Medina, E. (eds). *Seasonally dry tropical forests.* Cambridge University Press, Cambridge, U.K.

46 Sarmiento, G. 1992. A conceptual model relating environmental factors and vegetation formations in the lowlands of tropical South America. Pp. 583-601. In, Furley, P.A., Proctor, J. and Ratter, J.A. (eds). *Nature and dynamics of forest-savanna boundaries.* Chapman & Hall, London.

47 Rundel, P.W. and Boonpragob, K. 1995. Dry forest ecosystems of Thailand. Pp. 93-123. In, Bullock, S.H., Mooney, H.A. and Medina, E. (eds). *Seasonally dry tropical forests.* Cambridge University Press, Cambridge, U.K.

48 Menaut, J.C., Lepage, M. and Abbadie, L. 1995. Savannas, woodlands and dry forests in Africa. Pp. 64-92. In, Bullock, S.H., Mooney, H.A. and Medina, E. (eds). *Seasonally dry tropical forests.* Cambridge University Press, Cambridge, U.K.

49 Gentry, A.H. 1995. Diversity and floristic composition of neotropical dry forests. Pp. 146-194. In, Bullock, S.H., Mooney, H.A. and Medina, E. (eds). *Seasonally dry tropical forests.* Cambridge University Press, Cambridge, U.K.

50 Murphy, P.G. and Lugo, A.E. 1986. Ecology of tropical dry forest. *Annual Review of Ecology and Systematics* 17: 67-88.

51 Ceballos, G. 1995. Vertebrate diversity, ecology and conservation in neotropical dry forests. Pp. 195-220. In, Bullock, S.H., Mooney, H.A. and Medina, E. (eds). *Seasonally dry tropical forests.* Cambridge University Press, Cambridge, U.K.

52 Dodson, C.H. and Gentry, A.H. 1991. Biological extinction in western Ecuador. *Annals of the Missouri Botanical Garden* 78: 273-295.

53 Janzen, D.H. 1988. Tropical dry forests, the most endangered tropical ecosystem. Pp. 13-137 In, Wilson, E.O. 1988. (ed). *Biodiversity.* National Academy Press, Washington, D.C.

55 Bye, R. 1995. Ethnobotany of the Mexican tropical dry forests. Pp. 423-438. In, Bullock, S.H., Mooney, H.A. and Medina, E. (eds). *Seasonally dry tropical forests.* Cambridge University Press, Cambridge, U.K.

56 Maass, J.M. 1995. Conversion of tropical dry forest to pasture and agriculture. Pp. 399-422. In, Bullock, S.H., Mooney, H.A. and Medina, E. (eds). *Seasonally dry tropical forests.* Cambridge University Press, Cambridge, U.K.

57 Bullock, S.H. 1995. Plant reproduction in neotropical dry forests. Pp. 277-304. In, Bullock, S.H., Mooney, H.A. and Medina, E. (eds). *Seasonally dry tropical forests.* Cambridge University Press, Cambridge, U.K.

58 Tukhanen, S. 1999. The northern timberline in relation to climate. Pp. 29-62. In, Kankaapää, Tasanen, S.T. and M.-L. Sutinen (eds). *Sustainable development in northern timberline forests.* Finnish Forest Research Institute Research papers 734.

59 Solbrig, O.T., Medina, E. and Silva, J.F. 1996. Determinants of tropical savannas. Pp. 31-41. In, Solbrig, O.T., Medina, E. and Silva, J.F. (eds), 1996. *Biodiversity and savanna ecosystem processes: a global perspective.* Springer, Berlin.

60 Newsome, A.E. 1983. The grazing Australian marsupials. Pp 441-459. In, Bourlière, F. (ed). *Ecosystems of the world 13: tropical savannas.* Elsevier, Amsterdam.

61 Lavelle, P. 1983 The soil fauna of tropical savannas. I. The community structure. Pp. 477-484. In, Bourlière, F. (ed). *Ecosystems of the world 13: tropical savannas.* Elsevier, Amsterdam.

62 Solbrig, O.T. 1996. The diversity of the savanna ecosystem. Pp. 1-27. In, Solbrig, O.T., Medina, E. and Silva, J.F. (eds), 1996. *Biodiversity and savanna ecosystem processes: a global perspective.* Springer, Berlin.

63 Murphy, P.G. and Lugo, A.E. 1995. Dry forests of Central America and the Caribbean. Pp. 9-28. In, Bullock, S.H., Mooney, H.A. and Medina, E. (eds). *Seasonally dry tropical forests.* Cambridge University Press, Cambridge, U.K.

64 Beltrán, J and Esser, J. 1999. Analysis of the contribution of the non-public sector to in-situ biodiversity conservation in Costa Rica, Honduras and Nicaragua, Central America. GTZ, Frankfurt.

65 Holdsworth, A.R and Uhl, C. 1997. Fire in Amazonian selectively logged rain forest and the potential for fire reduction. *Ecological Applications* 7: 713-725.

66 Laurance, W.F. and Bierregaard, R.O. (eds). 1997. *Tropical forest remnants: ecology, management and conservation of fragmented communities.* University of Chicago Press, Chicago, USA.

67 Smith, T.M., Leemans, R. and Shugart, H.H. 1992. Sensitivity of terrestrial carbon storage to CO_2-induced climate change: comparison of four scenarios based on general circulation models. *Climatic Change* 21: 367-384.

68 Upton, C. and Bass, S. 1995. *The forest certification handbook.* Earthscan Publications Ltd., London.

69 World Conservation Monitoring Centre. 1999. Contribution of protected areas to global forest conservation. In, *International forest conservation: protected areas and beyond. A discussion paper for the intergovernmental forum on forests*. Commonwealth of Australia, Canberra.

70 Williams, M. 1989. Deforestation: past and present. *Progress in Human Geography* 13: 176-208.

71 Williams, M. 1991. Forests. In, Turner II, B.L., Clark, W.C., Kates, R.W., Richards, J.F., Mathews, J.T. and Meyer, W.B. (eds). *The Earth as transformed by human action*. Pp. 179-201. Cambridge University Press with Clark University. Cambridge.

72 Pounds, J.A., Fogden, M.P.L., and Campbell, J.H., 1999. Biological response to climate change on a tropical mountain. *Nature* 398:611-615.

73 Bliss, L.C. 1997. Arctic ecosystems of North America. Pp. 551-683. In, Wielgolaski, F.E. (ed.) *Ecosystems of the world 3: polar and alpine tundra*. Elsevier, Amsterdam.

74 Wielgolaski, F.E. 1997. Introduction. Pp. 1-6. In, Wielgolaski, F.E. (ed.) *Ecosystems of the World 3: polar and alpine tundra*. Elsevier, Amsterdam.

75 Chernov, Yu. I. and Matveyeva, N.V. 1997. Arctic ecosystems in Russia. Pp. 361-507 In, Wielgolaski, F.E. (ed.) *Ecosystems of the world 3: polar and alpine tundra*. Elsevier, Amsterdam.

76 Diaz, A., Péfaur, J.E. and Durant, P. 1997. Ecology of South American páramos with emphasis on the fauna of the Venezuelan páramos. Pp. 263-310. In, Wielgolaski, F.E. (ed.) *Ecosystems of the World 3: polar and alpine tundra*. Elsevier, Amsterdam.

77 Campbell, J.S. 1997. North American alpine ecosystems. Pp. 211-261. In, Wielgolaski, F.E. (ed.) *Ecosystems of the world 3: polar and alpine tundra*. Elsevier, Amsterdam.

78 Grabherr, G. 1997. The high-mountain ecosystems of the Alps. Pp. 97-121. In, Wielgolaski, F.E. (ed.) *Ecosystems of the World 3: polar and alpine tundra*. Elsevier, Amsterdam.

79 Hedberg, O. 1997. High-mountain areas of tropical Africa. Pp. 185-197. In, Wielgolaski, F.E. (ed.) *Ecosystems of the World 3: polar and alpine tundra*. Elsevier, Amsterdam.

80 Bazilevich, N.I. and Tishkov, A.A. 1997. Live and dead reserves and primary production in polar desert, tundra and forest tundra of the former Soviet Union. Pp. 509-539. In, Wielgolaski, F.E. (ed.) *Ecosystems of the World 3: polar and alpine tundra*. Elsevier, Amsterdam.

81 UNEP. 1995. Heywood, V. (ed.) *Global biodiversity assessment. United Nations Environment Programme*. Cambridge University Press, Cambridge.

82 Risser, P.G. 1988. Diversity in and among grasslands. Pp. 176-180. In, Wilson, E.O. and Peter, F.M. (eds). *Biodiversity*. National Academy Press, Washington, D.C.

83 Coupland, R.T. 1992. Overview of the grasslands of North America. Pp. 147-150. In, Coupland, R.T. (ed.) *Ecosystems of the World 8A: natural grasslands: introduction and western hemisphere*. Elsevier, Amsterdam.

84 Coupland, R.T. 1993. Review. Pp. 471-482. In, Coupland, R.T. (ed.). *Ecosystems of the World 8B: natural grasslands: eastern hemisphere and resumé*. Elsevier, Amsterdam.

85 Lavrenko, E.M. and Karamysheva, Z.V. 1993. Steppes of the former Soviet Union and Mongolia. Pp. 3-60. In, Coupland, R.T. (ed.) *Ecosystems of the World 8B: natural grasslands: eastern hemisphere and resumé*. Elsevier, Amsterdam.

86 Cowling, R.M. and Holmes, P.M. 1992. Flora and vegetation. Pp. 23-61. In, Cowling, R.M. (ed.) *The ecology of fynbos: nutrients, fire and diversity*. Oxford University Press, Capetown.

87 Myers, N. 1990. The biodiversity challenge: expanded hot-spots analysis. *Environmentalist* 10:243-256.

88 Davis, G.W., Richardson, D.M., Keeley, J.E. and Hobbs, R.J. 1996. Mediterranean-type ecosystems: the influence of biodiversity on their functioning. Pp. 151-183. In, Mooney, H.A., Cushman, J.H., Medina, E., Sala, O.E. and Schulze, E-D. (eds). *Functional roles of biodiversity: a global perspective*. John Wiley & Sons Ltd., Chichester, UK.

89 Keeley, J. F. 1992. A Californian's view of fynbos. Pp. 372-388. In, Cowling, R.M. (ed.) *The ecology of fynbos: nutrients, fire and diversity*. Oxford University Press, Capetown.

90 Richardson, D.M., Macdonald, I.A.W., Holmes, P.M. and Cowling, R.M. 1992. Plant and animal invasions. Pp. 271-308. In, Cowling, R.M. (ed.) *The ecology of fynbos: nutrients, fire and diversity*. Oxford University Press, Capetown.

91 Middleton, N. and Thomas, D. (eds), 1997. *World atlas of desertification*. Second Edition. United Nations Environment Programme. Arnold, London.

92 West, N.E. 1983. Approach. Pp. 1-2. In, West, N.E. (ed.) *Ecosystems of the World 5 temperate deserts and semi-deserts*. Elsevier, Amsterdam.

93 Huenneke, L.F. and Noble, I. 1996. Ecosystem function of biodiversity in arid ecosystems. Pp. 99-128. In, Mooney, H.A., Cushman, J.H., Medina, E., Sala, O.E. and Schultze, E.-D. (eds). *Functional roles of biodiversity; a global perspective*. John Wiley & Sons, Chichester, U.K.

94 Loh, J., Randers, J., MacGillivray, A., Kapos, V., Jenkins, M., Groombridge, B., Cox, N. and Warren, B. 1999. *Living planet report 1999*. WWF International, Gland, Switzerland.

95 World Conservation Monitoring Centre. 1999. Natural capital indicators for OECD countries. Unpublished report for RIVM.

Suggested introductory source

Huston, M.A. 1994. *Biological diversity: the coexistence of species on changing landscapes*. Cambridge University Press, Cambridge, U.K.

Turner II, B.L., Clark, W.,C., Kates, R.W., Richards, J.F., Matthews, J.T. and Meyer, W.B. (eds). 1990. *The Earth as transformed by human action*. Pp. 179-201. Cambridge University Press with Clark University, Cambridge.

8

INLAND WATER BIODIVERSITY

Inland waters comprise a minute proportion (less than one-hundredth of one percent) of the world's water resource. Despite this, they encompass a very wide range of habitat types and contain a disproportionately high fraction of the world's biodiversity.

Freshwater is also a vital resource for human survival. In consequence, inland water ecosystems are placed under many, often conflicting, pressures, with increasingly adverse consequences for their biodiversity. There are indications that, overall, inland water species are declining faster than marine or terrestrial ones.

INLAND WATER HABITATS

Despite their vastly smaller extent, inland aquatic habitats show far more variety in their physical and chemical characteristics than marine habitats. They encompass systems as varied as the world's great lakes and rivers, small streams and ponds, temporary puddles, thermal springs and even the minute pools of water that collect in the leaf axils of certain plants, such as bromeliads. Chemically they range from almost pure water to highly concentrated solutions of mineral salts, toxic to all but a few specialised organisms.

Inland water habitats can be divided into running or *lotic* and standing or *lentic* systems. They may also be divided into permanent water bodies, periodically (usually seasonally) inundated, and ephemeral or transient. Each of these has its own distinct set of ecological characteristics.

There is no rigid dividing line between an inland aquatic habitat on the one hand and a terrestrial or marine habitat on the other. In the former instance, any temporarily inundated area, such as a river floodplain, is effectively a hybrid or transitional system, being at some times essentially aquatic, at other times terrestrial. Similarly there are many areas that consist of shifting mosaics of land and shallow water, or areas of saturated vegetation, such as sphagnum moss bogs that are neither strictly land nor water. These transitional areas are often collectively termed 'wetlands'. Similarly, estuarine areas are transitional areas between inland and marine systems.

Although the terms 'inland water' and 'freshwater' are often used more or less interchangeably they are not equivalent. A considerable number of inland waters are saline, some much more so than seawater. Conversely, waters of the deltaic regions of some major river systems (most notably the Amazon) may be fresh a considerable distance out to sea.

A highly simplified representation of the global distribution of major river basins, lakes and wetlands is provided in Map 8.1.

LOTIC SYSTEMS: RIVERS AND CATCHMENT BASINS

A river system is a complex but essentially linear body of water draining under the influence of gravity from elevated areas of land toward sea level. The typical drainage system consists of a large number of smaller channels (streams, rills, etc.) at higher elevation merging as altitude falls into progressively fewer but larger channels, which in simplest form discharge by a single large watercourse. Most such systems discharge into the coastal marine environment; some discharge into lakes within enclosed inland basins; a few watercourses in arid regions enter inland basins where no permanent lake exists.

The source area of all the water passing through any given point in the drainage system is the *catchment* area for that part of the system. In parallel with the hierarchical aggregation of tributaries of the major river system, sub-catchments aggregate into a single major catchment basin; this is the entire area from which all water at the final discharge point of the system – ie. usually the sea – is derived. Strictly, the *watershed* is the line of higher elevation dividing one catchment basin from another, but this term is increasingly used as a synonym of catchment.

The speed and internal motion of river water depends largely on water volume and the shape of its channel. These factors typically differ greatly through the river system, from narrow, steep and fast upland feeder streams, to broad level and slow downstream reaches. In virtually all river systems water volume also varies seasonally. Some rivers with arid or semi-arid catchments only flow for part of the year, or in extreme cases only once every several years.

Large rivers may span many degrees of latitude and pass through a wide range of climatic conditions within their catchments. Variations in water flow and underlying geology also create a wide range of habitats within any river and often within a short distance. Different organisms are typically adapted to different parts of any given river system.

Table 8.1
Physical and biodiversity features of major long-lived lakes

Notes: A few other lakes have notable endemism among fishes, molluscs, crustaceans or other groups – among these are lakes Inle (Myanmar), Lanao (Phillippines); Malili (Indonesia) and the Cuatro Cienegas basin (Mexico) – but their ages are not yet firmly established. Qualitative remarks (eg. "very high", "low") in the 'biodiversity' column are related to long-lived lakes, not to lake systems in general.
Source: collated from data in Martens et al.[44]

Lake	Country	age (mill. yrs)	max depth (m)	vol. (km³)	Biodiversity
Baikal *Largest, deepest, oldest extant freshwater lake (20% of all liquid surface fresh water on Earth)*	Russia	25-30	1 637	23 000	very high spp richness, exceptional endemism in fishes and several invertebrate groups total animal spp: 1 825 endemic: 982 fishes: 56 spp., 27 endemic
Tanganyika	Burundi, Tanzania, Zambia, Zaire	20	1 470	18 880	very high spp richness, high endemism, especially high among cichlid fishes total animal spp: 1 470 endemic: 632 fishes: 330 spp., 241 endemic
Victoria World's second largest freshwater lake (area)	Kenya, Tanzania, Uganda	>4 ??	70	2 760	high spp richness, especially of fishes exceptional endemism among cichlid fishes *many fish endemics depleted or extirpated following introduction of Nile Perch* fishes: ca 290 spp., ca 270 endemic
Malawi	Malawi, Mozambique Tanzania	>>2	780	8 400	very high spp richness, high endemism, especially high among cichlid fishes fishes: ca 640 spp., >600 endemic *more fish species than any other lake*
Titicaca One of world's highest altitude lakes	Bolivia, Peru	3	280	890	moderate species richness and endemism (highest among fishes) total animal spp: 533 endemic: 61 fishes: 29 spp., 23 endemic
Biwa	Japan	4	104	674	moderate species richness and endemism (highest in gastropod molluscs and fishes) total animal spp: 595 endemic: 54 fishes: 57 spp., 11 endemic
Ohrid Fed mainly by subterranean karst waters	Albania, Macedonia (FYR)	3	295	50	moderate species richness, exceptional endemism in several groups (planarians, oligochaetes, gastropod molluscs, ostracod crustaceans) fishes: 17 spp., 2 endemic

River systems can change course radically as a result of deposition and erosion of their channel, and the uplift and erosion of watershed uplands. Despite the dynamic physical state of these systems, large rivers rarely disappear, and although direct evidence is scarce, indications are that some have been in continuous existence for tens of millions of years. This is consistent with the fact that running waters include representatives of almost all taxonomic groups found in freshwaters, and that several invertebrate taxa occur only in running waters or attain greatest diversity there. By far the largest river catchment in the world is the Amazon (with the Ucayali) in South America which covers just under 6 million km² and is nearly 60% larger than the next largest, the Congo in central Africa. Unsurprisingly the former is the major repository of the world's freshwater biodiversity. Between them the 20 largest river catchments cover around 45 million km², or about one third of the world's ice-free land surface[1].

LENTIC SYSTEMS

Lakes and ponds

The great majority of existing lakes, of which around 10 000 exceed 1 km² in extent, are formed as a result of glacial activity with most of the rest a result of tectonic activity. Tectonic lakes are formed either as a result of faults caused by deep crustal movements or by volcanism. In the case of the former, a lake may form in a depression caused by a single fault, or in a depressed area between two or more faults – these being graben lakes – or in a rift valley. Most volcanic lake form in craters or calderas of volcanoes while a few (usually short-lived) may form behind dams caused by lava flows. Glacial lakes occupy basins caused by the scouring action of ice masses. Most of the world's existing lakes are glacial and geologically very young, dating from the retreat of continental ice-sheets at the start of the Holocene, around 11 500 years before present. All such lakes are expected to fill slowly with sediment and plant biomass, and to disappear within perhaps the next 100 000 years along with any isolated biota. Lakes may also be caused by the dissolution of soluble rocks, mostly notably limestone in karst regions which is gradually dissolved by dilute acids in water running through it, and by changes in the course of rivers in floodplain regions, these resulting in ox-bow and scroll lakes.

Only about 10 existing lakes are known with certainty to have origins much before the Holocene (Table 8.1)[2], and most of these occupy basins formed by large scale subsidence of the Earth's crust, dating back to at most 20 million (Lake Tanganyika) or 30 million (Lake Baikal) years before present.

There is good evidence that some extinct lake systems in the geologic past were very large and very long-lived under different climatic and tectonic conditions. In general, the long-lived lakes are of particular interest in terms of biodiversity because these systems tend to be rich in species of several major groups of animals and many of these species are restricted to a single lake basin.

WETLANDS

As indicated above, the distinction between a wetland, an aquatic system and a terrestrial system may be essentially arbitrary. However, a number of mixed shallow-water and terrestrial habitat types share several characteristics and are habitually grouped as wetlands. Wetlands in this sense are typically heterogeneous habitats of permanent or seasonal shallow water dominated by large aquatic plants and broken into diverse microhabitats[3]. The four major broad habitat types are:

Bogs

Peat-producing wetlands in moist climates where organic matter has accumulated over long periods. Water and nutrient input is entirely through precipitation. Bogs are typically acid and deficient in nutrients and are often dominated by sphagnum moss.

Fens

Peat-producing wetlands that are influenced by soil nutrients flowing through the system and are typically supplied by mineral-rich ground water. Grasses and sedges, with mosses, are the dominant vegetation. Fens are typically more productive and less acidic than bogs.

Marshes

Inundated areas with herbaceous emergent vegetation, commonly dominated by grasses, sedges or reeds. They may be either permanent or seasonal and are fed by ground or river water, or both.

Swamps

Forested freshwater wetlands on waterlogged or inundated soils where little or no peat accumulation occurs. Like marshes, may be either permanent or seasonal.

DIVERSITY IN INLAND WATERS

At high taxonomic levels the diversity of freshwater organisms is considerably narrower than in the sea or on land. Only one extant eucaryote phylum (Gamophyta – green conjugating algae) is apparently confined to freshwater habitats. The number of species overall (species richness) is also low compared with marine and terrestrial groups. However, species richness in relation to habitat extent may be very high. For example, about 10 000 (40%) of the 25 000 known fish species are freshwater forms[1]. Given the distribution of water on the Earth's surface this is equivalent to one fish species for every 15 km[3] in freshwaters compared with one for every 100 000 km[3] of sea water. This high diversity of freshwater fishes relative to habitat extent is undoubtedly promoted by the extent of isolation between freshwater systems. Many lineages of fishes and invertebrates have evolved high diversity in certain water systems, and in some cases, species richness and endemism tend to be positively correlated between different taxonomic groups[4].

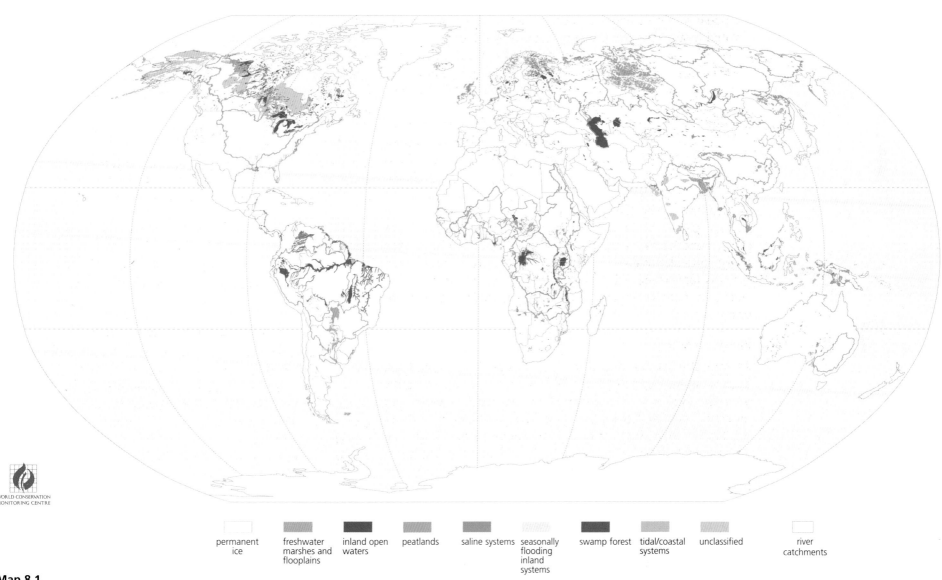

**WORLD CONSERVATION
MONITORING CENTRE**

permanent
ice

freshwater
marshes and
flooplains

inland open
waters

peatlands

saline systems

seasonally
flooding
inland
systems

swamp forest

tidal/coastal
systems

unclassified

river
catchments

Map 8.1

Major inland water areas of the world.

The 20 largest river catchments are shown in outline

As is the case with terrestrial habitats, species richness increases strongly toward the equator; so that in most groups of organisms, there are many more species in the tropics than in temperate regions, although in a few specific cases (eg. freshwater crayfish) this appears to be reversed.

Protoctists

The larger algae comprise some 5 000 species in three major groups (the green, brown and red algae), the great majority of which are marine or brackish water forms ('seaweeds'). The green algae Chlorophyta include one order of around 80 species (Ulotrichales) that is mainly freshwater. However, one major group sometimes associated with the green algae – the stoneworts (Charophyta) – is almost entirely freshwater. The stoneworts include some 440 species, most of which are endemic at continent level or below; they tend to be very sensitive to nutrient enrichment and have declined in many areas[5].

Fungi

There are more than 600 species of freshwater fungi known, currently more from temperate regions than from the tropics, although probably only a small fraction of existing species have been described, and the tropics have been little sampled[6]. Virtually all described freshwater fungi are ascomycotes with few basidiomycotes and zygomycotes having been identified. They occur wherever vascular plant material is available as a substrate. They appear to be important as parasites, endophytes and saprophytes of emergent aquatic macrophytes, as decomposers of submerged allochthonous woody debris, and as a food resource for invertebrates[6-8]. Most are very small, with fruiting bodies less than 0.5 mm in diameter.

Plants

Wetland or aquatic species occur with some frequency in the non-vascular plant phyla, which generally prefer moist habitats, and among the ferns and allies. Mosses in the order Sphagnales (a single family Sphagnaceae and genus *Sphagnum*) often grow submerged and are key components of peat bogs. Many hygrophilous terrestrial mosses (eg. *Thamnium*, *Bryum*, *Mnium*) have aquatic forms. Several genera of Bryales are aquatic or have aquatic species. A number of liverwort species growing otherwise in wet terrestrial situations may also live submerged, sometimes at considerable depths. Truly aquatic liverworts include several species of *Riccia* and *Ricciocarpus natans* (Ricciaceae; Marchantiales) live free-floating on the surface of eutrophic lakes. At least 16 species of *Riella* in the Riellaceae (Jungermanniales) are aquatic, characteristic of temporary waters of semiarid regions, reaching highest diversity in northern and southern Africa. Among the lycophytes, most of the 60 or so species of *Isoetes* (family Isoeteacae) are aquatic, some of great limnological importance, and *Stylites* is an endemic member of the littoral community of Andean lakes. Sphenophytes (horsetails) often occur in moist situations, including around water margins, eg. *Equisetum fluviatile* is a notable emergent littoral form of north temperate lakes.

The Filicinophyta include several aquatic forms. The genus *Ceratopteris* (4 spp. family Pteridaceae) are the only truly aquatic (floating) homosporous ferns; some are cultivated ornamentals, others are edible. A few other species, eg. *Microsorum pteropus* (Polypodiaceae) and *Microlepia speluncae* (Dennstaedtiaceae) can grow in water. Among heterosporous ferns, the family Marseliaceae comprises 3 genera and 55–75 species which are either amphibious or fully aquatic. All members of the families Salviniaceae (one genus and *c* 10 spp.) and Azollaceae (one genus and *c* 6 spp.) are floating aquatic ferns. The latter supports the symbiotic nitrogen-fixing *Anabaena azollae* (Phylum Cyanobacteria).

Vascular plants are essentially terrestrial forms, and existing aquatic species are derived from terrestrial ancestors; several different lineages include aquatic species and this transition has therefore occurred several times. Most inland water plant species are relatively widespread, ranging over more than one continental land mass; many are cosmopolitan, occurring around the world and on remote islands. Of the widespread forms, some are essentially northern temperate species extending to a great or lesser extent into the tropics; some are mainly tropical[9].

The Podostemaceae is particularly noteworthy for its many monotypic genera, and a large number of narrowly endemic species, in at least one instance with several forms restricted to different stretches of a single river; tropical South America, Madagascar, Sri Lanka, India, Myanmar, and Indonesia hold such localised species[9].

It has been estimated that at most 2 percent of ferns and 1 percent of angiosperms, ie. up to 250 and 2 500 species respectively, are aquatic forms[9]. These groups together comprise around 400 families; only some 33 of these include aquatic species, and most of these are not rich in species.

INLAND WATER ANIMALS

Animal species are considerably more diverse and numerous in inland waters than plants. Most of the major groups include terrestrial or marine species as well as freshwater forms. Apart from fishes, important groups with inland water species include crustacea (crabs, crayfishes and many smaller organisms), molluscs (including mussels and snails), insects (including stoneflies Plecoptera, caddisflies Trichoptera, mayflies Ephemoptera), sponges, flatworms, polychaete worms, oligochaete worms, numerous parasitic species in various groups, and numerous microscopic forms. The diversity and ecological rôle of microorganisms and microinvertebrates in freshwater sediments have recently been reviewed[10].

Insects with an aquatic larval phase but a winged adult phase are often restricted to particular river basins (even if adults disperse widely, they may not find suitable habitat), but in general are much less restricted in this way than entirely aquatic species. A relatively large number of species, particularly of crustaceans, occupy temporary pools and have a desiccation-resistant stage that can undergo long-range passive dispersal between drainage basins; some such species are thus widely distributed.

Insects

As on land, insects (Phylum Mandibulata) are as far as is known by far the most diverse group of organisms in inland waters. The true number of aquatic insects remains unknown; data for three relatively well known areas (Europe, Australia and North America) and extrapolations for possible global totals are included in Table 8.2. In contrast to terrestrial faunas, where beetles (Order Coleoptera) are the most diverse, flies and their relatives (Order Diptera) appear to be by far the most abundant group, although also one of the less well known.

In terms of life histories, there are two main groups of aquatic insects: those in which the adult stage and the active immature stages are passed in water (in some cases with a terrestrial pupal stage); and those in which, after a nymphal or larval stage in water, the adult stage is spent on land or in the air. The great majority of Diptera, and therefore most aquatic insects, form part of the latter group. Included amongst their number are several of enormous economic importance to man, of which the most significant are almost certainly mosquitoes of the genus *Anopheles*, intermediary hosts of the malaria parasite.

Most aquatic insects are benthic, living in or on the bottom; a small number are planktonic and live suspended in the water column; around half of the aquatic Hemiptera and a few other insects and non-insect invertebrates, and live on the water surface (epipleuston).

Table 8.2

Insects of inland waters

Source: collated from data in Hutchinson[45]

Order	Australia	North America	Europe	World
Ephemeroptera	84	614	224	2250
Odonata	302	415	127	4875
Plecoptera	196	578	387	2140
Orthoptera[1]	–	c 20	–	c 20
Blattodea[1]	–	0	–	c 10
Hemiptera	236	404	129 (81)[2]	3200
Megaloptera	26	43	6	300
Neuroptera	58	6	9	c 100
Coleoptera	730	1655	1072	5000
Hymenopterac[3]	–	55	74	c 100
Diptera	1300	5547	4050	>20 000
Trichoptera	478	1340	895	7000
Lepidoptera	–	–	5	c 100

[1] Partially aquatic as adult and sometimes as nymph
[2] Number in parentheses refers to fully aquatic Nepomorpha
[3] All these species are parasitoids as larvae

Fishes

Around 45% of known fish species occur in freshwater: almost exactly 10 000 species are confined to freshwater, and a further 1100 or so occur in freshwater but are not confined to it (Table 8.3). The latter includes catadromous and anadromous species. Freshwater fishes are taxonomically diverse, although not as diverse as marine ones. Thirty-four of the 57 extant orders of fishes have at least one freshwater species, while a further two, the sawfishes (Pristiophoriformes) and tarpons (Elopiformes) have one species that occurs in freshwater but is not confined to it[11]. This compares with 48 orders that have at least one marine species.

Of the orders of fishes with freshwater species, ten are entirely freshwater and another five are very largely so (more than 90% of known species are freshwater ones). A further 12 are very largely marine with a small proportion of freshwater species (<10%) while the remainder have significant numbers of both marine and freshwater species. Over 90% of freshwater species are confined to just four orders: the carps and their relatives (Cypriniformes); the characins (Characiniformes); the catfishes (Siluriformes); and the perches and their relatives (Perciformes). The first three of these are wholly or almost entirely freshwater, while the last, the largest order of fishes with nearly 40% of known species, is unusual in having significant numbers of both marine and freshwater species.

Amphibians

The great majority of the 5000 or so living amphibian species have aquatic larval stages and, as none is known to occur in seawater, all these are dependent on inland waters of various kinds for continued survival of populations. In some cases such water bodies may be temporary pools or puddles, or water in the leaf-axils of plants. Relatively few species are fully aquatic (Table 8.4). Although the number of fully aquatic species in each of the three extant orders is roughly similar (c 20–30 in each), these represent very different proportions of each order, being less than 1% of anurans, around 5% of caudate amphibians and over 10% of caecilians.

Aquatic caudate amphibians are neotenic, that is retain features of the larval stage, most notably external gills. In addition to the fully aquatic amphibians (several of which can survive for short periods in damp conditions out of water), many other species may lead largely aquatic lives or may, as in the case of the Mexican axolotl *Ambyostoma mexicanum*, have completely aquatic neotenic populations.

Reptiles

Very few completely aquatic inland water reptiles are known. The three file-snakes in the family Acrochordidae are live-bearing and may pass their entire lives in water, often in coastal and estuarine areas as well as freshwaters. Virtually all other reptiles of inland waters are egg-laying and return to land at least to

Table 8.3

Fish diversity in inland waters, by order

Note: strictly freshwater orders in bold

Source: Nelson[11]

Order	Common fishes	No. of families	No. of genera	No. of species	No. of freshwater species	No. of species using freshwater	% freshwater species	% species using freshwater
Petromyzontiformes	Lampreys	1	6	41	32	41	78	100
Carcharhiniformes	Ground sharks	7	47	208	1	8	0	4
Pristiophoriformes	Sawfishes	1	2	5	0	1	0	20
Rajiformes	Rays	12	62	456	24	28	5	6
Ceratodontiformes	Australian lungfish	1	1	1	1	1	100	100
Lepidosireniformes	Lungfishes	2	2	5	5	5	100	100
Polypteriformes	Bichirs	1	2	10	10	10	100	100
Acipensiformes	Sturgeons	2	6	26	14	26	54	100
Semionotiformes	Gars	1	2	7	6	7	86	100
Amiiformes	Bowfin	1	1	1	1	1	100	100
Osteoglossiformes	Bonytongues	6	29	217	217	217	100	100
Elopiformes	Ladyfishes and tarpons	2	2	8	0	7	0	88
Anguiliformes	Eels	15	141	738	6	26	1	4
Clupeiformes	Herrings and anchovies	5	83	357	72	80	20	22
Gonorhynchiformes	Milkfish and beaked sandfishes	4	7	35	28	29	80	83
Cypriniformes	Carp, minnows, loaches	5	279	2662	2662	2662	100	100
Characiformes	Characins	10	237	1343	1343	1343	100	100
Siluriformes	Catfishes	34	412	2405	2280	2287	95	95
Gymnotiformes	Knifefishes	6	23	62	62	62	100	100
Esociformes	Pikes and mudminnows	2	4	10	10	10	100	100
Osmeriformes	Smelts	13	74	236	42	71	18	30
Salmoniformes	Salmonids	1	11	66	45	66	68	100
Percopsiformes	Trout-perches, pirate perch, cavefishes	3	6	9	9	9	100	100
Ophidiiformes	Pearlfishes, cusk-eels, brotulas	5	92	355	5	6	1	2
Gadiformes	Cods, hakes, rattails	12	85	482	1	2	0	0
Batrachoidiformes	Toadfishes	1	19	69	5	6	7	9
Mugiliformes	Mullets	1	17	66	1	7	2	11
Atheriniformes	Silversides	8	47	285	146	171	51	60
Beloniformes	Needlefishes, sauries, flyingfishes, halfbeaks	5	38	191	51	56	27	29
Cyprinodontiformes	Rivulines, killifishes, pupfishes, four-eyed fishes, poeciliids, goodeids	8	88	807	794	805	98	100
Gasterosteiformes	Pipefishes, seahorses, stickle backs, sandeels, seamoths, snipefishes, shrimpfishes, trumpetfishes	11	71	257	19	41	7	16
Synbranchiformes	Swamp-eels	3	12	87	84	87	97	100
Scorpaeniformes	Gurnards, scorpionfishes, velvetfishes, flatheads, sablefishes, greenlings, sculpins, oilfishes, poachers, snailfishes, lumpfishes	25	266	1271	52	62	4	5
Perciformes	Perches, basses, sunfishes, whitings, remoras, jacks, dolphinfishes, snappers, grunts, damselfishes, dragonfishes, wrasses, butterflyfishes etc	148	1496	9293	1922	2815	21	30
Pleuronectiformes	Plaice, flounders, soles	11	123	570	6	20	1	4
Tetraodontiformes	Triggerfishes, puffers, boxfishes, filefishes, molas	9	100	339	12	20	4	6

Table 8.5.

Partial list of global hotspots of freshwater biodiversity

Notes: this table includes sites and areas documented in greater detail in WCMC[1].

Sources: See sources cited at end of Table 17 in WCMC[1].

continent	area name	taxonomic group		
Africa	L Malawi	fishes	molluscs	
Africa	L Tanganyika	fishes	molluscs	crabs
Africa	L Victoria	fishes	molluscs	
Africa	Madagascar	fishes	molluscs	crabs
Africa	Niger-Gabon	fishes		crabs
Africa	Upper Guinea	fishes	molluscs	crabs
Africa	lower Congo	fishes		crabs
Australia	SE Australia & Tasmania	fishes	molluscs	crayfish
Australia	SW Australia	fishes		fairy shrimp
Eurasia	SE Asia and Lower Mekong River	fishes	molluscs	crabs
Eurasia	Balkans (southwest)	fishes	molluscs	
Eurasia	L Baikal	fishes	molluscs	
Eurasia	L Biwa	fishes	molluscs	
Eurasia	L Inle	fishes	molluscs	
Eurasia	L Poso	fishes	molluscs	
Eurasia	Malili Lakes	fishes	molluscs	
Eurasia	Sri Lanka	fishes		crabs
Eurasia	Western Ghats	fishes	molluscs	crabs
North America	East Mississippii drainage (Ohio, Cumberland, Tennessee rivers)	fishes	molluscs	crayfish
North America	Mobile Bay drainage	fishes	molluscs	crayfish
North America	western USA	fishes	molluscs	fairy shrimp
South America	L Titicaca	fishes	molluscs	
South America	La Plata drainage	fishes	molluscs	
South America	Amazon basin	fishes	?	crabs

Biogeography and hotspots

Freshwater lineages that originated within continental water systems may show general patterns of distribution similar to terrestrial groups, corresponding more or less to broad biogeographic realms. Lineages of marine origin may remain restricted to peripheral systems corresponding to the area where the ancestral forms moved into freshwater.

Unlike many terrestrial species, that can disperse widely in suitable habitat, the spatial extent of the range of strictly freshwater species tends to correspond to present or formerly continuous river basins or lakes. These species include fishes and most molluscs and crustaceans. Watersheds between river basins are the principal barriers to their dispersal between systems, and their ranges are extended mainly by physical changes to the drainage pattern (eg. river capture following erosion or uplift can allow species formerly restricted to one system to move into another), or by accidental transport of eggs by waterbirds, or by flooding.

In many instances, the range within a system will also be restricted by particular habitat requirements (variations in water turbulence or speed, shelter, substrate, etc). These frequently differ between different stages in the life cycle (eg. in fishes, different conditions and different sites are often required for egg deposition and development, for early growth of fry, and for feeding and breeding of adults).

Many cave or subterranean freshwater aquatic species (eg. of fishes, amphibians and crustaceans) have very restricted ranges, perhaps consisting of a single cave or aquifer, and very limited opportunities for dispersal, depending on the surrounding geology and the consequent morphology of the water system occupied.

A recent analysis of areas important for maintenance of global freshwater biodiversity was based on the expert view of a number of regional and taxonomic specialists. The analysis was designed to make effective use of readily available information, and although preliminary, has yielded the first global overview of freshwater biodiversity hotspots[1]. Maps 8.2 and 8.3 show, respectively, important areas for freshwater fishes and for molluscs, crabs, crayfish, fairy shrimp. Further details of all these areas can be found in Table 17 in the WCMC analysis[1]. Table 8.5 lists the sites and areas that have been identified as of special importance for more than one of the above groups. It is not a comprehensive global listing because it omits several large but imprecisely defined areas of known high diversity, and it omits diverse taxa not covered in the assessment (eg. amphipods, copepods), nor does it mention sites of key importance mainly for one group of animals. The Amazon basin is included because, although it is a vast region rather than an identifiable site, it has such an exceptional diversity of fishes that it could not reasonably be excluded from a draft list of global hotspots.

Continental reviews are now also available for Asia, including discussion of taxonomy, hotspots and policy[12], for Latin America[13] and North America[14].

Analysis of data from some 151 river basins indicates that, as would be expected, there is a strong correlation between the extent of a river catchment and the number of fish species therein. The 'size' of a river can be represented by the area of the basin, or by the volume of water flowing through the river system in any given period; the latter is a better predictor of fish species richness than is basin area. When area is taken into account, there is also a strong relationship between species richness and the latitude of the basin. Recent analysis appears to indicate that latitude may be a surrogate measure for energy availability and productivity within the basin[15,16].

INLAND WATERS AND HUMANKIND

Freshwater – as precipitation, groundwater or in inland water ecosystems – is essential for human survival, chiefly because humans must drink and also because it is needed, in far greater quantity, to produce food. It also has a wide range of subsidiary uses – for transport, industrial production, cleaning, waste disposal, generation of hydroelectric power, recreation, aesthetic purposes and in the form of inland water ecosystems as sites for the production of food.

Many of these demands conflict with each other, so that for example the use of water for disposal of noxious wastes is incompatible with the provision of safe drinking water. Moreover, while the amount of freshwater available is limited

demands on it continue to grow remorselessly as the global human population continues to expand. This problem is exacerbated by the fact that freshwater is very unevenly distributed around the world, so that it is often not available where and when needed, nor in the appropriate amounts, nor with the necessary quality. The two last are particularly important to the maintenance of freshwater biodiversity.

Freshwater systems are therefore under growing pressure, as flow patterns are disrupted and the load of waste substances increases. Inevitably *per capita* shares of water for human use are decreasing and water stress is becoming more widespread[17].

Agriculture consumes around 70% of all water withdrawn from the world's rivers, lakes and groundwater[18]. In places, more than half the water diverted or pumped for irrigation does not actually reach the crop, and problems of waterlogging and salinisation (deposition in soil of salts left by evaporation of pumped groundwater) are increasing. However, irrigated agriculture produces nearly 40% of world food and other agricultural commodities on only 17% of the total agricultural land area, and is thus disproportionately important to global food security[18].

USES OF INLAND WATER SPECIES

The principal use of freshwater species, not considering properties of aquatic systems themselves, is as food. Subsidiary uses include the aquarium trade, materials for medicinal or ornamental use, and as fertilizer. For many human communities, particularly in countries less-developed industrially, capture fisheries provide a major portion of the diet.

Inland water fishery production has two components: capture fisheries and aquaculture, although as discussed below the distinction between the two is becoming increasing blurred. Traditionally such production has been regarded as far less important than marine fisheries. It is true that with one or two exceptions, where countries have access to both marine and inland aquatic resources, reported yield from inland waters is a small fraction of marine yield. Even in land-locked countries, the recorded inland harvest is often, but not always, low both in absolute size and in relation to consumption of meat and other agricultural produce.

Globally, reported inland water capture fishery for 1997 was around 7.7 million tonnes, with around 17 million tonnes of aquaculture production recorded; over 85% of the former and over 95% of the latter comprised finfishes, virtually all the remainder being molluscs and freshwater crustaceans[19]. This compares with reported marine capture fisheries of some 87 million tonnes and marine and brackish water aquaculture animal production of just under 12 million tonnes (see Chapter 6). Reported global inland water capture fishery has increased slowly in the period 1984–1997, by around 1.7% per year, although this masks considerable regional variation, with declines in some areas (eg. Europe and North America) and more marked increases elsewhere (notably Asia)[19].

However, national statistics do not adequately reflect the actual magnitude, location or importance of inland fisheries. The reported inland capture production is certainly a gross under-estimate because much of the catch is made far from recognised landing places where catches are monitored, and is consumed directly by fishers or marketed locally without ever being reported. The evidence suggests that actual capture fisheries catch may be twice or conceivably even three times the reported total, ie. around 15–23 million tonnes per year[20]. Because a far higher proportion of inland fisheries than marine fisheries harvest is apparently used directly for human consumption (rather than production of oils and meals, often used for livestock feed), and because discards are believed to be negligible, it has been argued by some that the provision of foodfish from inland waters is not that much less than that from recorded marine catch[21].

Inland water capture fisheries, particularly in countries less-developed industrially, certainly provide a staple part of the diet for many human communities. This is the case in West Africa generally, locally in East Africa, and in parts of Asia and Amazonia. In some land-locked countries inland fisheries are of crucial importance, providing more than 50% of animal protein consumed by humans in Zambia[22] and nearly 75% in Malawi[23]. In low income food deficit countries, fish protein may be particularly important in times of food stress.

It is impossible at the global level to carry out any meaningful analysis of the relative contribution of different species or species groups to inland capture fisheries because of the inadequacy of reporting. In FAO statistics, by far the largest group recorded is 'freshwater fishes not elsewhere included', that is those that are completely unclassified other than being identified as finfishes. These comprise just under half of all reported landing by weight with a further 15% comprising molluscs and crustaceans similarly classified. The majority of the remaining catch is classified into broad species groups (eg. cyprinids, characins, siluroids), with only three individual fish species having annual reported global landings of over 100 000 tonnes. These are the Nile perch *Lates niloticus* (c 330 000 tonnes reported in 1997), Nile tilapia *Oreochromis niloticus* (226 000 tonnes in 1997) and the common carp *Cyprinus carpio* (100 000 tonnes in 1997).

It is, however, evident that the importance of different species of freshwater finfishes varies considerably. In terms of food security for local subsistence or mixed market/subsistence communities, particularly in the tropics, there is increasing evidence that the diversity of species harvested is in itself a major factor in ensuring a continuous food supply. Many of the species which contribute to these fisheries are often small and would be considered 'trash' fishes in orthodox fisheries and far less valuable than larger (often non-native) species regarded as having market potential and which may be considered for

other
important
areas of fish
diversity

key areas
of fish
diversity

Map 8.2

Major areas of inland water fish diversity

Source: WCMC[1], based on data provided in part by relevant IUCN/SSC specialist groups

introduction. However, these small species are easy to preserve and keep under local conditions and moreover are eaten whole, providing a valuable source of calcium and other minerals. Larger species, such as the introduced Nile perch in Lake Victoria, cannot be easily preserved locally and are in any case not eaten whole, leading to danger of calcium deficiency. Fisheries for such species tend to become industrialised or semi-industrialised, producing fish products for commercial high-value markets, often for export. While these may improve balance of payments for the countries concerned, they may ultimately worsen the nutritional status of local people. Additionally, there are some indications that fish populations in mixed species fisheries are more stable over time, that is less susceptible to 'boom and bust' than those based on a small number of often introduced species.

It is difficult rigorously to assess the condition of inland fish stocks because they appear able to respond rapidly to changing environmental conditions. However, there is a consensus that, regionally, most stocks are fully exploited and in some cases over-exploited. Exploitation has become more efficient because of new technologies, and developing infrastructure has allowed easier access to freshwater resources. Some stocks, especially in river fisheries, appear to be in decline, but this is seemingly a result mainly of anthropogenic changes to the freshwater environment.

As well as food, in many parts of the world fishing is also of high recreational value. Locally, notably in the Amazon Basin and in parts of Southeast Asia, capture for the ornamental fish trade may be an important source of income, and potential impact on wild populations. Increasingly it is becoming difficult to distinguish between truly wild fish stocks and those which are artificially managed or enhanced in some way.

Other exploited animal groups in inland waters are far less important globally than finfishes, but may still be highly significant. These include: freshwater crustaceans, notably crayfishes and freshwater shrimps, both exploited for food; freshwater bivalve molluscs, taken for pearls and for food; frogs (chiefly family Ranidae), exploited for food; crocodilians, hunted mainly for leather; freshwater chelonians, taken for food and to a lesser extent for medicinal purposes, particularly in eastern Asia; waterfowl which are hunted for recreation and for food; fur-bearing mammals, such as beavers *Castor* spp., otters (subfamily Lutrinae) and muskrats (*Ondatra zibethicus* and *Neofiber alleni*), taken for their skins; manatees (family Trichechidae), mostly for food, although also used non-consumptively on a small scale for biological control of weeds.

Relatively few plants associated with inland waters are heavily exploited in the wild state; most are marginal or wetland species. Some (eg. *Aponogeton* spp. in Madagascar) are collected for use as ornamentals; reeds are used as building materials (eg. thatch); and some are collected for food or as medicines (eg. *Spirulina* algae). Rhizomes, tubers and seeds (rarely leaves) of aquatic and wetland plants are used as a food source, mainly in less developed regions where they can be important to food security in times of shortage, but globally they make a relatively minor contribution to human nutrition. Most important are some forms of edible aroid (Araceae), notably some cultivars of *Colocasia* (taro) and the giant swamp taro *Cyrtosperma chamissonis* which grow in flooded conditions and are important food crops in the Caribbean, West Africa and the Pacific islands. Conservation and collection of wild forms of these is considered a high priority. Sago Palms *Metroxylon* spp. in southeast Asia and the Pacific and Watercress *Rorippa nasturtium-aquaticum* in Europe are other examples of cultivated aquatic plants whose wild relatives merit conservation.

Rice is the major cultivated wetland plant and is discussed in detail in Chapter 4. Aquatic plants have been widely used for medicinal purposes, documented for at least two millennia, but such use appears at present to be minor and probably of real significance in few areas. However, interest in ornamental or

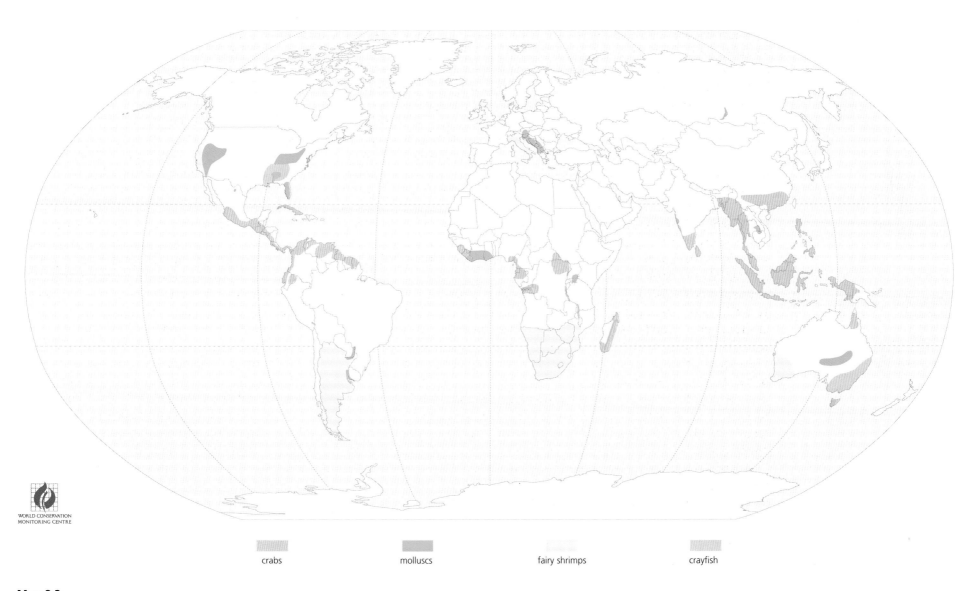

| crabs | molluscs | fairy shrimps | crayfish |

Map 8.3

Major areas of diversity of selected inland water invertebrate groups

Source: WCMC[1], based on data provided by relevant IUCN/SSC specialist groups.

aquarium water plants is very widespread and of some economic importance.

Inland aquaculture

Reported inland aquaculture production has been rising, to around 18 million metric tonnes in 1997, ie. over twice the reported production of inland capture fisheries. Most inland aquaculture production, and most of the recent increase in this sector, is attributable to China. In this particular instance the dividing line between aquaculture and capture fisheries is indistinct; no husbandry is involved beyond release of hatchery stock, and the fishery operates as a capture fishery[19].

OTHER MAJOR FACTORS INFLUENCING INLAND WATER ECOSYSTEMS

Physical alteration and destruction of habitat

Destruction of inland water ecosystems is most simply effected by the removal of water. Although humans have always made use of freshwater systems, the last 200 years (the Industrial Revolution, the growth of cities, the spread of high-input agriculture) have brought about transformations on an unprecedented scale. The rate of water withdrawal rose steeply at the start of the present century, and further after mid-century. Major changes in the distribution of water has resulted mainly from withdrawals for irrigation and secondarily from domestic and industrial use. It has been estimated that humans use 26% of the total evapotranspiration from land surfaces and 54% of the accessible runoff[24]. Unregulated withdrawal can lead to the wholesale destruction of inland water ecosystems, as has occurred with the Aral Sea in central Asia. Similarly, many wetlands have been completely destroyed by drainage, often for conversion to agriculture. Other factors can modify or destroy particular habitats within inland water ecosystems. For example, canalisation, usually to improve navigability, generally destroys riparian (shoreline) habitats while flood-control systems drastically alter regimes in floodplains[25].

Dams and reservoirs

Dams, particularly large dams, have a major impact on the rivers they are built on. They affect flow regimes, often dramatically, destroy large areas of existing habitat (while at the same time creating new ones) and can catastrophically disrupt the life-cycles of species that migrate up and down rivers[26]. Large dams are very unevenly distributed across the world's major catchments, with an enormous concentration in North America, particularly within the contiguous states of the USA, where at least eight catchments have over one hundred large dams in each. In contrast, small and medium-sized dams, which may cumulatively have as much impact, are concentrated in eastern Asia, particularly China[27]. Dams may be primarily for the generation of hydro-electric power, or to create reservoirs for the storage of water, or both. The size of a dam is not necessarily directly related to the area or volume of the impoundment created or to its downstream impact.

Pollution and water quality

Assessment of anthropogenic changes in water quality is not always easy as such changes are invariably superimposed on natural background variations. Historically, a similar sequence of water quality issues has became apparent in both Europe and North America during rapid socio-economic development over the past 150 years. Problems of faecal and other organic pollution were evident in the mid-19th century, followed by salinisation, metal pollution, and eutrophication in the first half of the 20th century, with radioactive waste, nitrates and other organic micropollutants, and acid rain most prominent in recent decades. Newly-industrialising countries are likely to face these problems over a much more compressed period, and typically without the capacity to monitor and analyse water quality, or manage water use appropriately[28].

Different kinds of pollutants appear to affect different classes of water system to differing extent[29]. With regard to quality for human use, contamination by pathogens of faecal origin is the major problem in river systems, and eutrophication probably the most widespread problem affecting lake and reservoir waters[29-31].

Acid deposition through precipitation has been recognised as a regional transboundary phenomenon since the 1960s. Industrial emissions of sulphur and nitrogen oxides (SO_2, NO_x), mainly a result of fossil fuel combustion, are the principal source of acid rain. Most evidence of acid rain and its effects relates to North America and Europe, but emission rates are rising steeply in rapidly industrialising countries elsewhere. Acid rain in one country may be a consequence of compounds released into the atmosphere by industry in another country hundreds of kilometres distant. The geology, soil and vegetation of drainage basins strongly influences the acidification process. Acid rain has been shown to decrease species diversity in lakes and streams but has not been implicated in any recorded species extinction nor any major species decline. It has not yet been shown to be a significant issue in tropical freshwaters, where global freshwater diversity is concentrated[1].

Sedimentation

Removal or extension of forest cover, or any anthropogenic interference with soils and land cover (eg. agriculture, urbanisation, road construction, mining), modifies the rate of runoff from catchment slopes and also the density of particles carried in the drainage system. All moving waters carry some mass of suspended material, and there is considerable natural variation in this in space and time, but logging can increase sediment load by up to 100% for a short period, and 20-50% over the longer term. Sediment reaching lakes will be deposited and in effect enter long-term storage; depending on water velocity, sediment in rivers will settle out on floodplains or other parts of the course, or be carried into the coastal marine environment.

Increased sedimentation can have several effects on aquatic biodiversity: deposition can radically change the physical environment of species restricted to particular conditions of depth, light penetration and velocity; it is a major carrier of heavy metals, organic pollutants, pathogens and nutrient; and it can interfere mechanically with respiration in gill-breathing organisms[1].

Introduced species

Unplanned or poorly planned introduction of non-native species and genetic stocks is a major threat to freshwater biodiversity. Such introductions can have negative or positive effects on fishery production; it is a reasonable assumption that all successful introductions will have an impact on existing population levels and community structure, and many changes are likely to be undesirable[32]. A classic example of the effect of introduced species is the impact of the Nile perch *Lates niloticus* on the haplochromine cichlids of Lake Victoria discussed further below.

Several species of aquatic plant, particularly free-floating species able to spread rapidly by vegetative growth (most notoriously the South American water hyacinth *Eichhornia crassipes*), but also other forms, have dispersed widely over the globe and become major pest species. They block drainage channels, sluices and hydro-electric installations, impede boat traffic, and hinder fishing. In recent decades the question of how best to control or eradicate pest species has been the foremost issue in conservation and management of aquatic plants[1].

THE CURRENT STATUS OF INLAND WATER BIODIVERSITY

As with marine species, assessment of the status of wholly aquatic inland water species is hampered by difficulties of direct observation. However, because these species also in general have far more circumscribed ranges than marine species, it is easier to infer their status from assessment of habitat condition

Table 8.6

Numbers of threatened freshwater
fishes in select countries

Notes: These are the 20 countries whose fish
faunas have been evaluated completely, or
nearly so, and which have the greatest
number of globally threatened freshwater fish
species.

Source: WCMC[1], the threatened species data
in this table were collated for the IUCN Red
List[37]. The estimates of total fish species
present are all approximations.

	total species	threatened species	percent threatened
USA	822	120	15
Mexico	384	77	20
Australia	216	27	13
South Africa	94	25	27
Croatia	64	20	31
Turkey	174	18	11
Greece	98	16	16
Madagascar	41	13	32
Papua New Guinea	195	12	6
Hungary	79	11	14
Canada	177	11	6
Spain	50	11	22
Romania	87	11	13
Italy	45	9	20
Moldova	82	9	11
Portugal	28	9	32
Bulgaria	72	8	11
Sri Lanka	90	8	9
Germany	68	7	10
Slovakia	62	7	11
Japan	150	7	4

and from sampling efforts. Amphibious or surface dwelling species may be relatively easier to monitor. Where such species are of economic importance – as for example with those European and North American waterfowl that play a role in the recreational hunting industry – they may be among the best monitored of all wild species.

Threatened and extinct species

In the few cases where particular inland water faunas – usually fishes – have been studied in any detail it has generally been found that more species than suspected turn out to be threatened, or cannot be re-recorded at all[33-36]. Of the 20 or so countries that have been reasonably comprehensively assessed, on average just under 20% of the inland water fish fauna has been found to be threatened (Table 8.6), using the IUCN threatened species categorisation[37]. The proportion of inland water chelonians believed threatened is even higher, with some 35% of species having been assessed as either critically endangered, endangered or vulnerable. Amongst mammals and birds the proportions are considerably lower, probably because many semi-aquatic species are able to disperse from one inland water body to another relatively easily. Nevertheless, somewhat more species than average for the groups as a whole are regarded as threatened. Table 8.7 shows the category and taxonomic distribution of threatened inland water vertebrates.

Two groups of species, the Lake Victoria cichlid fishes and the Mobile Bay drainage gastropod molluscs, serve as exemplary case studies illustrating the major threats faced by inland water biota worldwide.

Lake Victoria, the largest tropical lake in the world, provides a classic example of the potential negative impacts of species introductions. Until some 30 years ago, when the large top predator, the Nile perch *Lates niloticus*, was introduced, the lake supported an exceptional 'species flock' of around 300 species of haplochromine cichlid fishes as well as smaller numbers from other families. The cichlids are of enormous interest in the study of evolutionary biology. Not all the species have yet been formally described; many of these are known among aquarists and others only by informal common names. At least half and up to two-thirds of the native species are believed to be extinct or so severely depleted that too few individuals exist for the species to be harvested or recorded by scientists. Predation by the Nile Perch is believed to be the major cause of this decline. Additional factors include excess fishing pressure, already evident before introduction of Nile Perch, and possible competition from tilapiine cichlids that were also introduced. The lake itself has now become depleted of oxygen, and a

Table 8.7

Taxonomic distribution and status of threatened inland water vertebrates

Source: WCMC database, data compiled in part for IUCN Red List[37].

Phylum and class or order	Family	Common name	Critically Endangered	Endangered	Vulnerable
Craniata – Cephalaspidomorphi					
Petromyzontiformes		Lampreys		1	2
Craniata – Elasmobranchii					
Carchariniformes		Ground sharks	1		
Rajiformes		Rays		3	
Craniata – Pisces					
Acipenseriformes		Sturgeons	6	11	8
Osteoglossiformes		Bonytongues		1	
Clupeiformes		Herrings and anchovies		1	2
Cypriniformes		Carp, minnow, loaches	41	36	114
Characiformes		Characins		1	
Siluriformes		Catfishes	8	6	22
Esociformes		Pikes and mudminnows			1
Osmeriformes		Smelts		1	
Salmoniformes		Salmonids	8	3	20
Percopsiformes		Trout-perches, cavefishes etc	1	3	
Ophidiiformes		Pearlfishes, cusk-eels, brotulas			6
Mugiliformes		Mullets		1	
Atheriniformes		Silversides	6	5	31
Cyprinodontiformes		Rivulines, killifish etc	18	20	25
Beloniformes		Needlefishes, sauries etc	2	3	8
Gasterosteiformes		Pipefishes, seahorses etc	2		1
Synbranchiformes		Swamp eels		1	
Scorpaeniformes		Gurnards, scorpionfishes etc	2		4
Perciformes		Perches etc	53	25	105
Craniata – Amphibia					
Caudata	Cryptobranchidae	Giant salamanders and hellbenders			1
	Pletodontidae	Lungless salamanders			1
	Proteidae	Mudpuppies and olm			1
Anura	Pipidae	Clawed frogs and pipid toads			1
Craniata – Reptilia					
Testudinides	Carettochelidae	Pig-nosed soft-shelled turtle			1
	Trionychidae	Soft-shelled turtles	3	1	5
	Chelydridae	Snapping turtles			1
	Dermatemydidae	Central American river		1	
	Chelidae	Austro-american side-necked turtles	1	3	8
	Kinosternidae	Mud and musk turtles			3
	Pelomedusidae	Side-necked turtles		2	5
	Emydidae	Pond and river turtles	2	11	9
Crocodilia	Alligatoridae	Caimans and alligators	1	1	
	Crocodylidae	Crocodiles	3	1	3
	Gavialidae	Gharial and false gharial		1	
Craniata – Aves					
Anseriformes	Dendrocygnidae	Whistling-ducks			1
	Anatidae	Ducks, swans and geese	4	2	18
Gruiformes	Heliornithidae	Limpkin and sungrebes			1
	Rallidae	Rails, gallinules and coots			2
Ciconiiformes	Laridae	Gulls, terns, skua, auks, skimmers			1
	Podicipedidae	Grebes	2	1	1
	Phoenicopteridae	Flamingos			2
	Pelecanidae	Pelicans and shoebill			2
Passeriformes	Cinclidae	Dippers			1
Craniata – Mammalia					
Insectivora	Tenrecidae	Tenrecs and otter shrews		4	
	Soricidae	Shrews	2	1	
	Talpidae	Moles and desmans			2
Rodentia	Muridae	Mice			3
Cetacea	Platanistidae	River dolphins	1	2	1
Sirenia	Trichechidae	Manatees			3
Carnivora	Mustelidae	Mustelids		1	4
	Viverridae	Viverrids		1	

shrimp tolerant of oxygen-poor waters provides a major food source for the Nile Perch. In recent years the Nile Perch, and one of the introduced tilapiines have formed the basis of a high-yielding fishery, and an important national and export trade. However, it is thought unlikely that such high yields will be maintained because of continued overfishing and the suspected instability of the already highly disturbed lake ecosystem[1].

Dam construction is the prime cause of extinction in the gastropod fauna of the Mobile Bay drainage in USA. Historically, the freshwater snail fauna of Mobile Bay basin was probably the most diverse in the world, followed by that of the Mekong River. Nine families and about 118 species were known at the turn of the century to occur in the Mobile Bay drainage. Several genera and many species were endemic, particularly in the Pleuroceridae. Recent surveys suggest at least 38 species are extinct (32%); decline in species richness ranges between 33% and 84% in the main river systems. The richest fauna was in the Coosa River and this system has undergone the greatest decline (from 82 to 30 species). Almost all the snail species presumed extinct were members of the Pleuroceridae and grazed on plants growing on rocks in shallow oxygen-rich riffle and shoal zones. The system has 33 major hydroelectric dams and many smaller impoundments, as well as locks and flood control structures. A combination of siltation behind dams, and submergence of shallow water shoals has removed the snails' former habitat. Where habitat remains it has diminished in area and become fragmented[1].

The inland water living planet index

Application of the WCMC/WWF Living Planet Index methodology (see Chapter 5) to over 100 animal species of inland water ecosystems indicates that on average monitored populations have declined by 45% in the period 1970-95 (Figure 8.1). This compares with a decline over the same period of some 30% in marine and coastal species (see Chapter 6). Equivalent data are not as yet obtainable for terrestrial ecosystems, although a surrogate measure (natural forest cover) has only declined by around 10% over the same period. Although not conclusive, these provide strong indications that inland water ecosystems may overall be those suffering the greatest deleterious impact from human activities at present[38].

ASSESSMENT OF THE STATUS OF INLAND WATER ECOSYSTEMS

Indicators of habitat condition in river catchments

The Living Planet Index methodology can provide a clear indication of global trends in inland water biodiversity. Equally important, however, are methodologies to assess the overall condition of particular inland water ecosystems. One global approach involves combining two high-order indicators of likely habitat condition in different river catchments, these being a measure of wilderness and of water resource vulnerability.

A wilderness measure for each river catchment can be calculated using the Wilderness Index methodology developed by

Fig. 8.1

Inland water living planet index 1999

Source: Loh et al.[38]

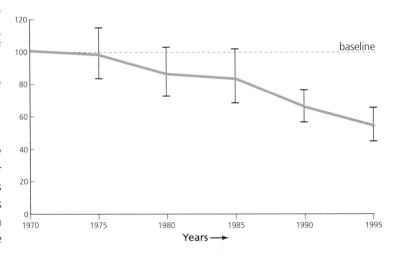

the Australian Heritage Commission[39,40] (see Chapter 4 and particularly Map 4.3). This provides a measure of the amount of wilderness in the land area of the catchment as a whole and does not directly reflect the condition of riverine ecosystems.

A Water Resource Vulnerability Index (WRVI) has been developed at country level by Raskin *et al.* on the basis of three water resource stress indices: reliability; use-to-resource; and coping capacity. For each country and for each index four classes from 'no stress' to 'high stress' have been designated. A compound WRVI for each country can be derived by giving equal weightings to each of the three stress indices. From this a measure of vulnerability for each catchment can be calculated by measuring that proportion of each catchment which lies within any given country and weighting this proportion by the WRVI of that country. This indicator is a measure of the extent to which water supply is or may be expected to become a problem to the human population in a country and is therefore only indirectly a measure of vulnerability of or degree of threat to riverine ecosystems.

Map 8.4 shows an overall vulnerability or impact value for 151 major river catchments calculated by normalising and combining the two measures. Overall, the pattern mapped agrees well with what might intuitively be expected, with few evident anomalies. This is particularly noteworthy because neither of the two indices used to generate this measure includes direct measures of the state of the riverine ecosystems (although the use-to-resource ratio, used as part of the Water Resources Vulnerability Index provides some insight into this in that it estimates what proportion of river water is physically removed). Globally, the most stressed catchments are to be found in South Asia (the Indian subcontinent), the Middle East and western and north-central Europe. The least stressed are those in the north-western part of North America (see WCMC[1] for further discussion).

Further refinements of such an analysis would involve applying water resource vulnerability measures to individual catchments rather than to countries and incorporating measures of direct impacts on inland waters, most importantly water quality and number and kind of dams and impoundments.

Management of inland waters

Inland waters are subject to management and use decisions made within many different sectors, including agriculture and forestry, navigation, public utility supply (water and waste disposal), and recreation. Most inland waters or their catchments also typically intersect several subnational administration units (counties, provinces, etc). Although it has been recognised for some time that the catchment basin is the fundamental unit within which management must be formulated, reconciling the many different interests concerned and coordinating actions have proved difficult. This is particularly problematic because sources of adverse impacts on such systems often originate far distant from where they are felt: thus pollutants and other inputs that enter the top reaches of a river system may have an impact in all downstream parts of the system as far as the river mouth and beyond. Inland water capture fisheries (other than some lucrative sports fisheries) and maintenance of inland water biodiversity have generally been accorded low priority in management decisions so that it has proven difficult to impose catchment-wide regulations or remedial measures for their benefit. This problem is exacerbated by the fact that it is often difficult to pinpoint a distant source of a problem, or to unequivocally demonstrate that actions in one place are having an adverse effect somewhere else (eg. to convince farmers that application of large doses of nitrogenous fertiliser on upstream agricultural land is causing deleterious eutrophication of estuarine wetlands).

Although they cannot address catchment-wide problems, inland water protected areas may play a valuable rôle in safeguarding particular sites or populations of species from immediate threats. Protection may be most effective where sites are relatively small and thus manageable, and have a relatively low level of allochthonous inputs. Wetlands, with their often abundant and highly conspicuous avifauna, have in general received most attention in this regard. Notable wetland protected areas include the Moremi Game Reserve in the Okavango Delta, Botswana, Camargue National Reserve in France, Keoladeo (Bharatpur) National Park in India, Doñana National Park in Spain and Everglades National Park in the USA.

In the context of protected areas, inland water ecosystems are unusual in that an international convention is dedicated specifically to them. This is the Convention of Wetlands of International Importance especially as Waterfowl Habitat (the Ramsar Convention), which was signed in 1971 and came into force in 1975. It places general obligations on contracting State Parties relating to conservation of wetlands throughout their territories, with special obligations pertaining to those wetlands that have been inscribed on the "List of Wetlands of International Importance". Each State Party must designate at least one such site. Wetlands are specifically though broadly defined under the convention and include areas of marine waters, the depth of which at low tide does not exceed six metres. There are presently 116 Contracting Parties to the Convention, with 1006 wetland sites, totalling 71.8 million hectares, designated for inclusion in the Ramsar List of Wetlands of International Importance[42].

Transboundary inland waters

Waters that delineate or cross international boundaries present a special class of management issues. Such waters and the living resources they contain are shared by one or more countries, and require positive international collaboration for effective use and management.

Available water in any given country within an international basin (or other administrative unit within a basin more generally) can be divided into endogenous, ie. locally generated runoff available in national aquifers and surface water systems, and exogenous, ie. remotely generated runoff imported in flow from upstream. Some countries (eg. Canada, Norway) have an abundance of water from endogenous sources, others (eg. Egypt, Iraq) have a small endogenous supply but large exogenous volumes (others have small supplies from both sources). Use of exogenous water carries an increasing risk because of dependence on sufficient supply from upstream countries.

There are well over 200 major international rivers and a host of smaller ones[43]. As demands on inland water resources continue to grow through the 21st century, as they undoubtedly will, management of these and the biological resources they contain will also grow ever more challenging.

References

1 World Conservation Monitoring Centre. 1998. *Freshwater Biodiversity: a preliminary global assessment.* By Groombridge, B. and Jenkins, M. WCMC – World Conservation Press, UK. vii + 104 pp + 14 maps.

2 Gorthner, A. 1994. What is an ancient lake? Speciation in Ancient Lakes. In, Martens, K., Goddeeris, B. and Coulter, G. (eds). *Archiv fur Hydrobiologie. Ergebnisse der Limnologie.* 44: 97-100.

3 Horne, J. A. and Goldman, C.R. 1994. *Limnology.* McGraw-Hill Inc., New York.

4 Watter, G.T. 1992. Unionids, fishes, and the species-area curve. *Journal of Biogeography* 19: 481-490.

5 Tittley, I. 1992. Contribution to chapter 7, Lower plant diversity, In, *Global Biodiversity.* Chapman and Hall.

6 Goh, T.K. and Hyde, K.D. 1996. Biodiversity of freshwater fungi. *Journal of Industrial Microbiology.* 17 :97-100.

7 Shearer, C.A. 1993. The Freshwater Ascomycetes *Nova Hedwigia* 56: 1-33

8 Shearer, C.A. Freshwater ascomycetes and their anamorphs. http://lords.life.uiuc.edu:23523/ascomycete/ as of Nov. 1999.

9 Sculthorpe, C.D. 1967. *The biology of aquatic vascular plants.* Edward Arnold, London.

10 Palmer, M., Covich, A.P., Finlay, B.J., Gilbert, J., Hyde, K.D., Johnson, R.K., Kairesalo, T., Lake, S., Lovell, C.R., Naiman, R.J., Ricci, C., Sabater, F. and Strayer, D. 1997. Biodiversity and ecosystem processes in freshwater sediments. *Ambio* 26(8): 571-577.

11 Nelson, J.S. 1994. *Fishes of the World.* 3rd Ed. John Wiley & Sons, Inc., New York.

12 Kottelat, M. and Whitten, T. 1996. *Freshwater biodiversity in Asia with special reference to fish.* World Bank Technical Paper No. 343.

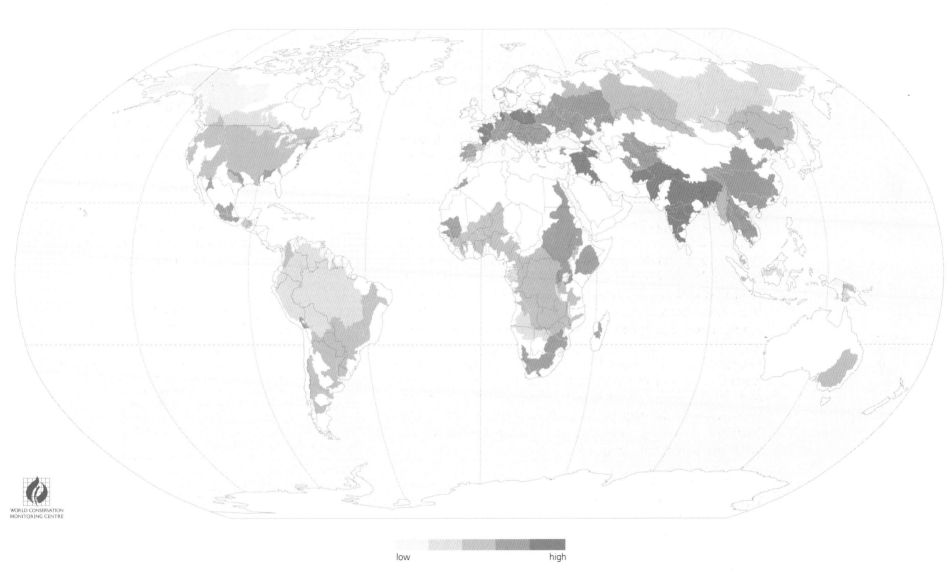

low   high

Map 8.4
River catchment vulnerability
Source: WCMC[1]

13 Olson, D.M., Chernoff, B. Burgess, G., Davidson, I., Canevari, P., Dinerstein, E., Castro, G., Morisset, V., Abell, R. and Toledo, E. (eds). 1997. *Freshwater biodiversity of Latin America and the Caribbean: a conservation assessment.* Proceedings of a workshop. World Wildlife Fund, Washington DC.

14 Abell, R., Olson, D., Dinerstein, E., Hurley, P., Eichbaum, W., Walters, S., Wetterngel, W., Allnutt, T. and Loucks, C. 1998. A *conservation assessment of the freshwater ecoregions of North America.* Final Report submitted to the U.S. EPA, April 1998. World Wildlife Fund-US.

15 Guégan, J.-F., Lek, S. and Oberdorff, T. 1998. Energy availability and habitat heterogeneity predict global riverine fish diversity. *Nature* 391, 22 January, 382-384.

16 Oberdorff, T., Guégan, J.-F. and Hugueny, B. 1995. Global scale patterns of fish species richness in rivers. *Ecography* 18: 345-352.

17 UN (CSD). 1997. Comprehensive Assessment of the Freshwater Resources of the World. Economic and Social Council. Fifth Session, 5-25 April. E/CN.17/1997/9. Available online through http://www.un.org/esa/sustdev/csd.htm.

18 FAO. 1996. *Food Production: the critical role of water.* Technical background document. World Food Summit. (available in .pdf from http://www.fao.org).

19 FAO. 1999. *Review of the state of the World fishery resources; inland fisheries.* FAO Fisheries Circular No. 942. FIRI/C942.

20 Coates, D. 1995. *Inland capture fisheries and enhancement: status, constraints and prospects for food security.* Paper presented at International conference on sustainable contribution of fisheries to food security. Kyoto, Japan 4-9 December 1995. FAO KC/FI/95/Tech/3.

21 Borgström, R. 1994. Freshwater Ecology and Fisheries. Chapter 3, pp. 41-69 In, Balakrishnam, M. Borgstrom, R. and Bie, S.W. *Tropical Ecosystems.* International Science Publisher. Oxford and IBH. New Dehli.

22 Scudder, T. and Conelly, T. 1985. Management systems for riverine fisheries. FAO Fisheries Technical Paper 263. FAO, Rome.

23 Munthali, S.M. 1997. Dwindling food-fish species and fishers' preference: problems of conserving Lake Malawi's biodiversity. *Biodiversity and Conservation* 6:253-261.

24 Postel, S.L., Daily, G.C. and Ehrlich, P.R. 1995.Human appropriation of renewable fresh water. *Science* 271: 785-788.

25 L'Vovich, M.I. *et al.* 1990. Use and transformation of terrestrial water systems. In, Turner II, B.L. (1990) *The earth as transformed by human action. Global and regional changes in the biosphere over the past 300 years.* P.235-252. Cambridge University Press with Clark University.

26 Dynesius, M. and Nilsson, C. 1994. Fragmentation and flow regulation of river systems in the northern third of the world. *Science* 266. 4 November, 735-762.

27 Avakyan, A.B. and Iakovleva, V.B. 1998. Status of global reservoirs: the position in the late twentieth century. *Lakes and Reservoirs: Research and Management*, 3: 45-52.

28 Meybeck, M. and Helmer, R. 1989. The quality of rivers: from pristine stage to global pollution. In, *Palaeogeography, Palaeoclimatology, Palaeoecology (Global and Planetary Change section)* 75: 283-309. Elsevier Science Publishers, B.V., Amsterdam.

29 Chapman, D.V. 1992. *Water quality assessment.* Chapman and Hall.

30 UNEP (United Nations Environment Programme). 1991. *Freshwater Pollution.* UNEP/GEMS Environmental Library No 6.

31 UNEP (United Nations Environment Programme). 1995. *Water Quality of World River Basins.* UNEP/GEMS Environmental Library No 6.

32 Moyle, P.B. 1996. Effects of invading species on freshwater and estuarine ecosystems. In, Sandlund, O.T. *et al.* (eds). *Proceedings of the Norway/UN Conference on Alien Species,* Trondheim, 1-5 July 1996. Directorate for Nature Management and Norwegian Institute for Nature Research. Trondheim, Norway, p.86-92.

33 Moyle, P.B. and Leidy, R.A. 1992. Loss of biodiversity in aquatic ecosystems: evidence from fish faunas. In, Fielder, P.L. (eds). *Conservation Biology, the theory and practice of nature conservation, preservation and management.* Pp. 129-169. Chapman and Hall, New York and London.

34 Stiassny, M. 1996. An overview of freshwater biodiversity with some lessons learned from African fishes. *Fisheries* 21(9): 7-13.

35 Reinthal P.N. and Stiassny, M.L.J. 1991. The freshwater fishes of Madagascar: a study of an endangered fauna with recommendations for a conservation strategy. *Conservation Biology* 5(2): 231-243

36 Kirchhofer, A. and Hefti, D. (eds), 1996. *Conservation of endangered freshwater fish in Europe.* Birkhäuser Verlag: Basel, Boston, Berlin. (Advances in Life Sciences).

37 IUCN. 1996. *The 1996 IUCN Red List of Threatened Animals.* IUCN, Gland, Switzerland.

38 Loh, J., Randers, J., MacGillivray, A., Kapos, V., Jenkins, M., Groombridge, B., Cox, N. and Warren, B. 1999. *Living planet report 1999.* WWF International, Gland, Switzerland.

39 R. Lesslie, *in litt.,* 30 May 1998.

40 Lesslie, R, and Maslen, M., 1995. *National wilderness inventory, handbook and procedures, content and usage.* Second Edition. Commonwealth Government Printer, Canberra

41 Raskin, P., Gleick, P., Kirshen, P., Pontius, G. and Strzepek, K. 1997. *Water Futures: Assessment of Long-range Patterns and Problems.* Background Report #3 of *Comprehensive Assessment of the Freshwater Resources of the World.* Stockholm Environment Institute.

42 The Convention on Wetlands (Ramsar) website: http://www.ramsar.org (as of 29 Nov. 99).

43 Anon. 1978. Centre for Natural Resources, Energy and Transport of the Department of Economic and Social Affairs, United Nations. Register of International Rivers. *Water Supply and Management.* 2: 1-58. (Special Issue). Pergamon Press.

44 Martens, K., Goddeeris, B. and Coulter, G. (eds). 1994 Speciation in ancient lakes. *Arciv für Hydrobiologie. Ergebnisse der Limnologie.* 44.

45 Hutchinson, G.E. 1993. A *Treatise on Limnology. vol. IV. The Zoobenthos.* John Wiley & Sons, Inc., NY.

Suggested introductory sources

Horne, A.J. and Goldman, C.R. 1994. *Limnology.* 2nd ed. McGraw-Hill, Inc. NY.

Hutchinson, G.E. A *treatise on limnology.* Vol. 1, 1957. Vol. 2, 1967. Vol. 3, 1975. Vol. 4, 1993. John Wiley & Sons, Inc., NY.